Perspectives on Ecological Degradation and Technological Progress

Veli Yilanci
Faculty of Political Sciences, Canakkale Onsekiz Mart University, Turkey

A volume in the Advances in Environmental Engineering and Green Technologies (AEEGT) Book Series

IGI Global
PUBLISHER of TIMELY KNOWLEDGE

Published in the United States of America by
IGI Global
Engineering Science Reference (an imprint of IGI Global)
701 E. Chocolate Avenue
Hershey PA, USA 17033
Tel: 717-533-8845
Fax: 717-533-8661
E-mail: cust@igi-global.com
Web site: http://www.igi-global.com

Library of Congress Cataloging-in-Publication Data

Names: Yilanci, Veli, 1982- editor.
Title: Perspectives on ecological degradation and technological progress / edited by Veli Yilanci.
Description: Hershey, PA : Engineering Science Reference, [2023] | Includes bibliographical references and index. | Summary: "The publication will explore the economic and social impacts of technological progress on environmental degradation from a multidisciplinary perspective. Besides, other factors that may have to affect the environmental degradation will be analyzed, and indicators that may be important for the environment will be determined"-- Provided by publisher.
Identifiers: LCCN 2022041295 (print) | LCCN 2022041296 (ebook) | ISBN 9781668467275 (hardcover) | ISBN 9781668467282 (paperback) | ISBN 9781668467299 (ebook)
Subjects: LCSH: Industries--Environmental aspects. | Technology--Environmental aspects. | Environmental degradation. | Nature--Effect of human beings on.
Classification: LCC TD194 .P47 2023 (print) | LCC TD194 (ebook) | DDC 333.7--dc23/eng/20221110
LC record available at https://lccn.loc.gov/2022041295
LC ebook record available at https://lccn.loc.gov/2022041296

This book is published in the IGI Global book series Advances in Environmental Engineering and Green Technologies (AEEGT) (ISSN: 2326-9162; eISSN: 2326-9170)

British Cataloguing in Publication Data
A Cataloguing in Publication record for this book is available from the British Library.

All work contributed to this book is new, previously-unpublished material.
The views expressed in this book are those of the authors, but not necessarily of the publisher.

For electronic access to this publication, please contact: eresources@igi-global.com.

Advances in Environmental Engineering and Green Technologies (AEEGT) Book Series

ISSN:2326-9162
EISSN:2326-9170

Editor-in-Chief: Sang-Bing Tsai, Zhongshan Institute, University of Electronic Science and Technology of China, China & Wuyi University, China; Ming-Lang Tseng, Lunghwa University of Science and Technology, Taiwan; Yuchi Wang, University of Electronic Science and Technology of China Zhongshan Institute, China

MISSION

Growing awareness and an increased focus on environmental issues such as climate change, energy use, and loss of non-renewable resources have brought about a greater need for research that provides potential solutions to these problems. Research in environmental science and engineering continues to play a vital role in uncovering new opportunities for a "green" future.

The **Advances in Environmental Engineering and Green Technologies (AEEGT)** book series is a mouthpiece for research in all aspects of environmental science, earth science, and green initiatives. This series supports the ongoing research in this field through publishing books that discuss topics within environmental engineering or that deal with the interdisciplinary field of green technologies.

COVERAGE

- Radioactive Waste Treatment
- Green Technology
- Biofilters and Biofiltration
- Electric Vehicles
- Cleantech
- Alternative Power Sources
- Renewable Energy
- Sustainable Communities
- Air Quality
- Green Transportation

IGI Global is currently accepting manuscripts for publication within this series. To submit a proposal for a volume in this series, please contact our Acquisition Editors at Acquisitions@igi-global.com or visit: http://www.igi-global.com/publish/.

The Advances in Environmental Engineering and Green Technologies (AEEGT) Book Series (ISSN 2326-9162) is published by IGI Global, 701 E. Chocolate Avenue, Hershey, PA 17033-1240, USA, www.igi-global.com. This series is composed of titles available for purchase individually; each title is edited to be contextually exclusive from any other title within the series. For pricing and ordering information please visit http://www.igi-global.com/book-series/advances-environmental-engineering-green-technologies/73679. Postmaster: Send all address changes to above address.

Titles in this Series

For a list of additional titles in this series, please visit: http://www.igi-global.com/book-series/

Nanopriming Approach to Sustainable Agriculture
Abhishek Singh (Sardar Vallabhbhai Patel Agriculture University, India) Vishnu D. Rajput (Academy of Biology and Biotechnology, Southern Federal University, Rostov-on-Don, Russia) Karen Ghazaryan (Biology Yerevan State University, Armenia) Santosh Kumar Gupta (NIPGR -National Institute of Plant Genome Research, India) and Tatiana Minkina (Academy of Biology and Biotechnology, Southern Federal University, Rostov-on-Don, Russia)
Engineering Science Reference • © 2023 • 300pp • H/C (ISBN: 9781668472323) • US $245.00

Contemporary Developments in Agricultural Cyber-Physical Systems
G.S. Karthick (Department of Software Systems, PSG College of Arts and Science, India)
Engineering Science Reference • © 2023 • 320pp • H/C (ISBN: 9781668478790) • US $240.00

Nematode-Plant Interactions and Controlling Infection
Waleed Fouad Abobatta (Horticulture Research Institute, Agriculture Research Center, Egypt) and Rehab Yasin Ghareeb (Plant Protection and Biomolecular diagnosis Dep., Arid Lands Cultivation Research Institute, City of Scientific Research and Technological Applications, Egypt)
Engineering Science Reference • © 2023 • 300pp • H/C (ISBN: 9781668480830) • US $240.00

Perspectives on Global Biodiversity Scenarios and Environmental Services in the 21st Century
Naveen Kumar Chourasia (Government College, Bichhua, India) and Kavita Chahal (Government College, Bichhua, Chhindwara, India)
Engineering Science Reference • © 2023 • 300pp • H/C (ISBN: 9781668490341) • US $250.00

Handbook of Research on AI-Equipped IoT Applications in High-Tech Agriculture
Alex Khang (Global Research Institute of Technology and Engineering, USA)
Engineering Science Reference • © 2023 • 450pp • H/C (ISBN: 9781668492314) • US $335.00

IGI Global
PUBLISHER of TIMELY KNOWLEDGE

701 East Chocolate Avenue, Hershey, PA 17033, USA
Tel: 717-533-8845 x100 • Fax: 717-533-8661
E-Mail: cust@igi-global.com • www.igi-global.com

Table of Contents

Detailed Table of Contents

chapter utilizes the panel stationarity test with sharp and smooth breaks introduced by Bahmani-Oskooee et al. There are two main advantages of this methodology over the conventional panel stationarity tests. First, it considers the cross-sectional dependency in the series. Second, the test combines both sharp and smooth structural breaks in the series. The empirical results of this study state that the energy efficiency index does not converge to the OECD average for 1971-2018. This result shows that the dissemination of the technology is limited across OECD countries. The empirical result of this study provides significant policy implications for energy and environmental issues.

Natural resources are components of the natural environment used in the production process to meet the material and cultural needs of society. Natural resources are inherently physical in origin, but in the process of their use, they become economic resources. Natural resources are divided into inexhaustible (agro-climatic, geothermal, hydropower) and exhaustible. In turn, exhaustible resources are divided into non-renewable (mineral) and renewable (land, water, biological, recreational). Based on this classification and development, this textbook distinguishes the following types of natural resources: mineral (minerals), energy, water, biological, land, agro-climatic, and recreational. When considering natural resources, it is important to assess the availability of resources (i.e., the ratio between explored reserves of resources and the volume of their use, also their impact on environmental conditions of the world).

This study investigated how income inequality is shaped according to energy-related variables. For this, income inequality is the independent variable; energy efficiency, energy import dependency, and energy efficiency are included in the model as independent variables. The data of the 17 founding countries of the OECD between 2011-2020 is used. According to the F Test results, the model has a one-way unit effect. According to the robust Hausman test, the random effect is valid. There are heteroscedasticity and autocorrelation in the context of deviations from the assumption. On the other hand, the data fit a normal distribution. Also, there is no multicollinearity and no inter-unit correlation. In light of all these, the model was estimated with Arellano, Froot, and Rogers Estimator. According to the model estimation results, the model is not statistically significant. This shows that the relationship between income distribution and energy, which Illich expressed qualitatively, is not quantitatively valid for the 17 OECD founding countries.

This chapter examines the employment efficiency in the Turkish labor market between 1988 and 2021 and measures the efficient level of Turkish employment by 2030, using the data obtained from TUİK (Turkish Statistical Institute), employing an alternative methodology named "Regressive Envelopment Analysis" or simply REA. The study uses a two-stage analysis where the first stage measures how efficiently the labor force is created from the pool of economically active population, while the second stage measures how efficiently the labor force is put to employment. The overall (1988-2021) employment efficiency average is found to be about 94%, and the bulk of the efficiency loss is the result of the first stage, which is fundamentally shaped by the structural unemployment. It is suggested that long-term unemployment rates, even at the efficient employment levels, will steadily rise from an average of 8% in 2000 to 12% in 2030 for both sexes, but especially for females, nearing 14%.

This study aims to test the tourism-induced EKC hypothesis, which claims the existence of an inverted U-shaped relationship between environmental pollution and tourism activities, for the period 1995-2018 in the most-visited countries in terms of the number of international tourists (Spain, United States, China, Italy, Turkey, Mexico, Thailand, Germany, and England). In this study, in which ecological footprint, which is a comprehensive and alternative variable, was used as an indicator of environmental pollution, the relations between the variables were analyzed using second-generation test techniques. Long-term relationships were examined by the LM and DH test, and the coefficients were estimated with the common correlated effects mean group (CCE–MG) and dynamic common correlated effects (DCCE) estimators. As a result of the empirical analysis, it was found that there is a U-shaped relationship between environmental pollution and tourism activity, and the validity of the tourism-induced EKC hypothesis could not be confirmed.

This study examines the evolution of ecological degradation within the context of consumption in the past three decades. The study used the R Studio program to conduct a bibliometric meta-analysis. The data set consists of 2183 articles selected from the Web of Science database by searching the keywords (ALL=("Ecological Degradation") AND ALL=("consum*")) OR (ALL=("environmental degradation") AND ALL=("consum*")). Graphics related to annual scientific production, corresponding author's country, most cited countries, most relevant sources, word cloud, word tree, trend topics, and thematic map, which are revealed using the visualization technique, will be included in the study. As a result, this research aims to create a comprehensive framework for articles on ecological degradation within the context of consumption published in the last thirty years.

This study investigates the convergence process of environmental-related technologies patents over the period 1991-2019 across 36 OECD countries. To do so, the club convergence methodology of Phillips and Sul is used, which allows the authors to analyze overall convergence and identify possible convergence clubs. The log-t test results show that OECD countries do not converge to the same equilibrium point in terms of environmental-related technologies patents per capita. However, the club clustering algorithm shows that while 33 OECD countries converge to a common point, Hungary, South Korea, and Lithuania are diverging countries.

The environmentalists argue that climate change is a real threat, and therefore, they believe both developed and developing countries should shift their energy sources from fossil-based to renewable alternatives, including solar, wind, geothermal, hydropower, bio-oils, and ocean power to sustain economic growth. In this context, this study investigates the causal relationship between two sets of pairs—(i) renewable energy consumption and economic growth and (ii) nonrenewable energy consumption—and economic growth for 16 OECD countries for the 1971-2015 period. To fulfill this objective, the study employs a recently introduced frequency domain approach in both single-country form and multi-country (panel) form. The study indicates some dissimilarities on renewable and nonrenewable energy consumption and economic growth nexus in OECD countries. Although the results of single country frequency domain test mostly support neutrality hypothesis, multi-country frequency domain test finds strong evidence supporting bidirectional causality among these sets of pairs.

The purpose of this chapter is to review the factors that have an impact on green consumption behavior of individual consumers, as they relate to environmental sustainability considerations. After a brief review of the evolution of environmentally related research and relevant terminology, factors expected to have an effect on pro-environmental behaviors of individuals, inclusive of various demographic, individual, social-cultural, and contextual/situational considerations are assessed. A distinction between high- and low-income countries with respect to their reasons for engaging in environmentally friendly behaviors is also made in the discussion section. The chapter ends with a number of suggestions for future research and some concluding remarks.

Environmental degradation will continue if the corporate continues to misbehave. This study offers evidence for whether environmental violators are more likely to disclose if they are in sensitive industries, faced higher penalties, and required to implement an environmental project. This chapter examines how environmental violations have shaped the mandatory and discretionary environmental disclosures of oil and gas companies in 10-k reports filed with the US Securities and Exchange Commission from 2011 to 2020. Using the Leximancer text mining tool, the violators' annual reports were examined before and after the violations. The results suggest that disclosure tended to be discretionary, and it is possible that selective information was traded off with other discretionary non-financial information for the sake of legitimacy. This chapter provides evidence of the effectiveness of regulators' efforts for higher mandated disclosures to encourage firms to deal with environmental matters proactively to prevent further environmental misbehaviour.

Preface

In the realm of economics, where the efficient allocation of scarce resources is paramount, scholars have long acknowledged the existence of limits to growth due to the finite nature of our natural resources. As early as 1972, the Club of Rome published a groundbreaking report titled "The Limits of Growth," predicting that global economic growth would encounter insurmountable barriers within a century. However, in the subsequent decades, the landscape has transformed with remarkable technological advancements and the discovery of new reserves of natural resources. As a result, the traditional boundaries on growth and development have begun to fade away. It is within this evolving context that the concept of a green economy has emerged, emphasizing policies and innovations that enhance human well-being, promote social equity, and effectively mitigate environmental risks by optimizing the use of scarce resources. Notably, new technologies promoting energy efficiency have unlocked growth opportunities in uncharted directions, and technological progress itself has exhibited a positive impact on the environment.

In this edited reference book, titled "Perspectives on Ecological Degradation and Technological Progress," we embark on a multidisciplinary exploration of the effects of technological development on environmental degradation. Adopting a comprehensive lens, we delve into the economic and social implications of technological progress on our environment. Moreover, we examine additional factors that may contribute to environmental degradation, analyzing their interplay with technological advancements. Throughout the book, we identify crucial indicators that hold significance for the environment and shed light on the complexities of achieving a sustainable future.

This publication is primarily intended to serve as a valuable resource for academicians, researchers, and professionals in the fields of ecology, economics, technology, and environmental economics. However, we also aspire to extend its relevance beyond the academic sphere. Policymakers, managers, and individuals concerned with shaping our collective future will find this book instrumental in comprehending the determinants of a green economy. By exploring the intricate relationship between ecological degradation and technological progress, we aim to facilitate informed decision-making and foster sustainable practices in a world poised for transformation.

The breadth of topics covered in this volume is broad, encompassing various aspects of the nexus between technological progress and environmental degradation. While the list is not exhaustive, it is indicative of the diverse themes addressed within the book. These topics include the impact of technology on ecological footprints, the convergence of indicators signaling ecological degradation, the interplay between technological development and renewable energy, and the pursuit of environmental sustainability. Additionally, we explore the correlation between green economic growth and new job opportunities, the intersection of a green economy with information and communication technologies, and the relationship between green economy and employment, finance, and ecological footprints. The examination of ecological degradation in both developing and developed countries is another crucial aspect of our analysis. Furthermore, we investigate how technological progress can be harnessed to foster sustainable development, while also exploring its implications for investments, as well as the emerging relationship between cryptocurrencies and clean energy stocks.

Chapter 1 focuses on the Index of Sustainable Economic Welfare (ISEW) as a crucial indicator of sustainable development. The primary objective is to calculate and analyze the ISEW in the Turkish economy, specifically examining its relationship with renewable energy generation. Through econometric tests, the chapter establishes a long-term relationship between these variables using the RALS-EG cointegration test. The dynamic OLS (DOLS) method is employed to estimate the long-term coefficient, revealing a negative impact of renewable energy production on sustainable development. This finding contrasts with most studies conducted in developed countries, challenging the common belief that renewable energy supports sustainable development.

Chapter 2 delves into the growing trend of studying energy efficiency in energy economics research. The aim is to investigate the stochastic convergence of energy efficiency across 27 OECD countries from 1971 to 2018. The chapter employs Fisher's Ideal index to calculate the energy efficiency index and utilizes a panel stationarity test with sharp and smooth breaks. This methodology considers cross-sectional dependency in the series and combines sharp and smooth structural breaks. The empirical results indicate that energy efficiency index does not converge to the OECD average during the observed period, suggesting limited dissemination of technology across OECD countries. These findings hold significant policy implications for energy and environmental issues.

Chapter 3 focuses on natural resources and their classification based on origin and exhaustibility. It examines the various types of natural resources, including mineral resources, energy resources, water resources, biological resources, land resources, agro-climatic resources, and recreational resources. The chapter emphasizes the importance of assessing resource availability, which involves evaluating the ratio between explored reserves and resource usage, as well as their impact on global environmental conditions. By providing a comprehensive overview of natural resources, this chapter contributes to understanding their significance and effects on the environment.

Chapter 4 explores the relationship between income inequality and energy-related variables. The study focuses on income inequality as the independent variable and includes energy efficiency, energy import dependency, and energy efficiency as the dependent variables. The data used in this chapter covers the 17 founding countries of the OECD from 2011 to 2020. The model estimation employs the Arellano, Froot, and Rogers Estimator, and various tests confirm the validity of the model. However, the results indicate that the model is not statistically significant, suggesting that the relationship between income distribution and energy, which was qualitatively expressed, lacks quantitative validity for the OECD founding countries.

Chapter 5 investigates employment efficiency in the Turkish Labor Market from 1988 to 2021 and predicts the efficient level of Turkish employment by 2030. Using data obtained from the Turkish Statistical Institute (TUİK) and employing the Regressive Envelopment Analysis (REA) methodology, this chapter conducts a two-stage analysis. The first stage assesses the efficiency of creating a labor force from the economically active population, while the second stage measures the efficiency of employing the labor force. The results reveal an average employment efficiency of approximately 94% during the observed period, with structural unemployment playing a significant role in efficiency loss. The chapter predicts a steady increase in long-term unemployment rates, especially for females, by 2030.

Chapter 6 aims to test the tourism-induced Environmental Kuznets Curve (EKC) hypothesis, which suggests an inverted U-shaped relationship between environmental pollution and tourism activities. Focusing on the most-visited countries in terms of international tourists, the study analyzes the relationship between environmental pollution (measured using the ecological footprint) and tourism activity. The empirical analysis utilizes second-generation test techniques, examines long-term relationships, and estimates coefficients with the Common Correlated Effects Mean Group (CCE-MG) and Dynamic Common Correlated Effects (DCCE) estimators. The findings indicate a U-shaped relationship between environmental pollution and tourism activity, failing to confirm the validity of the tourism-induced EKC hypothesis.

Chapter 7 conducts a bibliometric meta-analysis to examine the evolution of ecological degradation within the context of consumption over the past three decades. Using the R Studio program and data from the Web of Science database, the study analyzes 2183 articles based on specific keywords. The visualization technique is employed to present annual scientific production, corresponding author's country, most cited countries, most relevant sources, word cloud, word tree, trend topics, and thematic map. By creating a comprehensive framework for articles on ecological degradation and consumption, this chapter provides valuable insights into the research conducted in this field over the last thirty years.

Chapter 8 investigates the convergence process of environmental-related technologies patents across 36 OECD countries from 1991 to 2019. Using the club convergence methodology, the chapter analyzes overall convergence and identifies potential convergence clubs. The results show that OECD countries do not converge to the same equilibrium point in terms of environmental-related technologies patents per capita. While 33 countries converge to a common point, Hungary, South Korea, and Lithuania are identified as diverging countries. This chapter sheds light on the patterns and dynamics of technological innovation in environmental-related fields.

Chapter 9 explores the causal relationship between renewable and nonrenewable energy consumption and economic growth in 16 OECD countries from 1971 to 2015. By employing a frequency domain approach in both single-country and multi-country forms, the chapter examines dissimilarities in the nexus between energy consumption and economic growth. The results of the single-country frequency domain test generally support the neutrality hypothesis, while the multi-country frequency domain test reveals strong evidence of bidirectional causality between these sets of pairs. This chapter provides insights into the complex relationship between energy consumption, economic growth, and the choice of energy sources.

Chapter 10 reviews the factors that impact green consumption behavior, focusing on individual consumers' pro-environmental behaviors and their relation to environmental sustainability considerations. The chapter traces the evolution of environmentally related research and explores various demographic, individual, social-cultural, and contextual/situational factors that influence pro-environmental behaviors. Additionally, it distinguishes reasons for engaging in environmentally friendly behaviors between high and low-income countries. The chapter concludes by offering suggestions for future research and providing valuable insights into the factors shaping green consumption behavior.

Chapter 11 examines how environmental violations shape the mandatory and discretionary environmental disclosures of oil and gas companies. The study focuses on the 10-K reports filed with the US Securities and Exchange Commission from 2011 to 2020. Using the Leximancer text mining tool, the chapter analyzes the violators' annual reports before and after the violations to understand the impact on their disclosure practices. The results suggest that disclosures tend to be discretionary, and there is a possibility of trading selective information for legitimacy. This chapter sheds light on the effectiveness of regulators' efforts in encouraging firms to address environmental matters proactively and prevent further environmental misbehavior.

Through this collaborative effort, we aspire to illuminate the multifaceted dynamics between ecological degradation and technological progress. By embracing a multidisciplinary approach, we aim to offer valuable insights and facilitate an informed dialogue among scholars, practitioners, and policymakers alike. We express our deepest gratitude to the authors who have contributed their expertise, research, and perspectives to this volume, and we hope that readers find this book to be a source of inspiration and knowledge, ultimately contributing to the realization of a greener and more sustainable future.

Finally, I would like to express my gratitude to Muhlis Selman Saglam for his great effort and dedication during the creation of this book.

Veli Yilanci
Faculty of Political Sciences, Canakkale Onsekiz Mart University, Turkey

Chapter 1
The Relationship Between Index of Sustainable Economic Welfare and Renewable Energy Production in Turkey:
Findings From the RALS-EG Test

Muhammet Fatih Coşkun
İnönü Üniversitesi, Turkey

Gökhan Konat
Bolu Abant İzzet Baysal Üniversitesi, Turkey

ABSTRACT

ISEW (Index of Sustainable Economic Welfare), one of the important indicators of sustainable development, is an important tool that should be measured regularly and its relationship with other macroeconomic indicators should be closely monitored. In this context, the current calculation of ISEW in the Turkish economy and the analysis of its relations with renewable energy generation constitutes the purpose of this study. As a result of the econometric tests, it is found that there is a long-term relationship between the two variables considered according to the RALS-EG cointegration test. The long-term coefficient estimation of the co-integrated variables has been estimated by the dynamic OLS (DOLS) method, and it is seen that renewable energy production has a negative effect on sustainable development. This result differs from the majority of studies on developed countries. Theory and literature say that renewable energy will support sustainable development.

DOI: 10.4018/978-1-6684-6727-5.ch001

1. INTRODUCTION

After the Great Depression of 1929, different indices and methods were used to monitor macroeconomic trends closely. Before the Great Depression, the economics literature was largely shaped on the microeconomic base, but after this important event, the focus shifted to macroeconomics. Subsequently, important indicators such as gross domestic product (GDP) was calculated (Menegaki A., 2018, s. 1). However, GDP has some important shortcomings (Daly & Posner, 2012, s. 5).

Gross Domestic Product (GDP) is a widely-used indicator of a country's economic performance. However, it has several limitations in terms of measuring overall societal welfare. Firstly, it does not consider income inequality, which means that even if the GDP increases, the benefits may not be distributed equally. Secondly, GDP only represents the quantity of goods and services produced, without revealing their quality. Thirdly, defense expenditure is not taken into account, which can have a significant impact on a country's economic performance. Fourthly, GDP does not distinguish between value rises brought on by speculation and borrowing and growth in true economic value. Fifthly, natural resource degradation is not factored in, and the costs associated with it are not reflected in the GDP. Sixthly, goods and services that are not bought and sold in the market, such as housework, are not included in the GDP. Lastly, GDP offers limited information on social welfare, as it does not reflect the unemployment, poverty, and human development index indicators. Therefore, it is essential to complement the GDP with other indicators that provide a more comprehensive picture of a country's overall welfare.

These shortcomings of the GDP have been tried to be overcome by calculating new and other indices over time. However, these indices are mostly calculated as substitutes for GDP (Menegaki A., 2018, s. 2). ISEW (Index of Sustainable Economic Welfare) was developed not as a substitute for GDP, but as a new version of it with social welfare indices (Menegaki, 2018, 2). In this respect, this index, which is one of the important indicators of sustainable development, should be measured regularly and its relationship with other macroeconomic indicators should be closely monitored.

ISEW is calculated with three kinds of data (Menegaki A., 2018, s. 7-10). The first type of data is economic indicators and this dataset forms the Basic ISEW. Undoubtedly, sustainable welfare consists primarily of these activities. Data such as individual consumption expenditures, health expenditures, education expenditures and fixed capital formation are classified in this category. The second type of data consists of environmental data. Mineral depletion, energy depletion, carbon damage and particulate emission are some indicators that fall into this category. The sum of these forms the Solid ISEW. Finally, social data forms the Site-specific ISEW. Indicators falling into this category are mostly composed of willingness to pay index. However, it is difficult to measure and therefore not included in the calculation.

Figure 1. Types of index of sustainable economic welfare
Source: Menegaki A. (2018, s. 8)

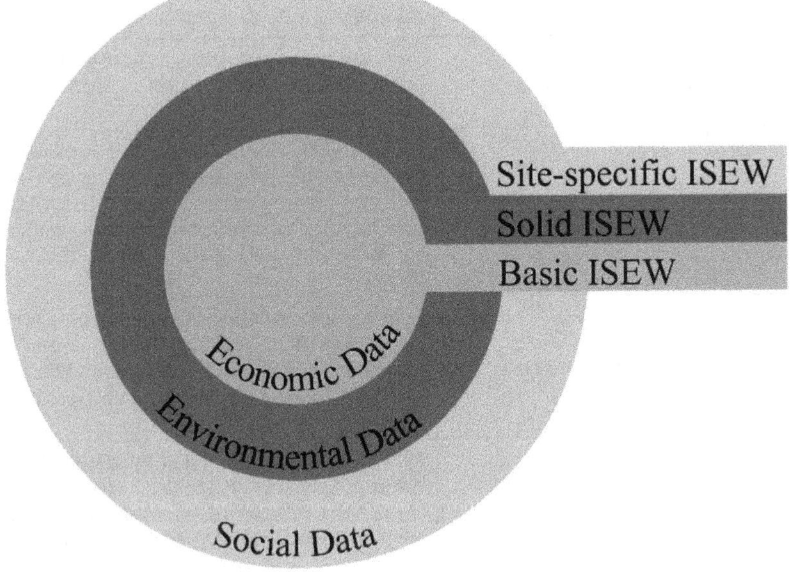

The continuation of the study consists of the related literature, obtaining the data set, econometric analysis and evaluation of the results, respectively.

2. RELATED LITERATURE

There is no single form of ISEW calculation. The researchers made different ISEW calculations according to the countries and time intervals, taking into account the availability of the data (Neumayer, 1999, s. 5). In addition, due to the lack of data, it is very difficult to obtain the Site-specific ISEW calculation, which is calculated by adding social variables on top of the Basic and Solid ISEW, and the Full ISEW, which is the sum of all ISEW varieties, and there are very few studies in the literature in which this calculation is made (Menegaki A., 2018, s. 1). In the light of this information, the relevant literature can be summarized as follows:

Table 1. Literature on ISEW

Author(s)	Interval	Country(ies)	Results
Stockhammer et al. (1997)	1955-1992	Austria	GDP correctly reflected welfare increase until the 1970s. However, the GDP did not provide the expected outcome afterwards. Although economic growth has expanded dramatically since the mid-1980s, welfare has decreased.
Castaneda (1999)	1965-1995	Chile	ISEW performed in parallel with GDP until 1985. GDP per capita doubled between 1985 to 1995, although ISEW plummeted.
Clarke (2004)	1975-1999	Thailand	Until the 1990s, fluctuations and differences were observed in GDP and ISEW. However, in the 1990s, the difference between these two variables gradually increased. In other words, while economic growth accelerated, the level of welfare did not increase or it decreased.
Clarke & Islam (2005)	1975-1999	Thailand	Until the 1990s, GDP and ISEW were close and parallel to each other. Both indices reached their peak in 1996. Despite the increase in economic growth in the 90s, welfare remained at a low level and the gap between the indices gradually increased.
Pulselli et al. (2006)	1999	Italy	Researchers took a year's ISEW value and GDP value and made a comparison. The study was conducted for the city of Siena, Italy. As a result, it was found that there is a large gap between GDP and ISEW, even at the local level..
Beça & Santos (2010)	1950-2005	USA	They developed a new calculation method by modifying ISEW. Results from the modified ISEW show that GDP has increased by 220% over 55 years, while ISEW has increased by only 17%.
Pulselli et al. (2012)	1971-2006	Italy	During the period under consideration, ISEW increased relatively slowly to GDP. The difference between the two indicators has steadily increased.
Menegaki & Tsagarakis (2015)	2000-2012	Greece	Two separate ISEWs were calculated with and without social indicators such as divorce and family problems. ISEW, for which social indicators are calculated, yielded a lower level result. It has been found that high indebtedness increases social problems in Greece, so the level of sustainable welfare does not increase significantly.
Menegaki & Tugcu (2018)	1990-2015	Selected Asian Countries	In the study, the effect of energy saving measures on welfare was analyzed. According to the findings, both GDP and ISEW responded negatively to energy saving measures. For these countries, energy saving negatively affects sustainable welfare.
Menegaki (2018)	2000-2012	Turkey	In the period under consideration, Turkey's per capita ISEW value increased from 2576 dollars to 8615 dollars. Although this situation shows that Turkey has come a long way in terms of sustainable welfare increase, a definite interpretation cannot be made because Full ISEW calculation cannot be made due to lack of data.

continues on following page

Table 1. Continued

Author(s)	Interval	Country(ies)	Results
Rugani et al. (2018)	1960-2010	Luxemburg	Due to the high spending on defensive measures, the growth rate of ISEW remained lower than the growth rate of GDP. In the period under consideration, the cost of growth in terms of sustainability has increased.
Armiento (2018)	1960-2014	Italy	Since 1991, the increase in ISEW has stopped, and decreases have been observed on occasion. The reasons for this are the increase in income inequality, the stagnation and decline in fixed capital formation and finally the decrease in unpaid household work.
Butnariu & Luca (2019)	1990-2017	Romania	During the time frame under consideration, GDP increased relatively faster than ISEW. Economic growth has been achieved with high costs in terms of sustainability.
Sánchez et al. (2020)	2001-2015	Ecuador	According to the findings, the difference between GDP and ISEW is high. Personal consumption was the item that contributed the most to ISEW. The biggest negative factor was natural resource degradation.
Azami (2020)	1995-2014	OPEC	The relationship between ISEW and energy consumption was analyzed. It has been found that energy saving policies has a sustainable welfare-enhancing effect.
O'Mahony (2021)	1970-2012	Spain	In the period under consideration, GDP per capita increased by more than one and a half times, while the ISEW per capita increased by only forty percent. The growth in the economy did not increase the level of welfare at the same level.
Liu et al. (2022)	2003-2019	China	ISEW prediction was made with artificial intelligence. According to the findings, ISEW shows an increasing trend. However, the rate of increase remains below the rate of increase of GDP. The most important contributor to ISEW was adjusted personal consumption expenditures.

3. DATASET AND ECONOMETRIC METHODOLOGY

In this study, the relationship between the ISEW (Index of Sustainable Economic Welfare) variable, which is one of the important indicators of sustainable development for Turkey, and renewable energy production is examined. The ISEW variable was calculated using the variable with 18 ISEW determinants obtained from the World Development Indicators, World Health Organization and Our World in Data sources. Renewable energy production was obtained from the official database of the World Bank and analyzes were carried out with the annual data set covering the period 2002-2019.

Table 2. Indicators and calculation method used in ISEW calculation

A- Households and NPISHs Final consumption expenditure (current US$)
B- Gini index (World Bank estimate)
C- $10.00 per day - share of population below poverty line
Item 1 = (A * (1-(B/100))*(1-(C/100))
D- Adjusted savings: education expenditure (current US$)
E- Public Spending on Education (% of GDP)
F- GDP (Current US$)
G- Public Spending on Education (Current US$)
Item 2 = G/2
H- Current health expenditure (% of GDP)
I- Ratio of general government health expenditure to total health expenditure (%)
Item 3 = [(F*(H/100))*(I/100)]/2
J- Vulnerable employment, total (% of total employment) (modeled ILO estimate)
K- Labor force, total
L- Real Minimum Wages (yearly) Constant prices at 2020 USD exchange rates
Item 4 = (J/100)*K*L
M- Gross capital formation (current US$)
N- Adjusted savings: consumption of fixed capital (current US$)
Item 5 = M-N
O- Adjusted savings: mineral depletion (current US$)
P- Adjusted savings: energy depletion (current US$)
R- Adjusted savings: carbon dioxide damage (current US$)
S- Adjusted savings: particulate emission damage (current US$)
Item 6 = O + P + R + S
ISEW
ISEW per capita = ISEW/Population

Source: Menegaki A. (2018, s. 9)

The calculation method of ISEW is shown in table 2. The calculation is exactly the same as Menegaki's (2018) calculation. First of all, the consumption expenditures are weighted by the Gini Index and poverty, and Item 1 is obtained. Then, the public expenditures for education are calculated, and Item 2 is obtained with fifty percent of the total expenditure, assuming that half is defensive expenditure (for maintaining the current level). Then, public expenditures for health services are calculated and Item 3 is obtained, assuming that fifty percent is defensive expenditure. Item 4 is calculated to calculate the value of jobs that cannot be calculated in GDP and that generate benefits even if they do not receive wages. Item 4 is computed to assess the amount of jobs that cannot be calculated in GDP yet provide value even if they are not paid. This item is calculated by multiplying the ratio of the vulnerable employment to the total employment by the annual minimum wage. Item 5 is calculated by subtracting the yearly contribution to total fixed capital from the current capital degradation. Finally, Item 6 is obtained, in which the damage to the environment is taken into account. The sum of all items gives the annual total ISEW value, and the ISEW value per capita is calculated by dividing this value by the population.

After the ISEW was obtained, its relationship with other variables was also analyzed with various econometric methods. First of all, the stationarity levels of the variables were investigated, then cointegration analysis was carried out, and in the last step, the long-term coefficients were estimated. For this purpose, the standard ADF unit root test for unit root research and the RALS-ADF unit root test recommended by Im & Schmidt (2008), in which RALS (Residual Augmented Least Squares) terms are added and which gives stronger results when the residuals are not normally distributed, were used. For the cointegration analysis, tests were carried out using the standard Engle-Granger (EG) cointegration test and RALS-EG cointegration techniques.

3.1. RALS-ADF Unit Root Test

Many unit root tests based on the Dickey-Fuller (1979) unit root test and its strengthened version, the Augmented Dickey-Fuller (1981) unit root test, were used in the literature after these two tests. The equations of the ADF unit root test are as follows:

$$\Delta Y_t = \beta_1 + \delta Y_{t-1} + \sum_{i=1}^{m} \alpha_i \Delta Y_{t-i} + e_t \tag{1}$$

$$\Delta Y_t = \beta_1 + \beta_2 t + \delta Y_{t-1} + \sum_{i=1}^{m} \alpha_i \Delta Y_{t-i} + e_t \tag{2}$$

Equations (1) and (2), respectively, show constant and constant and trended equation structures. Lagged values of the dependent variable were inserted to the model to overcome the autocorrelation problem. The test's main hypothesis is that the series contains a unit root. Using the ADF test, Im et al. (2014) proposed a new testing procedure. If the model's residuals are not normally distributed, the high moments of these residuals hold information about these residuals, which is included using this approach. They has proved in empirical studies that the test produces better results than standard unit root testing. The ADF model is augmented to the RALS-based structure with the term $\hat{w}_t$, with $\hat{e}_t$ being the residuals from the ADF models, and is as follows:

$$\hat{w}_t = h\left(\hat{e}_t\right) - \bar{K} - \hat{e}_t\hat{D}_2, \quad t = 1,2,\ldots,T \tag{3}$$

$$\bar{K} = \frac{1}{T}\sum_{t=1}^{T} h\left(\hat{e}_t\right) \text{ and } \hat{D}_2 = \frac{1}{T}\sum_{t=1}^{T} h'\left(\hat{e}_t\right).$$

It is defined as $h\left(\hat{e}_t\right) = \left[\hat{e}_t^2, \hat{e}_t^3\right]'$ to capture the information of non-normally distributed errors, which is indicated as the second and third moment of $\hat{e}_t^2$. For j=2,3, $m_j = T^{-1}\sum_{t=1}^{T}\hat{e}_t^j$, is obtained in the form of $\hat{w}_t = \left[\hat{e}_t^2 - m_2, \hat{e}_t^3 - m_3 - 3m_{2t}\hat{e}_t\right]'$. (Im vd., 2014: 321). The ADF testing process based on residuals augmented least squares (RALS) regression is shown again as follows:

$$\Delta y_t = \alpha_1 + \beta y_{t-1} + \sum_{j=1}^{p}\delta_j \Delta y_{t-j} + \hat{w}_t'\gamma + v_t, \; t = 1,2,\ldots,T \tag{4}$$

$$\Delta y_t = \alpha_1 + \alpha_1 t + \beta y_{t-1} + \sum_{j=1}^{p}\delta_j \Delta y_{t-j} + \hat{w}_t'\gamma + v_t, \quad t = 1,2,\ldots,T \tag{5}$$

The null hypothesis of the test expresses the unit rooted structure as in the standard ADF test. The long-term correlation is calculated for the appropriate critical values, and the test statistics obtained are compared with the critical values obtained by Hansen (1995), and it is decided whether the series contains a unit root.

3.2. RALS Engle-Granger Cointegration Test

The basic approach in the cointegration method proposed by Engle and Granger (1987) is that the error terms of a linear combination between two non-stationary time series have stationarity. Engle and Granger defined the cointegration model as follows:

$$y_t = \beta x_t + u \tag{6}$$

In order for the variables in the model to be cointegrated, it is assumed that both variables are difference stationary and the error term is stationary at level. To determine the long-term relationship between EG cointegration series, a linear equation was created as follows.

$$\Delta\hat{u}_t = \alpha_0 + \phi\hat{u}_{t-1} + \sum_{j=1}^{p}\alpha_j\Delta\hat{u}_{t-j} + e_t \tag{7}$$

In this equation, the parameters are estimated using the least squares method, and in the next part, the error terms obtained from the model are tested by applying the ADF test to see if they contain a unit root. The main hypothesis of the EG test is that there is no cointegration relationship between the variables.

Lee at al. (2015) suggested the RALS-EG cointegration test, in which the RALS method is used instead of OLS to increase the power of the EG test. As in the RALS-ADF unit root test, where $\hat{e}_t$ is the residuals obtained from the EG structure, the EG test is augmented to the RALS-based structure with the term $\hat{w}_t$ and is as follows:

$$\Delta\hat{u}_t = \alpha_0 + \phi\hat{u}_{t-1} + \sum_{j=1}^{p}\alpha_j\Delta\hat{u}_{t-j} + \hat{w}_t{}'\gamma + v_t \tag{8}$$

The hypothesis is as in the standard EG test and is tested with the t-statistic. The correlation between the test statistic of the RALS-EG method and the EG test statistic is calculated with the help of the following equation:

$$t^* \rightarrow pt + \sqrt{1-p^2}Z \tag{9}$$

t^* and t define RALS-EG and EG test statistics, respectively, and Z a standard normally distributed random variable. p shows the long-term correlation between the residuals obtained from the equations (7) and (8), respectively. Afterwards, the obtained test statistics are compared with the critical values and it is decided whether there is a long-term relationship between the series.

4. EMPIRICAL RESULTS

The long-term relationship between ISEW variable and renewable energy generation (REN) was investigated using the Residuals Augmented Least Squares Engle Granger (RALS-EG) cointegration test, which was introduced to the literature by Lee et al. (2015). Before proceeding to the cointegration test phase, whether the two variables discussed in the study contain a unit root was examined with standard Augmented Dickey-Fuller (ADF) and RALS-based ADF (RALS-ADF) unit root tests. Both variables have unit roots at levels and become stationary after taking the first difference, as shown in Table 3.

Table 3. ADF and RALS-ADF unit root test results

Variables	ADF Test St.	RALS-ADF Test St.	Rho	Critical Values %1 %5 %10
ISEW	-0.837 (0.940) [0]	-0.575	0.90	-4.526 -3.527 -3.159
Δ*ISEW*	-4.591(0.012) [1]**	-3.243*	0.79	-4.166 -3.423 -3.067
REN	0.703 (0.997) [2]	2.996	0.66	-4.041 -3.316 -2.579
Δ*REN*	-4.388 (0.017)**	-5.150***	0.79	-4.166 -3.423 -3.067

Note: ***, ** and ** denote significance at the 1%, 5% and 10% level, respectively. The values in brackets indicate the probability and the values in square brackets indicate the appropriate lag length.

According to these results, the long-term relationship between the variables with I(1) was investigated with both standard EG and RALS-EG tests, and the results are presented in Table 4 below.

Table 4. EG and RALS-EG cointegration test results

	Test Stat.	Rho	k	Critical Values %1 %5 %10
EG	-4.175***	-		4.32 3.67 3.28
RALS – EG	-3.910**	0.85		-3.917 -3.339 -3.004

Note: ***, ** and ** denote significance at the 1%, 5% and 10% level, respectively.

According to these results, both standard EG and RALS-EG analysis findings show that the series are related in the long term. The estimated long-term model results in order to determine the strength of the long-term relationship between the cointegrated series are presented in Table 5 below.

Table 5. Long-term model estimation results

	Variables	Coeff.	Standard Error	t-Stat.	Probability
Dynamic OLS	*REN*	-0.337	0.062	-5.432	0.000
	C	4440.530	548.682	8.093	0.000
Fully Modified OLS	*REN*	-0.409	0.0489	-8.377	0.000
	C	5198.704	390.068	13.327	0.000

According to the table 5, it is seen that both Dynamic OLS (DOLS) and Fully Modified OLS (FMOLS) results are statistically significant. In addition, it is seen that the change in renewable energy production in the long term negatively affects the ISEW variable both according to dynamic OLS and fully modified OLS estimation results.

5. CONCLUSION

In this study, the long-term relationship between the ISEW variable calculated for Turkey and renewable energy generation was investigated. The stationary analysis of both ISEW and renewable energy generation variables, which are primarily discussed with the data set covering the 2002-2019 periods, was investigated. Since the series are first-order integrated, the long-run relationship between the series was tested with the RALS-EG cointegration analysis proposed by Engle-Granger and Lee et al. (2015). According to the results of the two cointegration analyzes, it is seen that there is a long-term relationship between economic growth and oil prices for Turkey. It is seen that the long-term model estimation results performed afterward are statistically significant and that increasing renewable energy has a reducing effect on the ISEW variable.

REFERENCES

Armiento, M. (2018). The sustainable welfare index: Towards a threshold effect for Italy. *Ecological Economics*, *152*, 296–309. doi:10.1016/j.ecolecon.2018.06.014

Azami, S., & Almasi, S. (2020). Energy consumption and sustainable economic welfare: New evidence of organization of petroleum exporting countries. *International Journal of Energy Economics and Policy*, *10*(5), 31–40. doi:10.32479/ijeep.9435

Beça, P., & Santos, R. (2010). Measuring sustainable welfare: A new approach to the ISEW. *Ecological Economics*, *69*(4), 810–819. doi:10.1016/j.ecolecon.2009.11.031

Butnariu, A., & Luca, F. A. (2019). An index of sustainable economic welfare for Romania. *New Trends and Issues Proceedings on Humanities and Social Sciences*, *6*(8), 1–10. doi:10.18844/prosoc.v6i8.4543

Castaneda, B. E. (1999). An index of sustainable economic welfare (ISEW) for Chile. *Ecological Economics*, *28*(2), 231–244. doi:10.1016/S0921-8009(98)00037-8

Clarke, M. (2004). Widening development prescriptions: Policy implications of an Index of Sustainable Economic Welfare (ISEW) for Thailand. *International Journal of Environment and Sustainable Development*, *3*(3/4), 262–275. doi:10.1504/IJESD.2004.005075

Clarke, M., & Islam, S. M. (2005). Diminishing and negative welfare returns of economic growth: An index of sustainable economic welfare (ISEW) for Thailand. *Ecological Economics*, *54*(1), 81–93. doi:10.1016/j.ecolecon.2004.10.003

Daly, L., & Posner, S. (2012). *Beyond GDP new measures for a new economy*. Dēmos.

Dickey, D. A., & Fuller, W. A. (1979). Distribution of the estimators for autoregressive time series with a unit root. *Journal of the American Statistical Association*, *74*, 427–431.

Dickey, D. A., & Fuller, W. A. (1981). Likelihood ratio statistics for autoregressive time series with a unit root. *Econometrica*, *49*(4), 1057. doi:10.2307/1912517

Engle, R. F., & Granger, C. W. (1987). Co-integration and error correction: Representation, estimation, and testing. *Econometrica*, *55*(2), 251–276. doi:10.2307/1913236

Hansen, B. E. (1995). Rethinking the univariate approach to unit root testing: Using covariates to increase power. *Econometric Theory*, *11*(5), 1148–1171. doi:10.1017/S0266466600009993

Im, K. S., Lee, J., & Tieslau, M. A. (2014). More powerful unit root tests with non-normal errors. *Festschrift in honor of Peter Schmidt* (pp. 315–342). Springer.

Im, K. S., & Schmidt, P. (2008). More efficient estimation under non-normality when higher moments do not depend on the regressors, using residual augmented least squares. *Journal of Econometrics*, *144*(1), 219–233. doi:10.1016/j.jeconom.2008.01.003

Lee, H., Lee, J., & Im, K. (2015). More powerful cointegration tests with non-normal errors. *Studies in Nonlinear Dynamics and Econometrics*, *19*(4), 397–413. doi:10.1515nde-2013-0060

Liu, Y., Zhu, X., & Wang, Y. (2022). Revisiting and evaluation of the index of sustainable economic welfare based on artificial intelligence: Data from 30 Chinese provinces from 2003 to 2019. *Environment, Development and Sustainability*, 1–30. PMID:36618554

Menegaki, A. (2018). *The basic, the solid, the site-specific and the full or total index of Sustainable Economic Welfare (ISEW) for Turkey*. Academic Press.

Menegaki, A. N., & Tsagarakis, K. P. (2015). More indebted than we know? Informing fiscal policy with an index of sustainable welfare for Greece. *Ecological Indicators*, *57*, 159–163. doi:10.1016/j.ecolind.2015.04.037

Menegaki, A. N., & Tugcu, C. T. (2018). Two versions of the Index of Sustainable Economic Welfare (ISEW) in the energy-growth nexus for selected Asian countries. *Sustainable Production and Consumption*, *14*, 21–35. doi:10.1016/j.spc.2017.12.005

Neumayer, E. (1999). The ISEW: Not an index of sustainable economic welfare. *Social Indicators Research*, *48*(1), 77–101. doi:10.1023/A:1006914023227

O'Mahony, T. (2021). Spain's Index of Sustainable Economic Welfare. Are welfare and sustainability compatible? Cuadernos económicos de ICE, 159-183.

Pulselli, F. M., Bravi, M., & Tiezzi, E. (2012). Application and use of the ISEW for assessing the sustainability of a regional system: A case study in Italy. *Journal of Economic Behavior & Organization*, *81*(3), 766–778. doi:10.1016/j.jebo.2010.12.021

Pulselli, F. M., Ciampalini, F., Tiezzi, E., & Zappia, C. (2006). The Index of Sustainable Economic Welfare (ISEW) for a local authority: A case study in Italy. *Ecological Economics*, *60*(1), 271–281. doi:10.1016/j.ecolecon.2005.12.004

Rugani, B., Marvuglia, A., & Pulselli, F. M. (2018). Predicting sustainable economic welfare-analysis and perspectives for Luxembourg based on energy policy scenarios. *Technological Forecasting and Social Change*, *137*, 288–303. doi:10.1016/j.techfore.2018.08.005

Sánchez, M., Toledo, E., Ordóñez, J., & Associates. (2020). The relevance of Index of Sustainable Economic Wellbeing. Case study of Ecuador. Environmental and sustainability Indicators, 6, 100037.

Stockhammer, E., Hochreiter, H., Obermayr, B., & Steiner, K. (1997). The Index of Sustainable Economic Welfare (ISEW) as an alternative to GDP in measuring economic welfare. The results of the Austrian (revised) ISEW calculation 1955-1992. *Ecological Economics*, *21*(1), 19–34. doi:10.1016/S0921-8009(96)00088-2

Chapter 2
Convergence of Energy Efficiency in OECD Countries:
Consideration of Sharp and Smooth Breaks

Muhammed Sehid Gorus
https://orcid.org/0000-0002-7614-4567
Ankara Yıldırım Beyazıt University, Turkey

Erdal Tanas Karagol
Berlin Education Attaché of Turkey, Republic of Türkiye Ministry of National Education, Turkey

ABSTRACT

Studying energy efficiency is becoming a common trend in energy economics research recently. This chapter aims to explore the stochastic convergence of energy efficiency across 27 OECD countries from 1971 to 2018. For this aim, this chapter calculates the energy efficiency index based on the Fisher's Ideal index. Then, this chapter utilizes the panel stationarity test with sharp and smooth breaks introduced by Bahmani-Oskooee et al. There are two main advantages of this methodology over the conventional panel stationarity tests. First, it considers the cross-sectional dependency in the series. Second, the test combines both sharp and smooth structural breaks in the series. The empirical results of this study state that the energy efficiency index does not converge to the OECD average for 1971-2018. This result shows that the dissemination of the technology is limited across OECD countries. The empirical result of this study provides significant policy implications for energy and environmental issues.

DOI: 10.4018/978-1-6684-6727-5.ch002

INTRODUCTION

For decades, scientists and governments have been interested in the energy efficiency concept since economies may use less energy to generate the same quantity of goods and services when energy efficiency improves. Also, international initiatives have shown a particular interest in the energy efficiency issue. The United Nations' Sustainable Development Goals of Agenda 2030, for example, aim to double the worldwide rate of increase in energy efficiency by 2030 (United Nations, 2015).

Practitioners have analyzed the convergence behavior of energy variables such as energy consumption, energy productivity, and energy intensity for a few decades. However, most of the studies have failed to examine the convergence of energy efficiency. Investigating the convergence behavior of the energy efficiency index can provide some policy implications to policymakers. They can determine whether common (or differentiated) energy policies are proper for their countries, or not. Also, the convergent behavior of the energy efficiency index across OECD countries indicates that common energy policies are efficient and there is no need for reexamination of current policies (Yilanci et al., 2022), while the non-convergence behavior reveals that policymakers should design differentiated energy policies. In countries that have lower energy efficiency, policymakers should direct economic agents toward energy-saving innovations, while in countries that have higher energy efficiency, they should lead international technological cooperation (Yan et al., 2017). So, governments can design and implement proper energy policies based on the convergence results.

There are only a few studies considering the convergence of energy efficiency in the empirical literature; Menegaki and Tiwari (2019) for 35 European countries during 1995-2014 and Qi et al. (2019) for 60 Belt and Road Initiative (BRI) countries during 1995-2015. As seen, the examined time spans in these studies are very short. We believe that considering longer datasets can provide more reliable results regarding the convergence pattern of the series. Also, this is the first study that explores the convergence pattern of the energy efficiency index for OECD countries. Besides, the methods examined in the above studies fail to detect smooth structural breaks; Menegaki and Tiwari's (2019) study takes into account only sharp breaks. However, Leybourne et al. (1998) argued that series do not always display sharp breaks; structural shifts can take the form of gradual and smooth patterns in some cases.

In this study, there are two primary aims: a) calculating the energy efficiency index via the decomposition method for 27 OECD countries covering 1971-2018, and b) detecting the stochastic convergence behavior of the energy efficiency index through recent panel data techniques that consider both sharp and smooth breaks.

In this study, we chose the abovementioned economies as a sample because most of them lack rich energy resources. Thus, it would be rational to investigate their energy efficiency index levels and determine their convergence behavior.

This study employs Fisher's Ideal index to construct the energy efficiency index for the period examined. Since Fisher's Ideal index provides perfect decomposition and zero-value robustness, we employ this methodology in the study. Then, this study investigates the stochastic convergence behavior of the energy efficiency index through the panel stationarity test with sharp and smooth breaks introduced by Bahmani-Oskooee et al. (2014). The panel stationarity test with sharp and smooth breaks has two fundamental advantages compared to the conventional panel unit root tests such as LLC (Levin-Lin-Chu), IPS (Im-Pesaran-Shin), and Fisher-type tests. First, the methodology considers the cross-sectional dependency in the series; therefore, the empirical findings become more reliable. Second, the test combines both sharp and smooth structural breaks in the series.

This investigation is composed of six main sections. Section 2 overviews the energy structure in OECD countries, while Section 3 summarizes the empirical literature on the convergence of energy intensity, energy productivity, and energy efficiency, respectively. Section 4 introduces the dataset and the empirical methods utilized. Section 5 presents and discusses the empirical results, while Section 6 concludes this study.

ENERGY STRUCTURE IN OECD COUNTRIES

Energy resources have a great impact on the economic growth process of countries since it is one of the most significant factors of production. Especially in high- and upper-middle-income countries, energy consumption levels are higher than in low-income unindustrialized countries. For instance, the US primary energy consumption level was 94.65 exajoules in 2019, which was an enormous value after China. However, the high level of energy consumption in China can be attributed to its population level. Figure 1 exhibits that the share of primary energy consumption in OECD countries was 40% of the global level. Also, it can be stated that the European Union countries' share was around 11.8%. So, we can say that OECD countries have a great role in the energy markets in the world. In detail, we can also focus on the countries' per capita energy consumption level to eliminate the population effect. According to the British Petroleum (BP) data, Iceland, Canada, Norway, the US, Australia, and South Korea had very high per capita primary energy consumption in 2019 compared to developing countries except oil-rich countries.

Figure 1. Total primary energy consumption (exajoules) in OECD countries, 2019

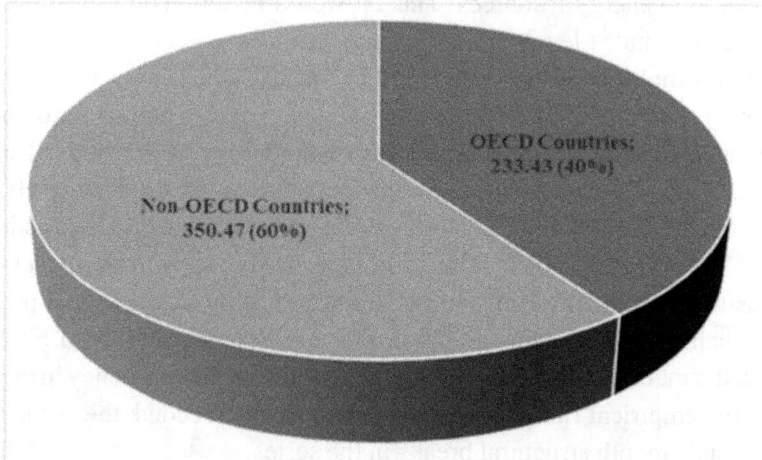

This study has revealed that OECD countries consume too many energy resources because of their industrialized economies throughout a long period. According to British Petroleum (2020) data, fossil fuels (oil, natural gas, and coal) still have the lion's share as 80% of primary energy consumption in this set of countries, while renewables (hydroelectric and other renewables) have only a 12% share. Among them, some economies are heavily dependent on fossil fuels, such as Poland (93.3%), Netherlands (92.5%), Mexico (91.5%), Australia (91.4%), Japan (87.4%), and Greece (87.3%). Besides, some developed countries meet their energy demand through nuclear energy, including France (36.8%), Sweden (26.1%), Finland (18.6%), and the Czech Republic (15.8%). Moreover, some countries diversified their energy sources through the advancement in renewable energy technologies, such as Sweden (42.2%), Austria (33.7%), Switzerland (31.3%), Canada (27.6%), Finland (26.1%), Portugal (24.5%), and Chile (22.9%).

So far, this study has emphasized the importance of energy sources and fossil fuels in OECD countries. However, it can be stated that most OECD countries lack energy sources except for North American countries (especially Canada and the US regarding oil), the US regarding natural gas, and the US, Australia, and Germany regarding coal. The proven oil and natural gas reserves in these countries were constant from 1999 to 2019, while the reserves for non-OECD countries increased in this period. On the one hand, at the end of 2019, the proved oil reserves were 260.1 thousand million barrels in OECD countries, while the remaining countries had 1473.7. In other words, only 15% of the proved oil reserves belong to OECD members. On the other hand, the proven natural gas reserves were estimated as 20.1 trillion cubic meters (10.1% of the total) in 2019 for OECD countries, while others had 178.7.

In addition to the above energy resources, OECD countries are rich in coal resources. They had 501,196 million tonnes of coal reserves in 2019, while non-OECD countries had 568,440 million tonnes. So, we can say that their reserve levels are quite similar; 46.9% and 53.1%, respectively. Besides, the reserves-to-production ratio in OECD countries was 308, while it was 88 in the remaining countries. In other words, the remaining reserves would last 308 years if the production rate would not change. The reserves-to-production ratio in OECD countries was 3.5 times higher than in non-OECD countries in 2019. It can be attributed to the environmental concerns of developed countries. Since the ecological damage of coal is very high, OECD members limit their coal production levels.

Since most OECD countries lack energy resources like oil, natural gas, and coal, they need to increase their energy efficiency through energy-saving technologies. The following part reviews the empirical literature related to energy intensity, energy efficiency, and energy productivity.

LITERATURE REVIEW

Recently, considerable literature has grown up around the theme of the convergence of energy variables such as aggregated or disaggregated levels of energy consumption including final energy, clean energy, or renewable energy. A detailed literature review considering these variables is provided by many researchers (see Yilanci and Gorus (2021)). However, in this study, we focus on the literature considering the convergence of energy intensity, energy productivity, and energy efficiency since these variables have received considerable scholarly attention in the last two-three decades.

Of the preliminary studies regarding examining the convergent behavior of energy intensity of the countries, Nilsson (1993) conducted a basic graphical analysis for 31 countries for the period 1950-1988. His empirical findings exhibited that energy intensity in most countries converges. Besides, Mielnik and Goldemberg (2000) utilized annual energy intensity data covering 41 countries for 1971-1992 and found that both developing and developed countries' energy intensity data converged to a common value. Markandya et al. (2006) aimed to test whether transition countries' energy intensity converges to European Union's average or not. Their empirical findings considering the beta-convergence backed up the validity of the convergence hypothesis of energy intensity for the period examined. In addition, the empirical results derived from Ezcurra (2007) demonstrated that energy intensity converged across 98 countries for the period 1971-2001.

Le Pen and Sévi (2010) rejected the validity of the convergence hypothesis through various time-series methods for the global setting (covering 97 countries); however, it was suggested that some of the region's energy intensity converge for the period 1971-2003. Liddle (2010) examined the convergence pattern of energy intensity for more than 100 countries through beta, sigma, and gamma-convergence. The empirical findings mainly supported the convergence behavior of the series for the periods examined. Also, Herreiras' (2012) findings based on the weighted distribution dynamics approach supported the validity of the convergence hypothesis across 83 countries. Also, Mulder and Groot's (2012) results showed a convergent behavior of energy intensity in OECD countries only after 1995.

Besides, Kiran (2013) utilized the fractional cointegration method to find the convergent pattern of energy intensity among OECD countries. The author found that the convergence of energy intensity is supported only for 14 out of 21 countries for the period 1980-2010. Apart from these studies, Yu et al. (2015) utilized the club convergence analysis to detect the convergent behavior of energy intensity across countries. Their dataset consisted of 109 countries for the period spanning from 1971 to 2010. They found that the series does not converge to a common value; instead, it was determined that there are four main convergence clubs. In detail, Club 1 included three countries, while Club 2 covered 18 countries. In addition, Club 3 and Club 4 consisted of 47 and 34 countries, respectively, while seven countries' data had non-convergent behavior for the period examined.

Burnett and Madariaga (2016) found the energy intensity converges across fifty US states and the Columbia district through the panel GMM during the period 1970-2013. Besides, Canel et al. (2017) conducted the Harvey linearity test to determine whether series have linear or nonlinear behavior for OECD countries. Then, they employed linear or nonlinear unit root tests according to the data properties. They found that the majority of the series does not converge to the OECD average for the period examined. Karimu et al. (2017) used the Fisher Ideal index to calculate energy intensity, energy efficiency, and activity indices for the Swedish economy covering the period 1990-2008. After, they found that energy intensity converges across the industrial sectors. Moreover, Bulut and Durusu-Ciftci (2018) conducted several unit root tests to detect the stochastic convergence behavior of energy intensity over 27 OECD countries covering 1980-2014. According to the findings, sharp and smooth breaks affected the unit root properties of the series significantly. The empirical results show that the convergence hypothesis was valid for 14 out of 27 countries for the period examined.

Recently, Santiago et al. (2020) found four convergence clubs regarding the energy intensity of 21 Latin American and Caribbean countries for the period spanning from 1970 to 2014. Similarly, Zhu and Lin (2020) utilized the club convergence analysis to determine the convergence of city-level energy intensity across 193

Chinese cities from 2005 to 2016. According to the empirical findings, there were four convergence clubs. Lastly, Shabani and Shahnazi (2021) employed the Markov chain method to check the beta-convergence pattern of the energy intensity in the Iranian provinces for the period spanning from 2002 to 2016. The authors found that the club convergence is supported for total energy intensity, natural gas intensity, electricity intensity, and oil consumption intensity.

Several studies investigate the convergence path of energy productivity in the empirical literature. Of the studies, Adhikari and Chen (2014) used spatial panel data analysis to determine the convergent pattern of energy productivity across 35 Asian countries for the period 1993-2010. The empirical results suggested a shred of weak evidence for sigma-convergence, while there was a convergent behavior of energy productivity according to the beta-convergence results. Besides, Apergis and Christou (2016) found no full convergence of energy productivity for 31 selected countries in the period 1972-2012. However, their methodology employed, the club convergence test, indicated that six convergence clubs converge different equilibria. It was also revealed that Iceland, Malta, and the US's data did not converge any equilibria.

Bhattacharya et al. (2018) stated that energy productivity does not display a stochastically convergent behavior (for most states) across Indian states and territories during 1988-2016. However, they also found that there are two convergence and one non-convergence club for the period examined. A similar analysis was conducted by Bhattacharya et al.'s (2020), and their empirical results showed that energy productivity in Australian states and territories converges to more than one equilibrium from 1990 to 2015.

To our knowledge, to date, only a limited number of empirical papers have been identified that test the validity of the convergence hypothesis of energy efficiency. Menegaki and Tiwari (2019) investigated the convergence behaviors of the final energy efficiency index, primary energy efficiency index, and energy productivity index across European countries from 1995 to 2014. For this purpose, they utilized a bunch of unit root tests (including time-series and panel data techniques) such as ADF, CADF, KSS (Kapetanios-Shin-Shell), Hadri-Rao, LM with one-break, LM with two-break, and club convergence analysis. Mainly, KSS and LM tests supported the convergent behavior of the variables examined. In addition, according to the club convergence analysis, there was only one convergent club in the European countries regarding energy productivity, and it consisted of only four countries. Besides, Qi et al. (2019) calculated the energy efficiency through the data envelopment analysis for 60 Belt and Road Initiative countries covering 1995-2015. The empirical results provided evidence for the convergent behavior of energy efficiency.

DATA AND METHODOLOGY

This paper constructs the energy efficiency index by following Metcalf (2008), Nillesen et al. (2013), and Gorus and Karagol (2022). Then, we investigate the convergence pattern of the energy efficiency index for OECD countries. For that purpose, we employ the panel stationarity test with sharp and smooth breaks developed by Bahmani-Oskooee et al. (2014) (hereafter, BCW). We use an annual dataset for 27 OECD member countries from 1971 to 2018 in this article. The economic activity measures used in this study are for four major sectors: residential, services, industrial, and transportation. The economic activity data are deflated and are from the United Nations database, while the sectoral energy consumption data come from the International Energy Agency.

The unit root and stationarity tests are beneficial techniques to determine the stochastic convergence behavior of the series in the empirical analyses. There are a bunch of conventional tests for econometric analyses in the empirical literature; ADF, PP (Phillips–Perron), KPSS (Kwiatkowski–Phillips–Schmidt–Shin) tests for the time-series analysis, and LLC, IPS, and Fisher-type (ADF and PP) tests for the panel data analysis. With the advancements in econometric techniques, the unit root tests that consider structural changes in the data have come into prominence for a few decades. For instance, Zivot and Andrews (1992) introduced a unit root test that allows one structural break, while Lumsdaine and Papell (1997) proposed a test that allows two structural breaks for the time series analysis. Besides, Carrion-i-Silvestre et al. (2005) (hereafter, CBL) developed a stationarity test that allows multiple structural shifts in the series for panel data.

Leybourne et al. (1998) argued that series do not always display sharp breaks; they asserted that structural shifts, in some cases, take the form of gradual and smooth patterns. For this reason, researchers have developed unit root or stationarity tests to capture gradual shifts using the Fourier approximation, see Becker et al. (2006), Enders and Lee (2012), and Rodrigues and Taylor (2012). Recently, Bahmani-Oskooee et al. (2014) extended the CBL stationarity test by incorporating the Fourier approximation. The methodology has a few fundamental advantages over conventional tests. First, the BCW test takes into consideration of the cross-sectional dependency in the series. Second, it combines both sharp and smooth structural changes in the data. Third, it provides country-specific stationarity test results based on the optimal number of breaks for each member (Yilanci et al., 2019). For these reasons, we conduct the BCW stationarity test to examine the convergent behavior of the energy efficiency index data for OECD countries.

The BCW stationarity test is based on the following specification (Yilanci et al., 2020):

$$y_{i,t} = \alpha_i + \sum_{t=1}^{m_i} \theta_{i,k} DU_{i,k,t} + \gamma_{1,i} \sin\left(\frac{2\pi k_i t}{T}\right) + \gamma_{2,i} \cos\left(\frac{2\pi k_i t}{T}\right) + \varepsilon_{i,t} \tag{1}$$

In the above equation, m_i denotes the number of sharp breaks, while DU denotes the dummy variable which can be represented as

$$DU_{i,k,t} = \begin{cases} 1 \; if & TB^i_{k-1} < t < TB^i_k \\ 0, & otherwise \end{cases}.$$

In this equation, TB^i_k shows the kth break date(s) for the ith panel member. In addition, the trigonometric terms are added to the equation to capture smooth structural shifts. In order, k, t, and T denote the number of frequency, time trend, and sample size. The null hypothesis of this test is *all members in the panel are stationary*, while the alternative is *there is a unit root in the panel* (Bahmani-Oskooee et al., 2014; Yilanci et al., 2019; Yilanci et al., 2020)[1].

EMPIRICAL RESULTS AND DISCUSSION

Energy Efficiency in OECD Countries

This paper presents the results of the decomposed energy indexes for OECD countries from 1971 to 2018 in Figure 2. According to the figure, it is seen that energy efficiency plays a significant role in the improvement of energy intensity in these countries.

Since energy efficiency is determined as the main driver for the improvements in energy intensity, we mainly focus on this index for the remaining parts of this study. Figure 3 exhibits the evolution of the energy efficiency index across different periods for each OECD country examined in this study. According to the graphs, all the countries benefit from energy-saving and energy-efficient technologies except for Portugal and Spain. In addition, this advancement in energy efficiency can also be attributed to the public consciousness and management capabilities on environmental issues.

In detail, it is seen that there was a significant decline in the energy efficiency index for several countries—Ireland (0.242), Sweden (0.361), the UK (0.352), and the US (0.305)—at the end of the period examined. Their index levels were below 0.40, and Ireland showed a remarkable performance among them with a 0.242 value. This means that the index value was 24.2% of its 1971 level. Besides, it can be expressed that if the structure of the Irish economy were fixed to its 1971 level, the energy intensity in Ireland would have been 24.2% of its base year value. In 2018, 15 countries' efficiency index was below the OECD average (0.597), while the remaining 12 countries' index value was higher than the average.

Figure 2. Time series plots of energy indexes (1971=1.00) of OECD average, 1971-2018

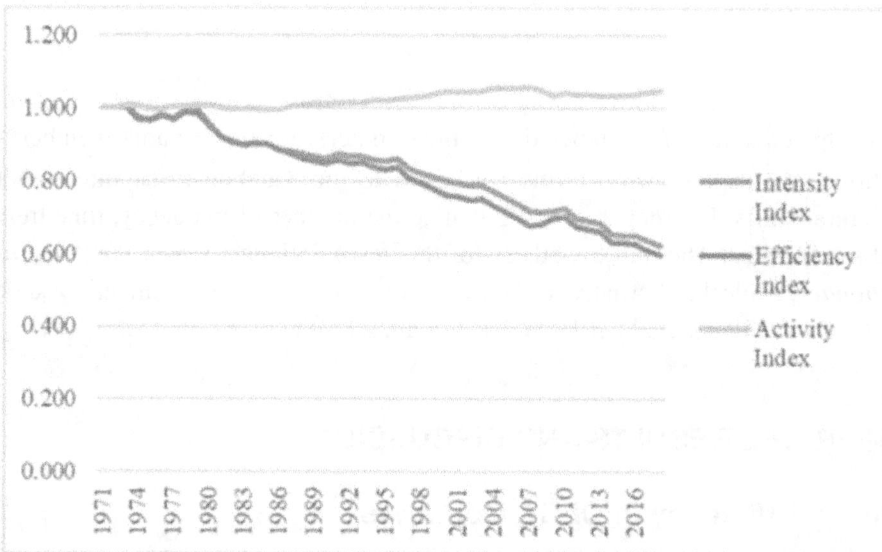

One can see that there are structural shifts in the series in Figure 3. There can be many reasons behind these changes. The energy efficiency index can be affected by international and domestic oil price shocks. So, an increase in energy prices can cause economic agents to increase energy efficiency (Metcalf, 2008). Also, Kilian (2015) argues that unexpected changes in energy prices force economic agents to use energy more efficiently by eliminating energy-inefficient equipment. Besides, the economic crises might affect the energy efficiency performance of the countries because their economic structure can be changed, and sectoral shifts can occur within the economy.

The United Nations, in particular, takes a keen interest in countries' efforts to enhance their energy efficiency. Goal 7.3 of the Sustainable Development Goals (SDGs) of the Agenda 2030 calls for doubling the worldwide rate of increase in energy efficiency by 2030 (United Nations, 2015). It is important to note that our energy efficiency calculations differ from those of the United Nations. In the context of this study, we measure energy efficiency through the Fisher Ideal index rather than the decrease in the ratio of global total primary energy supply per unit of gross domestic product. According to our calculations, the average growth rate of the energy efficiency index in OECD countries was -0.931 percent between 1990 and 2010, while the individual performances differ. Based on our estimates (considering the Exponential Smoothing, ETS, method), we believe that the target of doubling the energy efficiency index by 2030 would not be achieved by many OECD countries examined.

Figure 3. Plots of energy efficiency index across OECD countries, 1971-2018

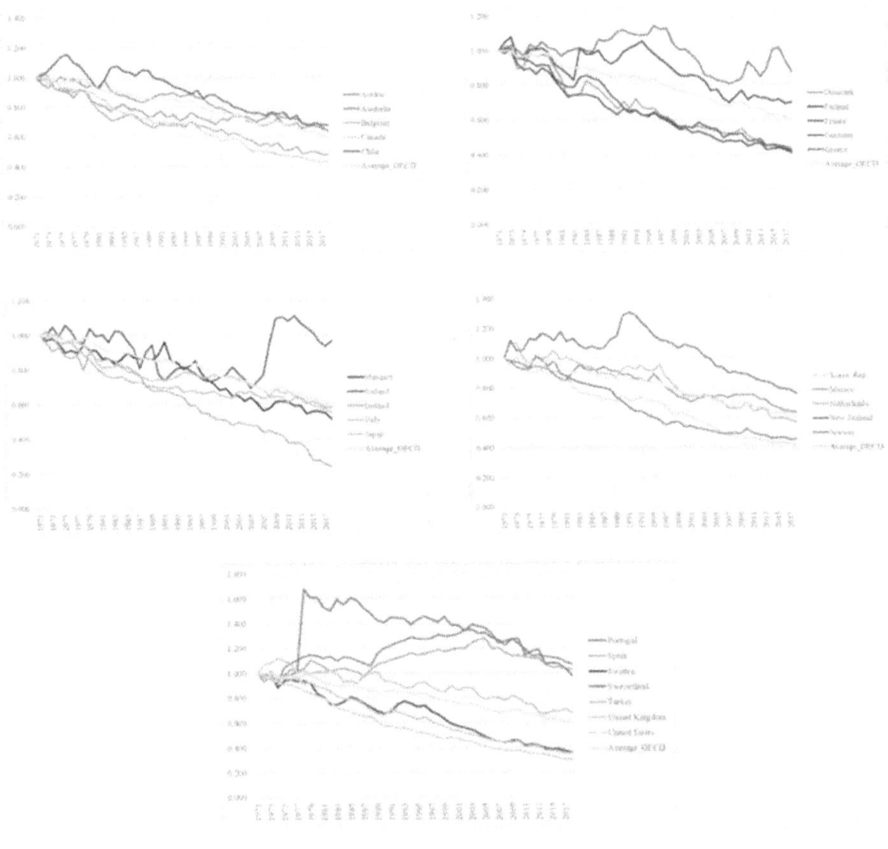

The forecast results exhibit that only Austria, Belgium, Denmark, Iceland, Ireland, Italy, Japan, Korea Republic, Netherlands, Portugal, Spain, Sweden, and Switzerland will reach their own 2030 targets. However, it is seen that the efficiency index level would be very high in Iceland, Portugal, and Spain compared to other members. Besides, it is expected that Canada, France, Germany, and New Zealand's performances considering energy efficiency will be remarkable. The rest of the countries (Australia, Chile, Finland, Greece, Hungary, Mexico, Norway, Turkey, the United Kingdom, and the United States) will mostly be far from their energy efficiency targets. Therefore, one can say these countries have to increase their energy efficiency levels significantly in the following years gradually.

In addition to the SDGs, governments have agreed to keep the increase in global average temperature to well below 2°C above pre-industrial levels in the Paris Agreement. For this purpose, countries try to decrease their greenhouse gas levels (European Commission 2021a) and improve their energy performances. For instance, according to the 2030 climate and energy framework, the European Union countries aim to improve energy efficiency by at least 32.5% in 2030 compared to 1990 levels (European Commission 2021b). According to the forecast results, it is expected that Belgium, Denmark, Finland, France, Germany, Hungary, Ireland, Netherlands, and Sweden will achieve this target, while Austria, Greece, Italy, Portugal, and Spain will not. Besides, the United Kingdom which ceased to be an EU member recently will show a successful performance; it will decrease its energy efficiency index from 0.663 in 1990 to 0.250 in 2030.

As discussed above, there have been significant improvements in energy efficiency across most of the OECD countries for the period examined. However, this study fails to examine the impact of energy efficiency advancements on the per capita energy consumption level. An increase in energy efficiency may not reduce energy consumption to the desired level, in some cases. One of the main reasons behind this issue is that these advancements may cause to decline in energy prices, and a decrease in prices can lead to an increase in the level of energy demanded. This increase in demand may offset the energy efficiency gains partially (Greening et al., 2000). This reaction of the economic agents to the increase in energy efficiency is called the *Rebound Effect* in the literature. This study does not engage with this issue (out of scope), and it can be accepted as one of the limitations of this study.

The Convergence of Energy Efficiency Index in OECD Countries

This paper also aims to investigate the convergence behavior of the energy efficiency index across OECD members covering 1971-2018. For this aim, we use the data which we calculated through the Fisher Ideal index in the previous subsection.

This study conducts the stochastic convergence analysis to find out whether the energy efficiency index of the countries converges to the OECD average or not. Therefore, we form and test the stationarity of the following series for each country:

$$eff_{i,t} = \ln\left(EFF_{i,t} \,/\, \overline{EFF_t}\right) \tag{2}$$

where $EFF_{i,t}$ and $\overline{EFF_t}$ denote the energy efficiency index for country i and its group average for OECD countries, respectively.

Next, this study conducts the first-generation panel unit root tests and cross-sectional dependency tests for the variables employed. The empirical results are displayed in Table 1. According to the empirical results, one can say that there is evidence for CD across the countries employed. Therefore, we should employ the second-generation panel unit root test results. Besides, we also report the first-generation test results below to enrich the study. According to the findings, it can be stated that the LLC test supports the validity of the convergence hypothesis for both models, while the Breitung and the IPS tests are in favor of non-stationarity. Besides, the remaining tests provide mixed results regarding the stationarity properties of the series. However, as CD test results, we should ignore the first-generation panel unit root test results because they do not take into account cross-sectional dependency in the series.

Table 1. First-generation panel unit root and CD tests results

Variable	Breusch-Pagan LM Test Statistics	Pesaran Scaled LM Test Statistics	Bias-Corrected Scaled LM Test Statistics	Pesaran CD Test Statistics
$ln(eff_{i,t})$	**5934.1300*** **	**210.7217*** **	**210.4344*** **	**-1.3860**
Methods	With Intercept Model		With Intercept and Trend Model	
	Test Statistics	Probability	Test Statistics	Probability
LLC	**-3.1506*** **	**0.0008**	**-2.3283*** **	**0.0099**
Breitung	—	—	**-0.8300**	**0.2033**
IPS	**-0.7356**	**0.2310**	**-1.1076**	**0.1340**
Fisher - ADF	**63.5255**	**0.1759**	**68.905* **	**0.0833**
Fisher - PP	**86.0411** **	**0.0036**	**66.7929**	**0.1135**

Note: ***, **, and * denote statistical significance at the 1%, 5%, and 10%, respectively.

Since we reject the null of no cross-sectional dependency in the series, this study employs the panel stationarity test with sharp and smooth breaks developed by Bahmani-Oskooee et al. (2014), the BCW test. This method has a few distinct advantages over the first-generation panel unit root tests. First, the BCW test considers the cross-sectional dependency among panel members. Second, it incorporates both sharp and smooth structural shifts in the series. Third, this methodology also provides country-specific stationarity test results in addition to the panel results. Thus, this study utilizes the BCW stationarity test to examine the convergent behavior of the energy efficiency index data for OECD countries. The empirical results are presented in Table 2 and Table 3.

Table 2 shows that, according to the panel results, the stochastic convergence hypothesis is valid in OECD countries, considering the significance level at 95%. However, the country-specific results differ from the panel results. On the one hand, the empirical findings exhibit that the series shows convergent behavior in 10 out of 27 countries; Austria, Belgium, Canada, Greece, Hungary, Iceland, Mexico, Netherlands, Spain, and the United Kingdom. On the other hand, it has non-convergent behavior in 17 countries, namely, Australia, Chile, Denmark, Finland, France, Germany, Ireland, Italy, Japan, South Korea (Korea, Rep.), New Zealand, Norway, Portugal, Sweden, Switzerland, Turkey, and the United States. In addition to the above findings, the stochastic convergence hypothesis is confirmed for seven countries for the significance level at 90%, while it is valid for 16 out of 27 OECD members for the significance level at 99%. Overall, it can be said that the common energy and environment policies have a significant impact on some of the OECD countries' energy efficiency advancements. However, the remaining countries need to develop country-specific policies in addition to the joint policies (see non-convergent countries) since each country has different geographical and political characteristics even though they share similar economic conditions.

Furthermore, these findings also show that the dissemination of technology regarding energy efficiency is limited across OECD countries. If technological diffusion was fully achieved between these countries, the difference between their energy efficiency levels would decrease through joint research or technology transfers, and they would converge to a common value. Overall, we can say that the non-convergence of the energy efficiency index is valid for the majority of the countries. Thus, policymakers in countries that have lower energy efficiency should direct economic agents to energy-saving innovations, while policymakers in countries that have higher energy efficiency should lead international technological cooperation (Yan et al., 2017). According to the authors, the most significant factors behind the non-convergence in technology (in this study, energy efficiency) are the low level of a) investment in innovation, b) human capital, and c) financial development.

Our findings are consistent with the claim of Yu et al. (2015); 'one size fits all' energy policies are improper. Therefore, governments should develop differentiated energy policies, but policymakers should also consider particular common energy policies since several countries benefit from them. Besides, the non-convergent behavior of energy efficiency across most of the countries provides significant insight into the future prospects of energy efficiency presented in the SDGs (Goal 7.3). In the previous section, we predicted the energy efficiency levels of OECD countries up to 2030 and found that it is hard to achieve targets of Goal 7.3. However, the stochastic convergence test results stress that countries not only should develop country-specific policies but also cooperate with each other to reach further improvements in energy efficiency. These actions may help them to reach the SDGs' target.

Table 2. Panel stationarity test with sharp and smooth breaks results

Stationarity Test Results	Bartlett	Critical Values		
		90%	95%	99%
Panel Results				
Homogenous long-run variance	**-0.1907**	-0.9467	-0.6511	-0.0614
Heterogeneous long-run variance	**0.0743**	0.1352	0.4656	1.1591
Univariate Results				
Australia	0.0518	0.0250	0.0288	0.0365
Austria	**0.0402**	0.0622	0.0785	0.1136
Belgium	**0.0282**	0.0523	0.0666	0.0958
Canada	**0.0510**	0.0433	0.0516	0.0700
Chile	0.0258	0.0200	0.0222	0.0268
Denmark	0.0282	0.0178	0.0205	0.0276
Finland	0.0519	0.0249	0.0284	0.0363
France	0.0707	0.0478	0.0571	0.0810
Germany	0.0549	0.0297	0.0352	0.0479
Greece	**0.0250**	0.0559	0.0719	0.1045
Hungary	**0.0428**	0.0502	0.0623	0.0894
Iceland	**0.0564**	0.1304	0.1662	0.2404
Ireland	0.0293	0.0227	0.0254	0.0301
Italy	0.0574	0.0222	0.0250	0.0297
Japan	0.0827	0.0298	0.0363	0.0519
Korea, Rep.	0.0599	0.0286	0.0334	0.0450
Mexico	**0.0200**	0.0190	0.0211	0.0247
Netherlands	**0.0756**	0.1625	0.2146	0.3292
New Zealand	0.0465	0.0334	0.0396	0.0560
Norway	0.0382	0.0258	0.0293	0.0353
Portugal	0.0278	0.0227	0.0253	0.0309
Spain	**0.0396**	0.0366	0.0447	0.0610
Sweden	0.0455	0.0202	0.0231	0.0292

continues on following page

Table 2. Continued

Stationarity Test Results	Bartlett	Critical Values		
		90%	95%	99%
Switzerland	0.0513	0.0342	0.0412	0.0571
Turkey	0.0425	0.0168	0.0193	0.0240
United Kingdom	**0.0235**	0.0324	0.0391	0.0553
United States	0.0450	0.0351	0.0423	0.0572

Note: The bold values show the validity of the convergence hypothesis. We take into consideration the critical values for 95%.

Table 3 provides optimal frequencies, F statistics for Fourier approximation, and structural break dates for each OECD member. It is seen that the trigonometric terms are statistically significant for all the countries except for Belgium and Greece for the period examined. Also, we find that the number of optimal frequencies varies across countries, from one to five. In addition, our methodology detects multiple structural breaks (up to five) for each country. It is seen that there are both sharp and smooth shifts in the series.

There are two main reasons behind these structural shifts. First, the oil price shocks affect the energy efficiency index significantly. The traditional approach to oil price shocks focuses on the production side of energy resources. According to this view, oil price shocks are caused by disruptions in oil production, and they are attributed to the political conflicts in the oil-producing countries, especially in the Organization of Petroleum Exporting Countries (OPEC) (e.g., Iranian Revolution in 1978-1979, the Iran-Iraq War in 1980-1988, the Gulf War in 1990-1991, the Venezuelan Civil Unrest in 2002-2003) (Kilian, 2008). However, it is recognized that the traditional interpretation of oil price shocks cannot explain the causes of the recent oil price movements. Therefore, a modern understanding of oil price shocks is developed recently. This approach argues that the shifts in the oil demand (the flow demand and the stock demand) are the most significant factor affecting oil prices (Kilian, 2015; Baumeister and Kilian, 2016). Second, economic crises can affect the energy efficiency performance of the economies since their economic structure can be changed, and sectoral shifts can occur within the economy. One can see that the detected break dates, based on our empirical results, are in line with the prior expectations regarding oil price shocks and economic fluctuations.

Figure 4 displays the plots of the energy efficiency index of OECD countries relative to the mean and fitted nonlinearities, based on the BCW test results. The blue lines show the actual series, while the red lines exhibit the predicted series. It is seen that there are several fluctuations in the series due to domestic or international shocks. We can say that our methodology captures most of the sharp and smooth breaks for each country.

Table 3. Results for optimum frequency and break dates

Countries	Optimal Frequency	F Statistics	Critical Values			Break Dates
			90%	95%	99%	
Australia	1	32.2321***	2.7481	3.3247	5.6587	1974, 1978, 1982, 2004
Austria	4	4.5341***	2.5876	3.2404	4.3694	1983, 1995, 2000
Belgium	4	2.6036	2.7345	3.0608	4.6186	1982, 1987, 1995, 2010
Canada	2	32.0666***	2.0989	2.9141	3.9155	1979, 2001
Chile	2	8.4967***	2.5819	3.0529	5.2877	1974, 1981, 1989, 2002
Denmark	1	19.4797***	2.6937	3.1344	5.4917	1975, 1981, 1985, 1992, 2012
Finland	5	5.6300**	2.8221	3.7733	6.2105	1976, 1983, 1993, 2007
France	4	4.5831**	2.8309	3.7171	4.6235	1981, 1987, 2011
Germany	2	4.4838**	2.6656	3.1777	4.6186	1983, 1987, 1992, 2002
Greece	3	2.2032	2.6552	3.2873	4.7692	1986, 1998, 2003, 2010, 2014
Hungary	4	10.9969***	2.6328	3.4391	4.1990	1982, 1991, 2006, 2010
Iceland	5	13.8916***	2.4979	3.1785	5.8775	1989, 2004, 2009
Ireland	1	17.5975***	2.6038	3.9070	5.5426	1978, 1987, 2006, 2014
Italy	1	45.7462***	2.6394	3.6250	6.2328	1974, 1987, 1996, 2002, 2011
Japan	4	24.4740***	3.3566	4.7623	7.7029	1979, 1984, 1989, 1995, 2007
Korea, Rep.	3	11.6338***	3.0263	3.6908	4.8678	1977, 1981, 1999, 2008
Mexico	3	20.1521***	2.7956	3.5704	5.4088	1974, 1981, 1994, 2000, 2010
Netherlands	5	11.2789***	2.4431	3.2167	5.0618	1995, 2000, 2006
New Zealand	1	13.1575***	2.4522	3.5661	5.7479	1985, 1989, 1999
Norway	2	11.5243***	2.3328	3.6721	5.6049	1975, 1988, 1992, 1997, 2011
Portugal	2	39.0225***	2.4976	3.2851	5.4451	1974, 1988, 1994, 2002, 2013
Spain	3	25.1353***	2.3808	3.6569	4.7020	1978, 1996, 2004, 2008
Sweden	1	49.7212***	2.6044	3.4611	5.4006	1974, 1981, 1986, 1991, 1996
Switzerland	3	29.7146***	2.3025	2.9727	3.6314	1977, 1982, 2003, 2010
Turkey	5	8.6667***	3.0087	3.7081	5.2700	1974, 1979, 1983, 1989, 1999
United Kingdom	1	5.4284***	2.6796	3.2748	4.6825	1977, 1981, 1990
United States	2	5.9489***	3.0809	3.7580	5.0029	1982, 1989, 1999

Note: *** and ** denote statistical significance at 1% and 5%, respectively. The number of bootstrap replications equals 10,000, while k_{max}=5 and m_{max}=5.

Figure 4. Plots of energy efficiency in OECD countries relative to the mean and fitted nonlinearities

Figure 5.

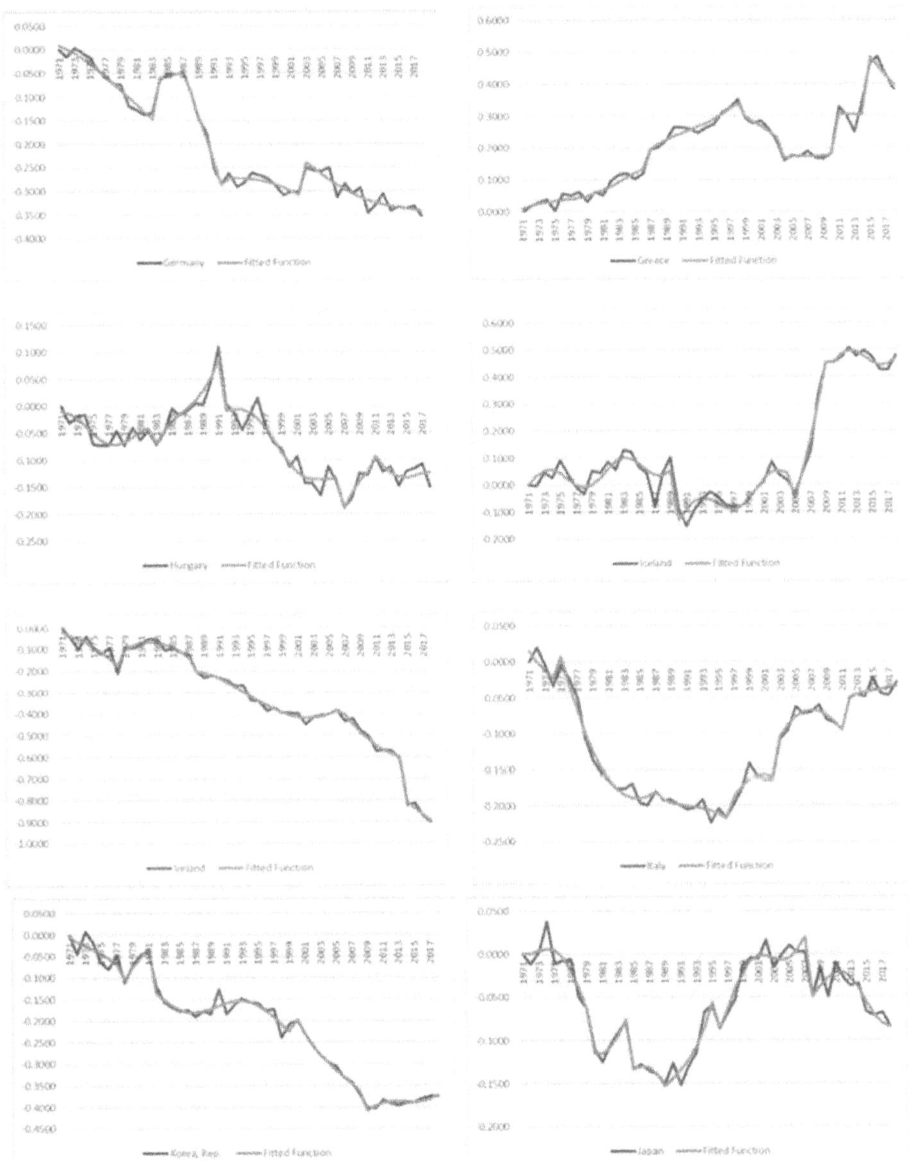

Figure 6.

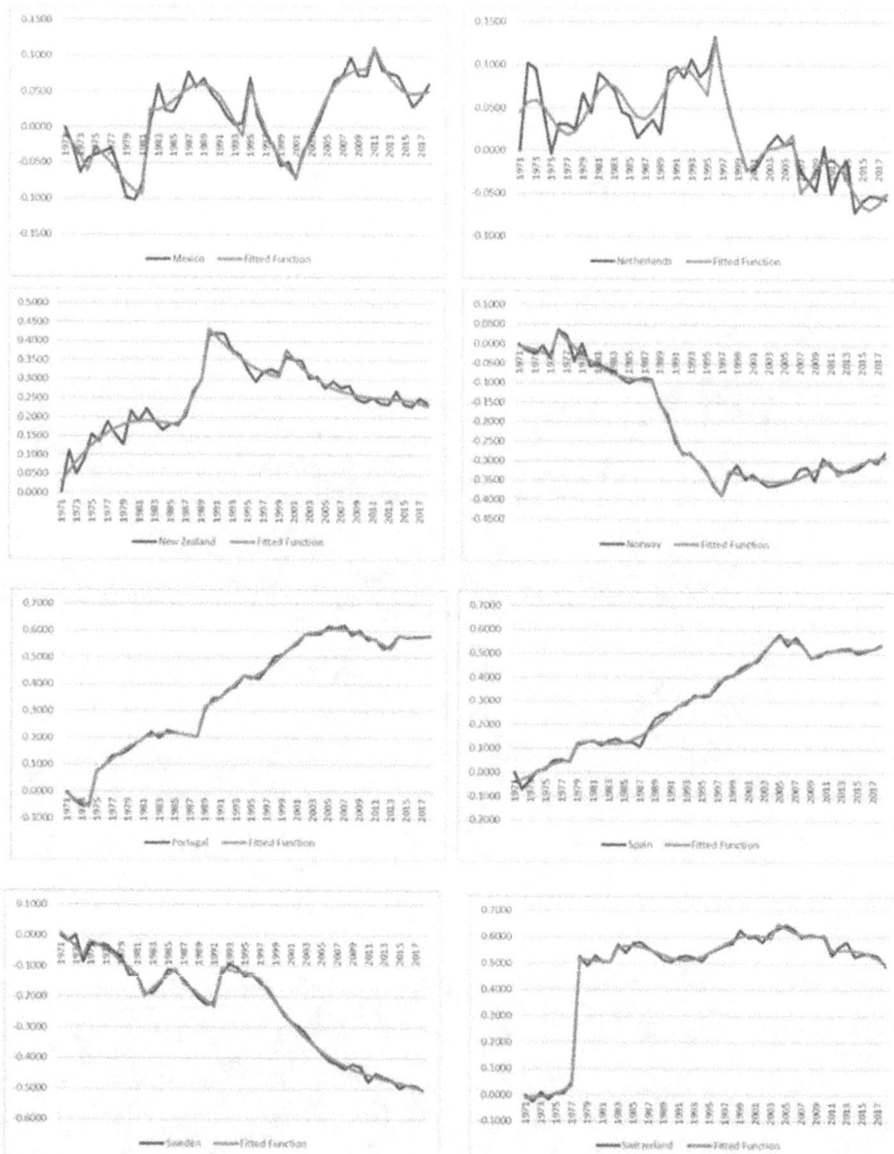

Figure 7.

CONCLUSION

This research aimed to construct an energy activity, energy efficiency, and activity index based on the Fisher Ideal index for 27 OECD countries covering 1971-2018. Another aim of the present investigation was to investigate the stochastic convergence behavior of the energy efficiency index across 27 OECD countries during 1971-2018. For this analysis, we utilized the panel stationarity test with sharp and smooth breaks proposed by Bahmani-Oskooee et al. (2014) that consider both sharp and smooth breaks in the series. The stochastic convergence test results confirmed the differences between energy efficiency indexes across countries. The findings showed that the country's energy efficiency index does not converge to the OECD average for 1971-2018. This result exhibits that the dissemination of the technology regarding energy efficiency is limited across OECD countries.

Our results support Yu et al.'s (2015) assertion that 'one size fits all' energy strategies are bad. Governments should thus create distinct energy policies, but they should also take into account certain shared energy policies because many different nations profit from them. In addition, politicians in countries with lower energy efficiency should encourage the development of energy-saving technology, while those in nations with better energy efficiency should spearhead global technical collaboration (Yan et al., 2017). They claimed that low levels of investment in innovation, human capital, and financial development are the leading causes of technological non-convergence.

The empirical findings of this paper exhibit that it is hard to reach the United Nations' SDGs regarding doubling the energy efficiency improvement at the end of 2030 for 10 OECD countries. The remaining 17 countries will most likely achieve their energy efficiency targets. Besides, according to the 2030 climate and energy framework target (aims to improve energy efficiency by at least 32.5% in 2030 compared to 1990 levels), most of the European Union countries will satisfy their own energy efficiency goals except for Austria, Greece, Italy, Portugal, and Spain. There should be differentiated energy policies across OECD members to reach SDG targets because each country has a different energy efficiency path. These countries should focus on energy-saving innovations and technological cooperation to enhance their energy efficiency performance.

REFERENCES

Adhikari, D., & Chen, Y. (2014). Energy productivity convergence in Asian countries: A spatial panel data approach. *International Journal of Economics and Finance*, *6*(7), 94–107. doi:10.5539/ijef.v6n7p94

Apergis, N., & Christou, C. (2016). Energy productivity convergence: New evidence from club converging. *Applied Economics Letters*, *23*(2), 142–145. doi:10.1080/13504851.2015.1058899

Bahmani-Oskooee, M., Chang, T., & Wu, T. (2014). Revisiting purchasing power parity in African countries: Panel stationary test with sharp and smooth breaks. *Applied Financial Economics*, *24*(22), 1429–1438. doi:10.1080/09603107.2014.925068

Baumeister, C., & Kilian, L. (2016). Forty years of oil price fluctuations: Why the price of oil may still surprise us. *Journal of Economic Perspectives, 30*(1), 139-60.

Becker, R., Enders, W., & Lee, J. (2006). A stationarity test in the presence of an unknown number of smooth breaks. *Journal of Time Series Analysis*, *27*(3), 381–409. doi:10.1111/j.1467-9892.2006.00478.x

Bhattacharya, M., Inekwe, J. N., & Sadorsky, P. (2020). Convergence of energy productivity in Australian states and territories: Determinants and forecasts. *Energy Economics*, *85*, 104538. doi:10.1016/j.eneco.2019.104538

Bhattacharya, M., Inekwe, J. N., Sadorsky, P., & Saha, A. (2018). Convergence of energy productivity across Indian states and territories. *Energy Economics*, *74*, 427–440. doi:10.1016/j.eneco.2018.07.002

British Petroleum. (2020). *Statistical Review of World Energy 2020*. Author.

Bulut, U., & Durusu-Ciftci, D. (2018). Revisiting energy intensity convergence: New evidence from OECD countries. *Environmental Science and Pollution Research International*, *25*(13), 12391–12397. doi:10.100711356-018-1551-z PMID:29460242

Burnett, J. W., & Madariaga, J. (2016). The convergence of US state-level energy intensity. *Energy Economics*, *62*, 357–370. doi:10.1016/j.eneco.2016.03.029

Canel, C., Guris, S., Guris, B., Öktem, B., & Oktem, R. (2017). Convergence of energy intensity in OECD countries. *Modern Economy*, *8*(7), 946–958. doi:10.4236/me.2017.87066

Carrion-i-Silvestre, J. L., Del Barrio-Castro, T., & Lopez-Bazo, E. (2005). Breaking the panels: An application to the GDP per capita. *The Econometrics Journal*, *8*(2), 159–175. doi:10.1111/j.1368-423X.2005.00158.x

Enders, W., & Lee, J. (2012). A unit root test using a Fourier series to approximate smooth breaks. *Oxford Bulletin of Economics and Statistics*, *74*(4), 574–599. doi:10.1111/j.1468-0084.2011.00662.x

European Commission. (2021a). *Paris Agreement*. https://ec.europa.eu/clima/policies/international/negotiations/paris_en

European Commission. (2021b). *2030 Climate & Energy Framework*. https://ec.europa.eu/clima/policies/strategies/2030_en

Ezcurra, R. (2007). Distribution dynamics of energy intensities: A cross-country analysis. *Energy Policy*, *35*(10), 5254–5259. doi:10.1016/j.enpol.2007.05.006

Gorus, M. S., & Karagol, E. T. (2022). Reactions of energy intensity, energy efficiency, and activity indexes to income and energy price changes: The panel data evidence from OECD countries. *Energy*, *254*, 124281. doi:10.1016/j.energy.2022.124281

Greening, L. A., Greene, D. L., & Difiglio, C. (2000). Energy efficiency and consumption—the rebound effect—A survey. *Energy Policy*, *28*(6-7), 389–401. doi:10.1016/S0301-4215(00)00021-5

Herrerias, M. J. (2012). World energy intensity convergence revisited: A weighted distribution dynamics approach. *Energy Policy*, *49*, 383–399. doi:10.1016/j.enpol.2012.06.044

Karimu, A., Brännlund, R., Lundgren, T., & Söderholm, P. (2017). Energy intensity and convergence in Swedish industry: A combined econometric and decomposition analysis. *Energy Economics*, *62*, 347–356. doi:10.1016/j.eneco.2016.07.017

Kilian, L. (2008). Exogenous oil supply shocks: How big are they and how much do they matter for the US economy? *The Review of Economics and Statistics*, *90*(2), 216–240. doi:10.1162/rest.90.2.216

Kilian, L. (2015). Energy Price Shocks. In S. N. Durlauf & L. E. Blume (Eds.), *The New Palgrave Dictionary of Economics*. Palgrave Macmillan. doi:10.1057/978-1-349-95121-5_3008-1

Kiran, B. (2013). Energy intensity convergence in OECD countries. *Energy Exploration & Exploitation*, *31*(2), 237–247. doi:10.1260/0144-5987.31.2.237

Le Pen, Y., & Sévi, B. (2010). On the non-convergence of energy intensities: Evidence from a pair-wise econometric approach. *Ecological Economics*, *69*(3), 641–650. doi:10.1016/j.ecolecon.2009.10.001

Leybourne, S., Newbold, P., & Vougas, D. (1998). Unit roots and smooth transitions. *Journal of Time Series Analysis*, *19*(1), 83–97. doi:10.1111/1467-9892.00078

Liddle, B. (2010). Revisiting world energy intensity convergence for regional differences. *Applied Energy*, *87*(10), 3218–3225. doi:10.1016/j.apenergy.2010.03.030

Lumsdaine, R. L., & Papell, D. H. (1997). Multiple trend breaks and the unit-root hypothesis. *The Review of Economics and Statistics*, *79*(2), 212–218. doi:10.1162/003465397556791

Markandya, A., Pedroso-Galinato, S., & Streimikiene, D. (2006). Energy intensity in transition economies: Is there convergence towards the EU average? *Energy Economics*, *28*(1), 121–145. doi:10.1016/j.eneco.2005.10.005

Menegaki, A., & Tiwari, A. K. (2019). Energy Efficiency in Europe; Stochastic-Convergent and Non-Convergent Countries. In M. Shahbaz & D. Balsalobre (Eds.), *Energy and Environmental Strategies in the Era of Globalization* (pp. 305–333). Springer. doi:10.1007/978-3-030-06001-5_12

Metcalf, G. E. (2008). An empirical analysis of energy intensity and its determinants at the state level. *The Energy Journal (Cambridge, Mass.)*, *29*(3), 1–26. doi:10.5547/ISSN0195-6574-EJ-Vol29-No3-1

Mielnik, O., & Goldemberg, J. (2000). Converging to a common pattern of energy use in developing and industrialized countries. *Energy Policy*, *28*(8), 503–508. doi:10.1016/S0301-4215(00)00015-X

Mulder, P., & De Groot, H. L. (2012). Structural change and convergence of energy intensity across OECD countries, 1970–2005. *Energy Economics*, *34*(6), 1910–1921. doi:10.1016/j.eneco.2012.07.023

Nillesen, H. L. P., Haffner, C. G. R., & Ozbugday, F. C. (2013). A Global Perspective on the Long-Term Impact of Increased Energy Efficiency. In F. P. Sioshansi (Ed.), *Energy Efficiency Towards the End of Demand Growth* (pp. 87–110). Academic Press. doi:10.1016/B978-0-12-397879-0.00003-7

Nilsson, L. J. (1993). Energy intensity trends in 31 industrial and developing countries 1950–1988. *Energy*, *18*(4), 309–322. doi:10.1016/0360-5442(93)90066-M

Qi, S., Peng, H., Zhang, X., & Tan, X. (2019). Is energy efficiency of Belt and Road Initiative countries catching up or falling behind? Evidence from a panel quantile regression approach. *Applied Energy*, *253*, 113581. doi:10.1016/j.apenergy.2019.113581

Rodrigues, P. M., & Taylor, A. M. R. (2012). The flexible Fourier form and local generalised least squares de-trended unit root tests. *Oxford Bulletin of Economics and Statistics*, *74*(5), 736–759. doi:10.1111/j.1468-0084.2011.00665.x

Santiago, R., Fuinhas, J. A., & Marques, A. C. (2020). An analysis of the energy intensity of Latin American and Caribbean countries: Empirical evidence on the role of public and private capital stock. *Energy*, *211*, 118925. doi:10.1016/j.energy.2020.118925

Shabani, Z. D., & Shahnazi, R. (2021). Energy intensity convergence in Iranian provinces: Evidence from energy carriers' consumption intensity. *Environmental Science and Pollution Research International*, 1–20. PMID:33495948

United Nations. (2015). *Resolution adopted by the General Assembly on 25 September 2015, Transforming our world: The 2030 Agenda for Sustainable Development (A/RES/70/1)*. The New York Declaration.

Yan, Z., Du, K., Yang, Z., & Deng, M. (2017). Convergence or divergence? Understanding the global development trend of low-carbon technologies. *Energy Policy*, *109*, 499–509. doi:10.1016/j.enpol.2017.07.024

Yilanci, V., Eryuzlu, H., & Hopoglu, S. (2020). Convergence of military burdens in the MENA region. *International Relations*, *17*(66), 41–59.

Yilanci, V., & Gorus, M. S. (2021). Considering nonlinearity and structural changes in the convergence of clean energy consumption: The case of OECD countries. *Management of Environmental Quality*, *32*(2), 243–259. doi:10.1108/MEQ-03-2020-0047

Yilanci, V., Gorus, M. S., & Aydin, M. (2019). Are shocks to ecological footprint in OECD countries permanent or temporary? *Journal of Cleaner Production*, *212*, 270–301. doi:10.1016/j.jclepro.2018.11.299

Yilanci, V., Gorus, M. S., & Solarin, S. A. (2022). Convergence in per capita carbon footprint and ecological footprint for G7 countries: Evidence from panel Fourier threshold unit root test. *Energy & Environment*, *33*(3), 527–545. doi:10.1177/0958305X211011461

Yu, Y., Zhang, Y., & Song, F. (2015). World energy intensity revisited: A cluster analysis. *Applied Economics Letters*, *22*(14), 1158–1169. doi:10.1080/13504851.2015.1013603

Zhu, J., & Lin, B. (2020). Convergence analysis of city-level energy intensity in China. *Energy Policy*, *139*, 111357. doi:10.1016/j.enpol.2020.111357

Zivot, E., & Andrews, D. (1992). Further evidence on the great crash, the oil price shock, and the unit root hypothesis. *Journal of Business & Economic Statistics*, *10*(3), 251–270.

KEY TERMS AND DEFINITIONS

Energy Efficiency: It means using fewer energy resources to produce a fixed level of goods or services.

Energy Intensity: The ratio of energy use to gross domestic product.

Fisher Ideal Index: It is a measure of change in volume from period to period. The geometric mean of the Paasche volume index and the Laspeyres volume index is used to calculate this index.

OECD: It is an intergovernmental organization with 38 member countries.

Stochastic Convergence: It exists when a series of a country *i* relative to the group average is mean-reverting.

ENDNOTE

This book chapter is derived from the Ph.D. dissertation titled "Three Essays on Energy and Environmental Economics" defended by the first author on August 17, 2021.

1 More details on the testing procedure can be found in Bahmani-Oskooee et al. (2014), Yilanci et al. (2019), and Yilanci et al. (2020).

Chapter 3
World Natural Resources and Their Impact on the Environmental Conditions of Our Planet

Nika Chitadze
International Black Sea University, Georgia

ABSTRACT

Natural resources are components of the natural environment used in the production process to meet the material and cultural needs of society. Natural resources are inherently physical in origin, but in the process of their use, they become economic resources. Natural resources are divided into inexhaustible (agro-climatic, geothermal, hydropower) and exhaustible. In turn, exhaustible resources are divided into non-renewable (mineral) and renewable (land, water, biological, recreational). Based on this classification and development, this textbook distinguishes the following types of natural resources: mineral (minerals), energy, water, biological, land, agro-climatic, and recreational. When considering natural resources, it is important to assess the availability of resources (i.e., the ratio between explored reserves of resources and the volume of their use, also their impact on environmental conditions of the world).

DOI: 10.4018/978-1-6684-6727-5.ch003

INTRODUCTION

Natural resources are regarded by us as environmental components which are used in the process of production for meeting the material and cultural needs of society.

At the same time, it should be known, that from the very first appearance, man began to develop natural resources for his needs, be it animal or plant. As man evolved, he moved from an appropriation economy to a production economy, that is, instead of hunting or gathering, he discovered some patterns and subsequently, after these events, began to create the means of his existence. Thus, man was able to domesticate several species of animals and domesticate many species of plants. From that moment on, people began to provide their food with their power. However, taking into account the exhaustion of the soil, man faced the need to create new fertile areas, which would provide him with the same yield and livestock feed as before. Due to the developed need, people began to move in search of favorable lands. Finding them presented him with another problem: now he had to somehow protect himself from external conditions and adapt to the environmental conditions for which he needed to build a dwelling. This led to the massive development of forest resources. Wooden houses began to be built, which were relatively protected from the outside world and warm. But as a result, the large-scale use of wood as a resource has also adversely affected the natural environment. It caused the problem of deforestation, which is still relevant today. However, the use of wood for construction gradually decreased due to the emergence of various new materials, for example, stone or brick. But at the same time, trees still found wide use as fuel for stoves. In rural areas, the purchase of wood is still the main source of heating, especially in winter. later during the 18th century. In connection with the intensive development of technology and transport system in society, it became necessary to switch to qualitatively new resources, such as coal, natural gas, and oil. This transitional circumstance was marked by the fact that with the development of these resources, the subsequent ecological situation in the world worsened, causing several problems in the environment. This happened because at that time the level of scientific development was not sufficient to organize production free of ecological waste, therefore industrial and agricultural resources were not yet fully developed and their processing was impossible. In addition, due to the lack of processing, the man had to constantly develop more and more new deposits and deposits. Thus, the unused resources accumulated over many years have a direct impact on the world around our planet. Man's continuous use of natural resources has often hurt the environment and has caused many environmental problems. In addition, for many centuries wildlife has suffered from pollution from raw and unprocessed surpluses.

And although scientific progress has already achieved sufficient development in the development of waste-free production, there are many difficulties in upgrading enterprises (Tltaudit, 2023).

The main reason for the implementation of high-tech ecological production is the lack of adequate financial resources, thanks to which it will be possible to update the factories. In addition, thanks to investments, it is already possible to notice how actively such industries are entering, which allows for rational and more efficient use of natural resources.

Classification of the Natural Resources

As for the issue of the classification of natural resources, the most important ones need to be addressed here.

First, it is the classification of natural resources according to the natural source of their origin or genesis, according to which they are divided into lithosphere resources (mineral, land, soil), hydrosphere (water, land and oceans, energy rivers and tides), atmosphere (climate, wind) and biosphere (flora and fauna). Natural resources divided into minerals, water, land, soil, flora and fauna, solar radiation, moving water, etc. serve as the variation of this classification.

Secondly, it is a classification of natural resources according to their possible use in human activities. This involves resources for industrial production (mineral, water, forest, etc.), agricultural production (agro-climatic, land, soil, water, etc.), transport, recreation, tourism, etc.

They can be considered as fractional - such as resources for the fuel and energy industry, metallurgy, chemical, timber, textile industry, for construction.

Third, the classification of natural resources according to their degree of depletion (Figure 1). Obviously, they are divided into two large groups - exhaustible and inexhaustible resources. In the group of exhaustible resources, non-renewable resources are considered the economic exploitation which could ultimately lead to their exhaustion, and renewable resources are located within the Biosphere cycle of matter and with the ability to heal themselves in terms of being commensurate with human activities. Inexhaustible resources belong to such types of resources the lack of which does not represent a threat to the future generations of the people. For example, in terms of solar energy resources, it is estimated that they exceed the current needs of humanity by 20 thousand times! (Chitadze, 2017). Additionally, to be classified into natural resources according to their degree of exploration, they are subdivided into traditional and nontraditional.

Figure 1. Classification of natural resources in the degree of depletion
Source: http://greenenergyhomedesign.tk/tag/green-energy/page/237/

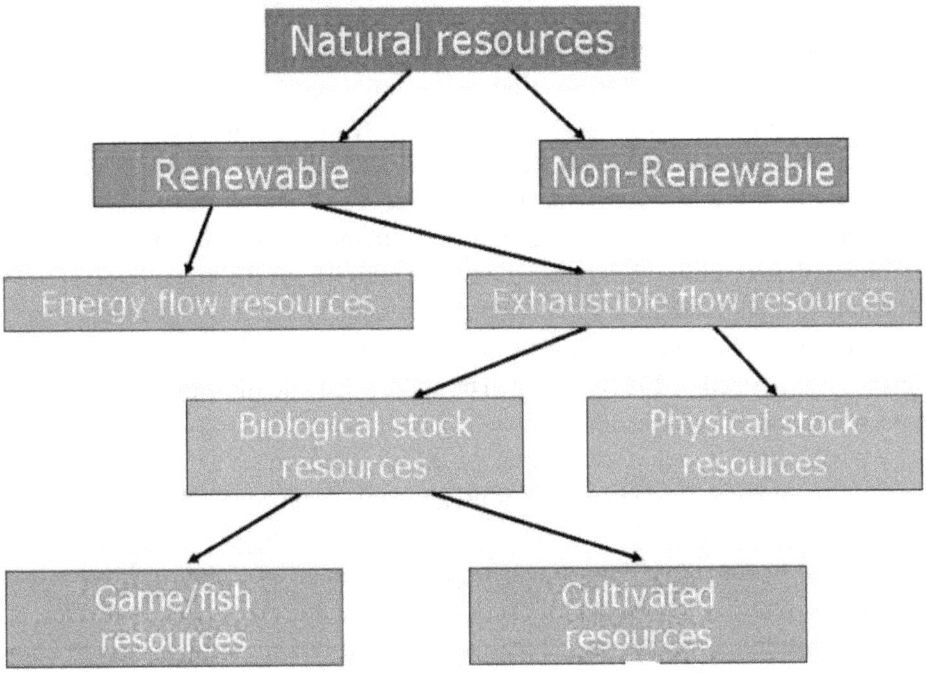

It is known, before the 60-70th of the XX Century, natural resources and environmental issues were paid comparatively less attention by scientific and governmental circles. But gradually, taking into account the actuality of the topic, the situation related to the environment and natural resources consumption studies has begun to be discussed and researched more deeply. Furthermore, the first international conference on environmental and natural resources consumption issues was held in Stockholm in 1972, and in the same year, at the UN, The United Nations Environment Program (UNEP) was founded. Within the 60-70th of the last century, several international non-governmental organizations in those fields – such as Greenpeace and WWF were established (UNEP, 2023).

In the research, opinions, and analysis of scholars from different countries on natural resources and environmental issues are reviewed and analyzed.

Geography of the World's Natural Resources: Mineral and Land Resources

In the socio-economic geography familiarity with natural resources usually starts with mineral resources (minerals) - the main "building blocks" of production. The fact is that people have learned to use minerals in ancient times, according to the names of these epochs of human development, such as the stone, bronze, and iron centuries. If in the Middle Ages from the Earth's crust were extracted only 18 types of minerals were, in the XIX century the number amounted to 47. Our days, according to the figurative expression of the geologist – A. Fersman "at the feet of humanity is composed all of the Periodic System of Mendeleev," or rather – more than 200 different kinds of mineral resources are used (University of Oxford. Science Encyclopedia. 2022). They can be subdivided into three groups: 1) fuel, 2) ore (metal) and 3) non-metallic.

We will not be wrong if we say that – It is reasonable to state that fuel resources have big importance. Their characteristic usually begins with a quantitative assessment, but you have to consider that we can talk about a wide variety of categories of such assessment. As for the geological data of the fuel resources of the world, now they are estimated at 5.5 trillion tons of fuel (in tons) and proven reserves, which we will continue to deal with - at 1.2 trillion tons. But it is more important to know the exact figures of proven reserves of coal, oil, and natural gas. Coal is one trillion tons, oil - is 192 billion tons, and gas - is 175 trillion m3 (Maksakovsky, 2009).

According to the quantitative importance, consideration of the geography of fuel resources is of particular interest to us. From the global analysis point of view, in general, these resources are widely distributed in the earth's crust. So, coal basins and deposits amount to 3.6 thousand and together they occupy 15% of the total land area, being located in more than 80 countries (Neidze, 2004). Oil and gas basins in the world amount to at least 600 deposits - up to 50 thousand (Neidze, 2004) whereas prospective oil and gas areas exist in more than 100 countries (Chitadze. 2004) and collectively occupy more territory than coal.

At the same time placing of fuel resources can not be considered equal. To understand its laws, you have to remember the geology and, in particular, the fact that deposits of fossil fuels are always associated with sediments and tectonics. So, maps of minerals should be of the tectonic framework which does not always take place. It is also important that fuel resources are typically spread waist, forming a vast belt of coal accumulation of oil and gas that formed in those geological eras when there were peaks of coal, oil, and gas.

For example, the major coal basins of Europe form a latitudinal belt stretching from the UK via Belgium, northern France, and western Germany, the southern part of Poland, and the northern part of the Czech Republic to the Donets Basin. The emergence of this belt of coal accumulation took place because the Carboniferous geological period was the one when the majority of geopolitical processes got reflected on the farthest north arch of the Epihercynian platform.

Therefore, this belt and pools - the Ruhr, Upper Silesia, Donetsk, etc. - exhibit certain geological similarities.

From the global, let's now proceed to the consideration of the regional level, using the data in

Table 1.

Region	Coal. Billion Tones	Oil, Billion Tones	Natural Gas, Trillion M3
Post-Soviet Space	230	20,3	56,0
Europe	125	2,7	6,0
Asia	215	106,2	82,5
Africa	55	15,1	13,0
North America	260	31,1	7,0
Latin America	30	16,7	7,5
Australia and Pacific	85	0,2	3,0
Whole World	1000	192,5	175,0

Source: Distribution of the proven energy reserves in major regions of the world at the beginning of the XXI Century (Chitadze, 2017)

With regard to the Proven reserves of coal in the world, there are highlighted in North America (26%), post-soviet space (23%) and Asia (21.5%), oil reserves in Asia (55%), and natural gas reserves in Asia (47%) and the post-soviet space (32%) (Maksakovsky. 2009). Therefore, during geological history best conditions for coal and oil and gas production evolved in these three regions of the world.

As for the third, country level, it can be assumed a priori that in these three regions the richest countries in the world possess fuel resources. They are considered in detail during laboratory exercises, based on the principle of "top ten."

Analyzing Table 2, we can be limited by the first three countries. A simple calculation shows that the share of the United States, Russia, and China accounted for more than half of all the world's known coal reserves whereas that of Saudi Arabia, Canada, and Iran for about 2/5 of the world's oil, and Russia, Iran, and

Qatar - almost half of the natural gas reserves. Those countries that occupy the first three places are - the United States, Saudi Arabia, and Russia should be especially singled out (Maksakovsky, 2009).

Table 2.

Country	Coal. Billion Tones	Country	Oil, Billion Tones	Country	Natural Gas, Trillion M3
USA	250	Saudi Arabia	35	Russia	48
Russia	195	Canada	28	Iran	27
China	115	Iran	18	Qatar	26
India	85	Iraq	16	Saudi Arabia	7
Australia	82	Russia	15	United Arab Emirates	6

Source: Distribution of the proven energy reserves by countries at the beginning of the 21st century (Maksakovsky, 2009)

There are a number of advantages of the above–mentioned countries. The most important one is the special richness of their fuel resources. Thus, the largest explored reserves of coal basins in the world are found in the United States (Illinois, Appalachia mountains), Russia (Kans -Achinsk, Kuznetsk), and China (Ordos) (Zgenti, 2010). The same applies to the richest petroleum provinces of the Persian Gulf. But in the case of oil and gas, a crucial role is not played by the total number of fields and the presence of these fields with giant and more unique stocks.

The number of unique non-competitive oil fields in the world occupy the Gulf countries, where they are genetically related to the sediments of the Arabian plate and the Mesopotamian basin. It is here that the major oil Ghawar (Saudi Arabia), Agha Jari (Iran), and the Greater Burgan (Kuwait) with initial reserves of more than 10 billion tons each appear (Leckie, 1993). According to a number of unique natural gas fields worldwide, the leader is the Russian northern part of Western Siberia. Qatar recently opened a unique gas field, Qatar -Nord, which immediately put forward this tiny country in the top four countries in the world according to the largest proven reserves (Karumidze, 2021).

With regard to metal reserves, they are more common in the earth's crust than in the fuel. This is explained by the fact that they are genetically linked not only with sediments but also crystalline rocks (remember Baltic shields or Canadian). For ore resources, it is a much typical zonal distribution. We should be aware of two major metallogenic belts of the earth - the Alpine- Himalayan and Pacific, stretching huge arcs at 30 thousand km (Chitadze, 2017). Both of these zones are associated with deep crustal faults, originated in the Alpine orogeny and that within them should first seek many ore minerals - whether iron ore in India, tin - in Malaysia, or copper in Chile.

In assessing the ferrous and nonferrous metals it is a need to consider some of their features. First, the fact that they have rarely explored reserves that amount to hundreds and tens of billions of tons and, usually, they are considered billions, tens of millions, and millions of tons. Secondly, this is due to the content of the useful component of the ores varying from less than 1% and up to 60-70% (Elizbabashvili, 2012). It is clear that the low composition of the metal in ore leads to assessing their deposits according to useful components rather than ore and, therefore, their amount is further reduced. Third, the collection of ore resources itself is much broader than that of fuel – their amount is about 35. We should think at least about ferrous metals - iron, manganese, chrome, alloy metals - titanium, vanadium, nickel, cobalt, ferrous and light metals - magnesium, copper, lead, zinc, bismuth, precious metals - gold, silver, platinum. Therefore, we can meet with them only on the separate examples – we can just discuss separate examples.

As a first example, consider iron ore widely distributed in the earth's crust. Their resources amount to 350 billion tons in the world and are mainly concentrated in the post-soviet space, North and Latin America, and Asia. Explored reserves are estimated at 165 billion tons and they are known in some 100 countries with a strong predominance of just a few of them. Here, the first five are Russia, Brazil, Australia, Ukraine, and China. At the same time, Russia has a non-competitive first place - 33 billion tons or 20% of world reserves (Davitashvili, 2012) which are concentrated primarily in the Kursk magnetic anomaly and several other unique and large-sized swimming pools. Pool Hammersley dominates in the stocks of Australia in the northwest of the country whereas in Ukraine – Kryvyi Rih.

Bauxite - the main raw material for aluminum production serves as a second example, which is also very widespread in the earth's crust. Explored reserves of bauxite amount to 20 billion tons (Maksakovsky, 2009). To understand the main pattern of their placement on the globe, we must remember that bauxite deposits are genetically related to primarily weathering crust sections located within the tropical and subtropical climatic zones.

That's why, one of the main provinces of the bauxite world include Guinea in Africa (more than one-third of all proven reserves), North Australia, Caribbean Central America, and Mediterranean Europe.

Uranium resources are widely distributed in the earth's crust. However, it is cost-effective to develop only those fields that contain at least 0.1% of the useful component: in this case, 1 kg of uranium concentrates cost is less than $ 80 of the explored reserves of uranium available for retrieval at a price of 3.5 million tons and Australia, Kazakhstan, Russia, Canada, and South Africa are included in the top five countries in this case. In Canada, the uranium content in the ore is 10%, and in Australia - 0.5% (Maksakovsky, 2009).

The third group, as already mentioned, is non-metal resources. We will not consider them in detail. We note only that, according to the volume of those resources, among them are sodium and potassium salts, phosphorus, and sulfur.

To conclude the description of mineral resources, two other issues need to be considered.

First, these resources are distributed between economically highly developed and developing countries. Economically advanced countries are ahead in proven reserves of coal, iron, manganese and chrome ores, poly-metallic, uranium, and gold. Developing countries lead in oil resources (more than 4 / 5 of the total reserves), natural gas, bauxite, copper ore, tin, tungsten, and Diamond (Bliadze, 2021).

Secondly, to what extent humanity is provided by critical - mineral resources? If we consider only the proven reserves, many kinds of minerals will not be enough for a long time. For example, oil, natural gas, copper, zinc, lead, tin, and tungsten will disappear in about 60 years (Bliadze, 2021). Notably, this happens taking into consideration the volume of their extraction at present. Another added factor is that the "appetite" of humanity is continuously growing!

Of course, differences between regions and countries according to mineral resources can be very large. For example, proven oil reserves are more provided in Canada (which will be enough for 230 years), wherein the Alberta region there are the world's largest deposits of tar sands and scientists have only recently begun to consider those reserves in international statistics. This country is followed by Iraq, Iran, Kuwait, United Arab Emirates, and Venezuela where reserves will be kept for about 100 to 150 years. But in Europe and Australia oil wealth is enough for only 9 years whereas in the U.S. – it is 11 years. In Russia, this figure amounts to 32 years (Maksakovski. 2009).

Turning to the characteristics of the land (soil) resources, several scientists call them territorial resources and we must first note that the earth is a kind of universal resource without which neither man's economic activity nor his life will be possible. However, with this multi-purpose land use at any given time, one or another piece of land can only be used by one for several purposes - granting, plowing, etc. It is important to note that although the land (soil) resources we are belonging to the category of exhaustible, renewable resources and renewal of them really requires a lot of time.

During the discussion about land resources, a fundamental concept represents the land fund. To get an idea of the size of the land fund of the planet, we should take into account a total of Earth's land area (149 million km2 or 14.9 billion hectares) and subtract the area of Antarctica and Greenland. The final result is 134 million km2 or 13.4 billion hectares, and it is the total amount of the land fund (Neidze,

2004). This is a tremendous resource and is admittedly very encouraging. However, familiarity with the structure of the land fund (Figure 2) leads to somewhat different conclusions.

Figure 2. Types of the global land fund
Source: https://www.ufz.de/export/data/global/54001_LSA_Map.jpg

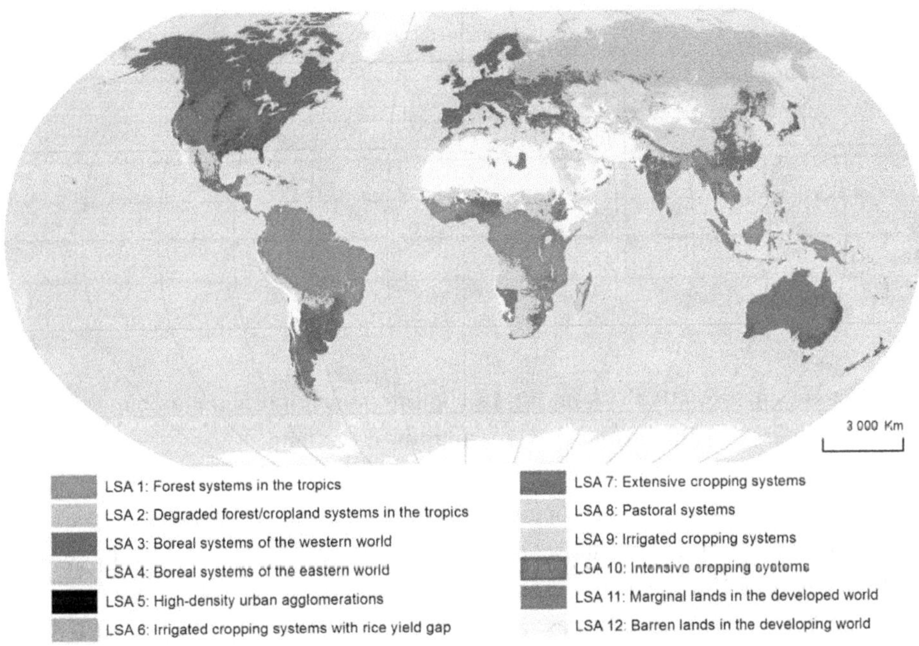

It turns out that the less or unproductive lands are wholly or partially unfit for life and carrying out business activities of people, occupying the first place in this structure. Forests and shrubs take second place. As for two types of agricultural land, arable and mowing ones, they share third and fourth positions, occupying only 1/3 of the total land fund, including, mainly cultivated ones, which yield nearly 9/10 of all necessary food products for people, accounting for only 11% (Neidze. 2004).

Admittedly, at the regional level, all these figures can vary greatly. The share of arable land is much higher in Europe and Asia than that of meadows and pastures - in Australia and Africa, the proportion of forests - in South America and in Russia, and the share of marginal and unproductive land - in Asia, North America, and Africa. Of course, there are even more differences between individual countries. For example, in Denmark, India, and Bangladesh plowed land prevails and amounts to 55%, in Mongolia 75% of the land is occupied by pastures, and in Libya, located mainly within the Sahara Desert, over 90% of the land is less productive and unproductive (Chitadze, 2017). Table 3 represents individual countries with the largest area of arable land.

Table 3. The first five countries according to the size of arable land at the beginning of the 21st century

Country	Area of the Land. Million Hectares	% to the Land Fund
USA	186	20,3
India	166	55,9
Russia	117	6,8
China	93	9,9
Australia	47	6,1

Source: Maksakovsky (2009)

Of course, historically the land fund structure does not remain unchanged. Two opposite processes Thus influence the structure of the land.

On the one hand, for hundreds, even thousands of years, people have sought to increase the area of land suitable for habitation and agriculture. Primarily, this means the offensive field on forest landscapes. It is not by accident, that XIV c. entered into the history of Europe as the "age of uprooting." Of course, the fields began to attack the landscapes and pastures. As a result, only in the twentieth century, the area of cultivated lands in the world has more than doubled. Remember though epic virgin lands that have been mastered in Canada, USA, Australia, Brazil, and China.

On the other hand, it was going on all the time and in the second half of the twentieth century, the process of the degradation of land (soil) resources accelerated. Now, in the world, the high and moderate degradation is exposed to 2/3 of all the arable land. The main cause of this degradation is the development of erosion due to which each year 6-7 million hectares of land falls from the agricultural turnover (Neidze, 2004).

As for the vast arid zone, anthropogenic desertification became the main reason for land degradation, which has already covered about 10 million km2, which is comparable to the territory of such giant countries as Canada, China, or the U.S. More than 1 billion people from about 100 countries around the world live under the conditions of anthropogenic desertification and soil degradation (Chitadze, 2017).

Figure 3. Areas for concern for soil degradation
Source: http://www.globalchange.umich.edu/globalchange2/current/lectures/land_deg/land_deg.html

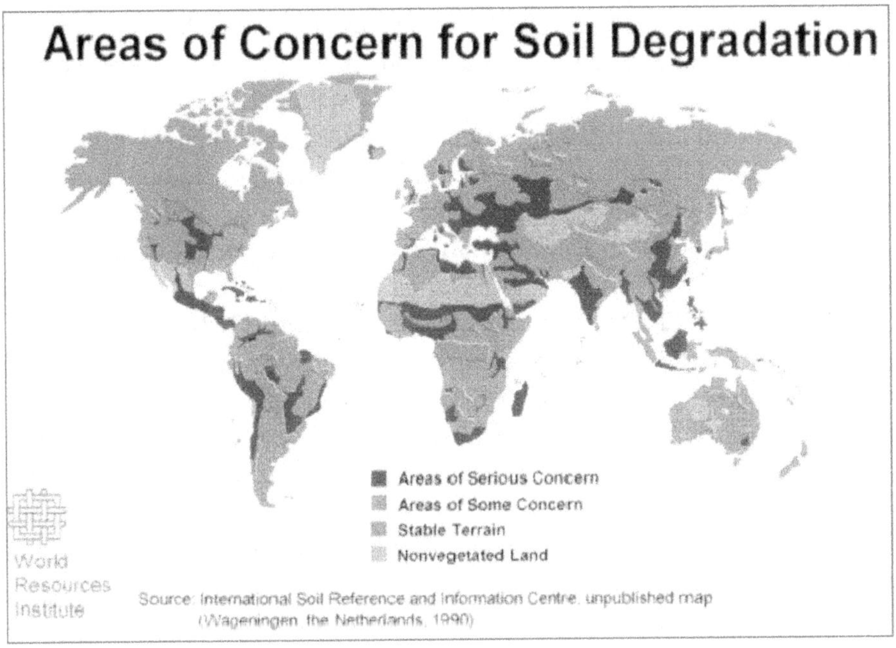

Despite the importance of the specific indicator of providing land resources, the more important index represents the availability of the most valuable cropland. Worldwide, it has declined from 0.5 hectares in the middle of the twentieth century to 0.2 ha in the early XXI century. Australia (2.6 hectares) and North America (0.6 ha) appear again the most successful in this regard and the lowest index is reported in East Asia (0.1 ha), South Asia, and Western Europe (0.2 hectares). From selected countries (besides Australia) Kazakhstan and Canada (1.5 hectares), Russia, Ukraine, and the United States (0.6-0.8 ha) occupy the leading positions, and the Netherlands, Japan, Egypt, Vietnam, Bangladesh, and China are at the end of the list with exponents from 0.03 to 0.07 hectares per person (Maksakovsky, 2009).

Geography of the World's Natural Resources: Water and Biological Resources of the Land

Water, like the earth, is an indispensable condition of human life, which satisfies their physiological and sanitary needs. Admittedly, people can stand much longer without food than without water. From this, by the way, comes some fundamentally new concept in medicine – This is where the new concept of medicine originates from. An almost equal amount of water is needed for a variety of businesses the people, which is largely based on the “wet” technology. It refers to the production of food, energy, and industrial products.

Let’s start with the introduction of land water resources according to their nature.

Although freshwater resources in the world amount to only 2.5% of the entire hydrosphere, this corresponds to 35 million km3 (Neidze, 2004). But the reason for optimism related to this fact is not so much – This does not quite give any reason for optimism. The fact is that almost 70% of this volume, is admittedly conserved in the ice sheets of Antarctica and Greenland, the Arctic ice, and mountain glaciers. Groundwater accounts for another 30% but they are used in relatively small quantities. It turns out that free fresh water is available in rivers, lakes, wetlands, atmosphere - it is only 0.3% of all fresh water on the Earth (Chitadze, 2017).

But even with this approach, the most reasonably available resources are considered to be the most dynamic part of fresh water - the river (river bed) water flowing into the oceans. Their lump sum amount in rivers is negligible - only 2.1 km3. But since this volume is renewed during the year on average 23 times, in fact, available resources of the river waters rise to 48 km3 (Bliadze, 2021). Apparently, this number characterizes the “water ration” of humanity that can (to some extent) withdraw from economic activity.

Now from the quantitative assessment of the freshwater resources of the world, let`s turn to the consideration of their geography. If we keep in mind the major regions of the world, according to common freshwater resources leading positions to have Russia (1/5 of the World`s supply), Latin, and North America (Kreselidze, 2021). If we consider only the resources of river flow, then go forward to Asia and Latin America, it is clear that here we have in mind first the all-river systems, the leading positions have Yangtze, Brahmaputra, Ganges, Mekong, and the second - the Amazon, Orinoco, Paraná. There are several changes and the order of the leading countries.

Table 4. The first five countries in the world in size resources stream flow at the beginning of the 21st century

Country	Resources, km³
Brazil	6950
Russia	4300
Canada	2900
China	2800
USA	2500

Source: Davitashvili (2012)

This is the situation related to the freshwater resources that nature made available to mankind. However, from the standpoint of socio-economic geography that is not enough – the principles of water consumption should also be taken into account, which is constantly increasing. It is enough to say that only in the twentieth century did global water consumption increase several times and now amounts to almost 3000 km3 per year (Bliadze, 2021). Some experts believe almost half of the total available amount of fresh water has been used on our planet. Moreover, almost 70% of it goes to agriculture and is lost forever (Neidze, 2004). With regard to the industry and utilities, where the water recycling principle is intensively used, those fields of economy are respectively in second and third places.

Undoubtedly, for large regions of the world, the water consumption rate also varies greatly. As expected, the Eastern part of Asia is leading in the first place. In this part of the world water consumption in agriculture predominates. Although, in some countries (China, Japan) an appreciable proportion of the industry is reported (The Water Footprint Network, 2014). Agricultural use is also prevalent in Africa, Australia, and Oceania, a large part of Latin America, and industrial and municipal - in North America, and Europe.

Figure 4. Consumption of freshwater per capita. M3 annually.
Source: https://www.mapsofworld.com/world-freshwater-resources.htm

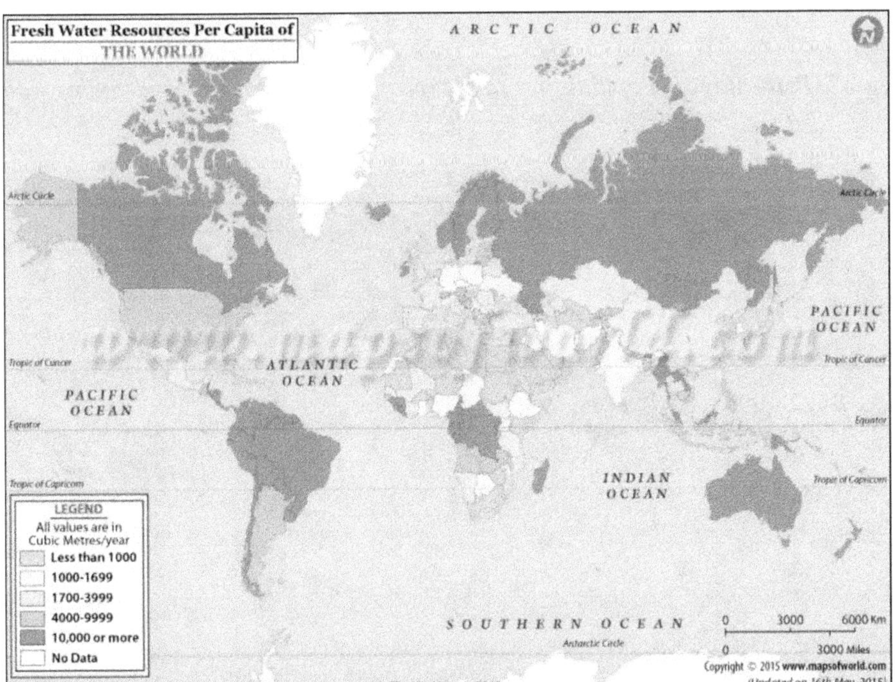

The main point to be noted in this regard is the characterization of provision with freshwater resources, which is calculated on the basis of per capita. The average per capita freshwater availability is constantly decreasing as though these resources are growing and, in any case, slower than the population. If this indicator is put at stake, the differences will be clear.

It was immediately struck by the presence of two well-defined zones of sufficient and excess moisture. The first of these is in the range of temperate and subtropical climate zones of the Northern Hemisphere and includes Canada, the United States, the Nordic countries, and Russia. However, countries of Europe, located in this zone, are already experiencing a lack of fresh water. The second zone extends within the equatorial and tropical climatic zones, mainly in the Southern Hemisphere. Between them extends the arid belt with the biggest shortage of fresh water.

Interestingly, if the analysis of this indicator applies to the principle of "most-most", it would appear that provision of renewable freshwater resources in a number of countries, and in this case, champions are French Guiana (over 800 thousand m3 per person!) and its neighboring Guyana and Suriname (300 thousand), DR of Congo (about 300 thousand) and Iceland (250 thousand), the minimum water supply indicators are in Kuwait (10 m3), the UAE and Qatar (less than 100 m3), Saudi Arabia and Libya (slightly more than 100 m3). This means that in French Guiana shower provision of such resources is higher than 8 thousand times (!) than in Kuwait (Chitadze, 2017).

Figure 5. Percentage of population without reasonable access to safe drinking water
Source: http://www.theglobaleducationproject.org/earth/human-conditions.php

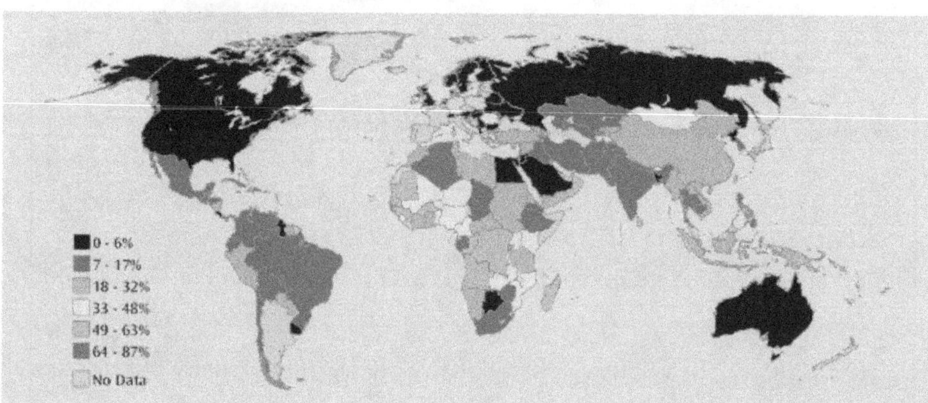

Reportedly, the problem of freshwater - as farmland - in fact, has already become global. In fact, according to the UN, at the beginning of the XXI Century, about 1.2 billion people lacked access to safe drinking water (Figure 5), and more than 2.4 billion to improved sanitation (UN, 2005). In these least developed countries in Asia and Africa, such as Nepal, Cambodia, Ethiopia, Chad, and Mauritania less than one-third of all residents have. secured access to drinking water. We must take into account the fact that in most developing countries there is not just a shortage of water, but also the quality of it is poor. It is the consumption of contaminated water in them that is the source of two-thirds of all diseases. According to UN projections, universal access to safe drinking water will be provided in Asia - until 2025, Latin America - up to 2040, and Africa – until 2050 (UN, 2005).

Here, we come to the important issue related to the resolution of the problem of water that humanity divided into major and minor factors.

To start with, the main focus should be made on principal factors. The use of water and reducing its wastage/deadweight loss of water during industrial processes serve as the key ones. Industry, is primarily concerned with the production of synthetic fibers during which approximately 3500 tons of water are consumed per ton of the product whereas, in case of nickel, it amounts to 800 tons and 200 for the production of iron, steel, and paper. As for agriculture, 10 000 tons account for cotton production and 7000 ones for rice growing. It is worth mentioning that in the process of cotton production the Aral Sea suffered significantly. Economical use of water is necessary for carrying out everyday activities too provided that in economically developed countries one city inhabitant does not consume less than 300-400 liters of water. 100 liters of water are used for several minutes while getting the shower (Neidze, 2004).

Sparingly, water should be consumed in everyday life. Indeed, in economically advanced countries an urban resident consumes at least 300-400 liters of water per day. Only receiving the shower for a few minutes required 100 liters (Maksakovsky, 2009).

The second most important measure is the construction of reservoirs for river regulation. It is estimated that with their help the global river flow could be increased by 1/4. Over the past half a century the number of reservoirs around the globe has increased by about 5 times. Now there are more than 60 million with a total useful volume of 6600 km^3. Together, they occupy 400 km^2 (Chitadze, 2017).

As in the case of land resources, many programs for the conservation and restoration of freshwater resources are directed by the UN, which in 2002 announced the "Decade of Water" and in 2003 proclaimed the "Year of Freshwater". This has stimulated the increased public attention to the problem of the water supply of the world, by its individual regions and countries, not only in the present but also in the future. Some politicians are already predicting the possibility of "water wars" - acute conflict over water resources (Simon, 2005).

To finish with water resources, they also have their hydro-energetic potential which has three grades. Potential resources of river flows and reservoirs are meant under the theoretical hydropower potential. Usually, it is estimated at 35-40 trillion kW/h. Next comes the technical hydropower potential, which is part of the theoretical capacity, which can be technically mastered. Most often it is estimated at 15 trillion kW/h. Finally, the economic hydropower potential is the total energy resources of the rivers the use of which is the Top of the Form, including, the cost of construction of the hydroelectric power station and the cost of electricity, which is economically feasible. It is estimated at 8 trillion W • h (Maksakovsky, 2009).

In economic geography literature, you often come across the indicator of economic hydropower. As the volume of river flow, here ahead of the other regions are Asia and Latin America.

There are five richest countries in hydropower resources (Table 5).

Table 5. The first five countries in the economic size of hydropower

Country	Hydro Energy Potential. Billion K/h	The Degree of Its Exploitation
China	1260	16
Russia	850	19
Brazil	765	37
Canada	540	65
India	500	16

Source: Maksakovsky (2009)

It is easy to calculate that these five countries account for almost half of all global economic hydropower. As for the extent of its development, it is the average for the world not reaching 1/3 whereas for Europe and North America it is 70% and 18% for Africa. (Chitadze, 2017).

France, Italy, and Switzerland serve as examples of countries where this potential is almost fully mastered whereas Japan and the United States are the ones with almost completely achieved potential.

Biological resources – there are resources of the Earth's biota, i. e. plants, and animals, which are measured in trillions of tons. The gene pool of such organisms is distinguished by exceptional biodiversity: according to various data, it comprises from 10 to 100 million different species. However, only 1.7 million are described among them (Kereselidze, 2021).

The biomass of planting resources on the earth's land is about 200 times more than the biomass of wildlife. It is represented by both cultural and wild plants. Including crop species, there are almost 6 thousand, but the amount of most common crops in the world equals only 80-90, and the most common - 15-20: wheat, rice, corn, potatoes, barley, sweet potatoes, soybeans, etc. (Neidze, 2004).

Forest vegetation, forming forest resources plays a major role among wild plants. With their characteristic, we must first remember that the forest as a part of the biosphere on Earth forms the largest ecosystem, which significantly affects photosynthesis, the oxygen balance of the atmosphere, and the preservation of the gene pool. In the same economic activity of people, wood is widely used for the production of 20 thousand different products, as well as fuel. Forest resources are exhausted as land ones but at the same time, renewable resources for multipurpose use emerge (Neidze, 2004).

For the forest`s resources assessment is typically measured by the forest or the forested (which is the main part of the forest) area. As we mentioned during the meeting with the structure of the land fund, the forest area of the world is 4.1 billion hectares, which corresponds to an average of 30.5% of the forest cover. According to various estimates, the stock of standing timber in the forests of the world reaches amounts to 330-380 billion m3 and annually rises by 5.5 billion m^3. (World Resources Institute, 2010). It would seem that it is quite a comforting figure. But they need a serious adjustment in light of two important factors.

The first is called the natural-geographic factor. It is a very uneven distribution of the forests on the land surface. If we consider the large regions of the world, it turns out that most wide areas of the forest are in Latin America. Here it is fixed at the highest percentage of forest and timber stock. The lowest rates are fixed in Australia and Oceania. But with regard to the characterization of such forests with geographical unevenness, forest reserves can be discussed in different ways based on the fact that the world's forests actually form two huge forest belts along the strike zones - north and south.

Obviously, the northern forest belt extends to a broad band across North America and Eurasia, taking up about half of the total forest area in the world. It is dominated by coniferous and mixed forests of the cold temperate and subtropical climatic zones of the earth. The Southern Forest belt consists of three parts - South American, African, and Asian - Australian. By the area, it is about the same, but almost entirely consists of deciduous forests and is a much more diverse forest, richer, and most importantly, updated much faster. Therefore, there are more stocks of forests in the southern belt. And between them is an almost completely treeless arid, torrid zone, which has been continuously mentioned above.

Figure 6. Schematic map of the world's forests
Source: https://www.mapsofworld.com/world-natural-forest.htm

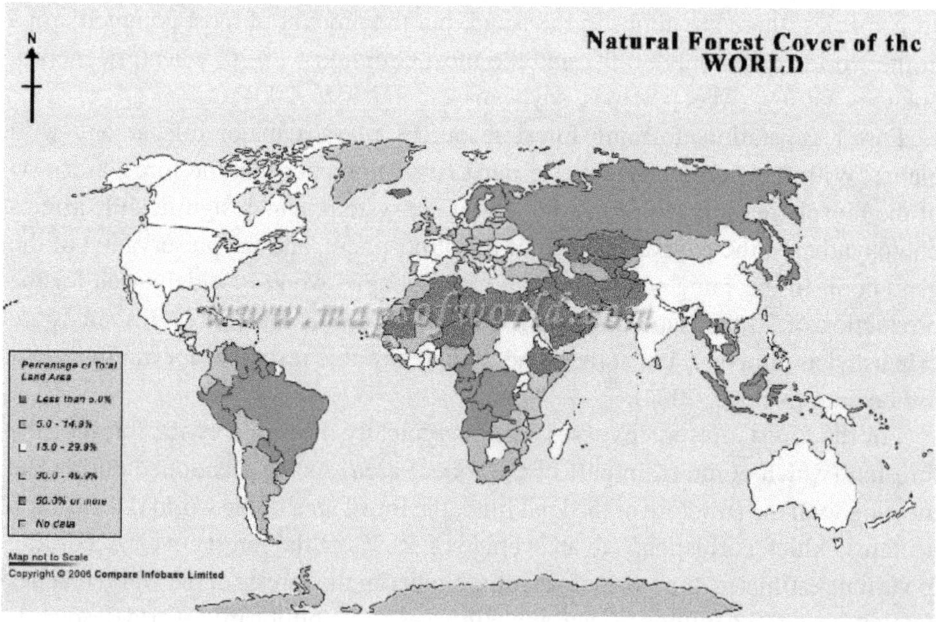

Such a belt approach is the "golden key" to understanding why some countries are very rich in forest resources, while others, on the contrary, are very poor. Clearly, the rich countries by the forest's resources can be founded within the northern and southern forest zones. In this case, it should be mentioned that in the northern zone it is Russia (810 million hectares of forest area), Canada (310), the United States (305), and partly China (195) and in the southern - Brazil (480), Australia (165), DR Congo (135) and some other countries (Chitadze, 2017).

Additionally, within these two zones, you will find a country with record levels of forest coverage, which in Finland, North Korea, the Democratic Republic of Congo, Gabon, more than 70% in Guyana and Papua - New Guinea - 80, and in Suriname, even 90%! The middle arid zone is distinguished from the most sparsely populated countries. Here, too Saudi Arabia, Jordan, Libya, Central African Republic are the "champions" among the countries where forests cover less than 1% of the area, excluding, Kuwait or Oman, where they are absent (UN, 2012).

This, again, first is the natural and geographical factor influencing the wealth of forest resources of the regions and countries. With regard to the second factor, it is only part of the natural essence. What it mainly represents is the "anthropogenic" factor, i.e., the impact of humanity on the process of reducing forest resources.

Man-made deforestation began in the Neolithic, when, as we already know, agriculture and animal husbandry were created. It continued in the era of the ancient civilizations of antiquity, in the Middle Ages, and in modern and contemporary times. Only in the last two centuries, the forest area of the world has halved and today, it continues to decline at a speed of 13 million hectares per year. But, the situation in the northern and southern forest zones is very different (FAO, 2012).

The Resources of the Oceans, Climate, and Space, Recreational Resources

If before we considered the resources of the Earth's land, now we turn to the resources of the oceans, atmosphere, and space, as well as a special kind of recreational resource.

Admittedly, the World Ocean is another pantry of various resources at the disposal of mankind. Resources of the oceans can be divided into 1) water; 2) mineral; 3) energy and 4) biological (Nellemann C. and Corcoran E. 2010).

From the materials in the physical geography, we should know about the fact that the World Ocean encompasses virtually inexhaustible water resources. Indeed, the total amount of its waters is 1.37 billion km3, which corresponds to 96.4% of the total terrestrial hydrosphere, and they occupy almost 71% of the surface of our planet (World Atlas. 2013).

Talking about the waters of the World Ocean, we must remember that they themselves have considerable economic importance since they contain about a hundred chemical elements. It's hard to imagine but one km3 of occan water can hold a huge amount of dissolved minerals. Sodium and chlorine are the most widespread ones in the water. So, even thousands of years B.C. Chinese learned to get table salt from it. It is also possible to get magnesium, bromine, iodine, potassium, hydrogen and oxygen, and other chemicals from marine water. Technologically, there are developed methods for recovering from their uranium and gold even though their content is 0.0001 mg / l (Christie, A and Brathwaite, R. 2012).

The same applies to obtaining deep, the so-called heavy water (it has a slightly different combination of hydrogen and oxygen isotopes) of deuterium required for thermonuclear fusion.

Mineral resources of the World Ocean are geological resources of raw materials and fuel, which are lying on the seabed or in the subsoil. Geographically and genetically, they are usually divided into resources of the continental shelf, the continental slope zone and Deepwater areas of the ocean floor (Figure 7). Resources of the continental shelf play a major role among them, which covers 31.2 million km2 or 8.6% of the total area of the ocean (Pinet, 1996).

Figure 7. Mineral resources of the ocean floor
Source: https://worldoceanreview.com/en/wor-1/energy/marine-minerals/

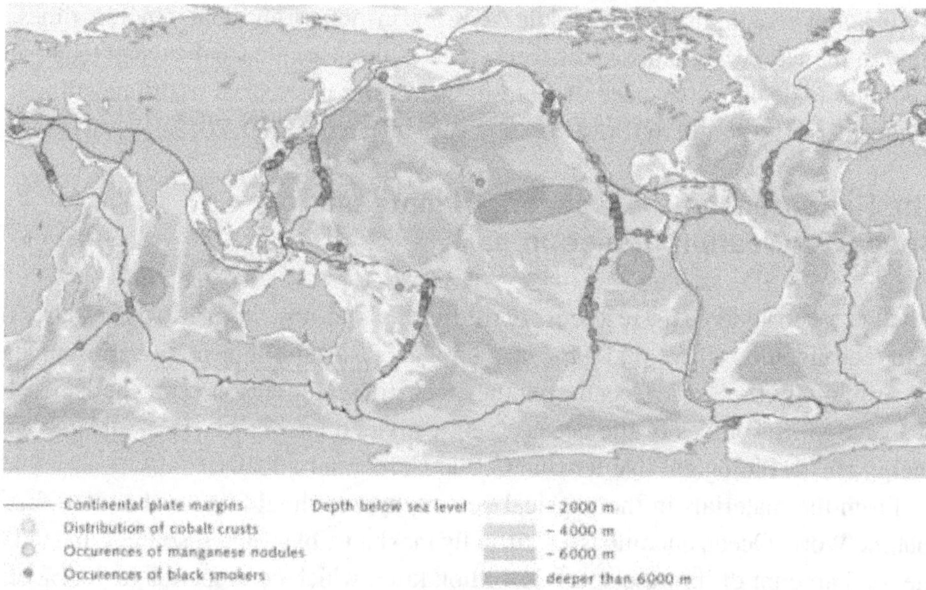

In its turn, among those resources, the basins of oil and natural gas are the most important together occupying almost more territory than oil and gas basins in the Earth's land area. It can, therefore, be concluded that the Atlantic Ocean is leading among the richest hydrocarbon resources, where they have already been explored in many marine spaces. Three of them need to be focused on: the Caribbean Sea and the Gulf of Mexico in Central America, the Gulf of Guinea in West Africa, and the North Sea in the northwestern part of Europe. Smaller deposits are discovered near the coast of Canada, Brazil, and Argentina. The Indian Ocean is in second place in those resources, where oil and gas are found on the shelves of India, Indonesia, and Australia, but most of all - in the Persian Gulf, where, incidentally, some offshore fields (Saffaniya, Qatar -Nord) are of the unique size. In the Pacific, the coast of Asia, North and South America, and Australia are famous for oil and gas resources, and in the Arctic Ocean - the coast of Alaska, Canada, and Russia (Barents and Kara Seas) dominate. The Caspian Sea should also be mentioned where near Baku oil production was conducted for a long time but the "big oil" was opened recently.

Besides oil and gas, many solid minerals are connected with the shelf of the oceans and related fields. Among them, there are indigenous deposits of coal, iron ore, and other salts that may be developed from the shore by means of inclined tunnels.

As for energy resources of the World Ocean, they are limitless similar to the water resources. Their main types are - tidal power, waves, temperature gradient, and ocean currents. Yet, they are mainly related to potential resources, but their use has already begun.

Speaking about tidal energy, we should note its advantages and disadvantages. The advantage is its inexhaustibility, regularity, and environmental cleanliness. In fact, the energy contained in only one tidal cycle reaches 8 trillion kW • h (which is only half of the world's electricity generation in the year) (Chitadze, 2017), and such cycles are repeated twice a day and they work "on schedule" to within a few minutes. As for disadvantages, tidal energy can be effectively used only in those places where the exceeding tide height is 5 meters, and around the globe, there are approximately 25-30. Admittedly, the highest tides are off the coast of the Atlantic Ocean: in the north-western part of the Bay of Fundy they reach 18 m, and in the northeastern part, near the shores of the English Channel and the Irish Sea - 10-13 m (Chitadze, 2017). Such tides are present in the north-western Pacific (e.g., in the Okhotsk Sea) and the Arctic Ocean (e.g., in the White Sea), i.e., in most cases in very remote and sparsely populated areas.

As for the biological resources of the world's ocean, their total biomass is often estimated at 35-40 billion tons, which is much less than the total biomass of the Earth's land (Ocean Resources. 2001).

Nevertheless, it has 180 thousand animal species and 10 thousand plant species (Ocean Resources Defense Council. 2014). This means that the water of the oceans is populated by the many living organisms of the world, from microscopic bacteria to the largest animals on earth – whales; while they live throughout the thickness of oceanic waters from the surface layer to the bottom of the deepest valleys.

As for the issue of climate and space resources, they belong to the category of inexhaustible resources and practices that are not removed from nature but nonetheless can significantly affect the living conditions of people and economic activity.

Climate resources are closely related to certain features of the climate. Undoubtedly, those are primarily agro-climatic resources, i.e., light, heat, and moisture, which determine the possibility of growing all crops.

Further, there are wind energy resources. Their use has started a long time ago with windmills and sailing ships. Although this energy is different by the scattering and inconstancy, the world still has a lot of places where the average wind speed exceeds 5 m/s, which makes this energy use, which is environmentally clean with the help of using economically viable wind turbines. Examples of such places only in Europe can serve the coast of the North, Baltic, and Black seas as well as mountain areas (Holttinen, Hannele. 2006).

When talking about cosmic resources, solar radiation - the largest energy source in the world is primarily meant. It forms the basic processes in the biosphere and ensures the existence of life. The power of the solar energy reaching the lower atmosphere and the earth's surface is measured by a huge quantity (1014 kW). It is by ten times superior to all of the energy contained in the proven reserves of fossil fuels and by thousand times - the current level of global energy consumption (Reference Solar Spectral Irradiance. 2013).

However, solar power is widely dispersed, so that its use is appropriate only in areas with low clouds, receiving it in quantities exceeding 200 W/m^2 (World Meteorological Organization. 2008).

All of them are hot climate zones of the Earth, and partly subtropical, within which mainly developing countries are located. But due to its economic and technical advantages, solar energy for commercial purposes is used in the United States, Japan, Israel, and Australia to the greatest extent.

In conclusion, we should analyze one more interesting question related to recreational resources. The word 'recreation' in Latin means "restoration" (The Free Dictionary, 2014). Respectively, those natural resources are involved with the help of which the people's health and their ability to work are maintained and restored. But, besides this, recreational resources are an important source of aesthetic pleasure, which is also worth mentioning.

Recreational resources serve as a basis for recreation and are closely associated with tourism. In other words, there are resources for recreation and tourism, though sometimes tourist resources are allocated separately. In recreational geography, there are four main types of recreational use of the territory. The first one is therapeutic, which uses healing waters, mud, comfortable climate. The second – is beaches of seas, rivers, lakes, reservoirs, forests, and parks. The third type represents sports activities, including, skiing, sailing, and mountain climbing. Fourthly, the recreational and educational types are based on natural and cultural heritage sites and cultural landscapes (Rechner. 2010).

According to another approach, all recreational resources can be grouped into two large classes.

The first of them form natural recreational resources, among which are the seashores, riverbanks and lakes, mountains and hills, forests, mineral springs outputs, and therapeutic mud. Here vacationers and tourists find natural diversity, scenic, and attractive, landscapes, the richness of vegetation, pleasant eye relief, healing climate. And often it is the combination of the above-mentioned attractiveness.

The second class is formed by cultural and historical attractions - monuments of history, archeology, urban planning, architecture, literature, and art, which are the main prerequisite for the organization of cultural and cognitive recreation and also largely determine the direction of recreational flows of people, in which you are involved too.

Obviously, the greatest interest among vacationers and tourists is expressed towards those countries, that possess a combination of natural, cultural, and historical attractions, such as Italy, Spain, France, Switzerland, Bulgaria, Egypt, and Mexico. The same applies to certain areas of many countries that are specialized in recreation and tourism for a long time.

The World Heritage Site needs to be mentioned here. This concept has been existing since 1972 when UNESCO adopted the Convention on the World Cultural and Natural Heritage. It lists the heritage objects, which are replenished every year. At the beginning of 2020, it already included 830 sites in 138 countries (Karumidze, 2021).

Of the total number of heritage sites, 644 are classified as cultural. The Forum and the Colosseum in Rome, Versailles near Paris, Westminster and the Tower of London, Hradcany in Prague, the great pyramids in Cairo, the Taj Mahal in Indian Agra, the historic imperial city in Beijing and the Great Wall of China, Statue of Liberty in New York, the ancient Mayan city in the Yucatan Peninsula in Mexico serve as examples of the most famous of them. The category includes 162 natural objects too. Many of them are well known: Bialowieza Forest in Poland and Belarus, Mount Everest in Nepal, Lake Victoria and Mount Kilimanjaro in East Africa, the Grand Canyon on the Colorado River in the United States, Galapagos Islands in South America, and the Great Barrier Reef in Australia. In addition, an additional 24 mixed, natural and cultural properties are allocated. World Heritage forms a huge recreational resource of universal significance, which is an important incentive stimulus for recreational activities.

Many geographers point out that among the city's millionaires, about 4/5 of all residents are eager to spend their holidays in nature, i.e., to be engaged precisely in such activities. The forms of such activities are extremely varied (Karumidze, 2021).

RAW MATERIAL PROBLEMS ON THE GLOBAL LEVEL

General Signs of Raw Material Problems

One of the serious problems of modernity is the maximum provision of humanity with raw materials. This problem has always existed, but it has acquired global significance in recent decades when unprecedented population growth and increasing production capacity have raised the raw material problem with all its severity.

Global raw material problems are associated with very rapid, often spontaneous growth in demand for raw materials (both energy and metallic and non-metallic minerals). Suffice it to say that in the last half-century the world has extracted and used more raw materials than in the entire history of mankind.

At the same time, this process is irreversible, and the supply of raw materials is steadily declining, which will pose an unresolved dilemma in the future.

The raw material crisis in recent decades has shown everyone that it is impossible to continue the growth of raw material consumption at such a pace and it is necessary to find a solution.

The raw material crisis is exacerbated by the fact that along with the reduction of raw material stocks, the mining-geological conditions of the deposits are deteriorating and, consequently, extraction is becoming more expensive.

In addition, the mining industry is associated with a whole range of environmental problems. Open-source mining reduces the cost but leads to land degradation and disruption. Pollution of the atmosphere, water, and soil is also an important problem. In recent years, the use of high-sulfur oil and coal has been increasingly avoided, and some countries (where the so-called acid rain is being actively fought) have banned the consumption of such raw materials altogether. All this creates new difficulties in providing raw materials.

Raw Material Problem: A Future Perspective

Given current consumption trends, which mean stabilizing or reducing demand for raw materials in developed countries and significant growth in the developing world, it is likely that the existing stock of some mineral wealth will already be depleted shortly. In particular, the depleted reserves of tin ore will be available to mankind only until 2002, by 2030 will run out of stocks of tungsten, silver, lead, and sure, by 2040 - cobalt, molybdenum, copper, nickel, by 2060 - aluminum (bauxite), 20 and iron - Manganese stock. Reusable metal materials (scrap recycling) can indeed be used repeatedly, which significantly increases the service life of metal raw materials, but the final introduction of other, containing materials is inevitable.

The situation is relatively better in terms of the supply of chemical and other non-metallic minerals. The threat of depletion of stocks of phosphorites, apatites, potassium salts, sulfur, and building materials does not threaten us in the 21st century, but unlike metal, most of these raw materials are impossible to use. Thus, it is impossible to consume those materials eternally, accordingly, sooner or later this problem will arise.

The situation is especially difficult in terms of fuel and energy raw materials (oil, natural gas, coal). The rate of their consumption is rapidly increasing, and stocks, despite the discovery of new deposits, are depleted and can disappear within 60-80 years (except for coal, the extraction of which will be possible for more than two centuries). That is why the provision of fuel and energy resources is the main problem of mankind related to raw materials. Due to their special severity and importance, energy problems are often singled out and attributed to several major global problems (Davitashvili, Elizbarashvili, 2012).

Fuel-Energy Problems

Despite the rapid growth of energy consumption in the past, no serious problems were encountered until the 1970s. The increase in energy production was provided by the growth of cheap oil and natural gas, also by the extraction and development of new fields. Those products were quite cheap and easy for transportation as well. Because of all those factors, oil and gas have become the main source of energy. Their share in the world energy balance in the 70s of the XX Century exceeded 70%, while in the 50s of the last century it was less than 40%. Accordingly, in the second half of the XX century, the role of other energy sources has consequently diminished and the economies of most states have become entirely dependent on oil and gas imports (Davitashvili, Elizbarashvili, 2012).

The unprecedented growth of demand for energy resources and the enormous scale of their use has made it clear that humanity is facing serious problems. It has become clear to everyone that the supply of energy resources can not continue eternally and a lot of effort is needed to avoid the current situation.

The above does not mean that the situation is catastrophic. Mankind is not facing an energy famine yet, but the fact that the era of cheap energy sources (oil and natural gas) is finally over, is beyond doubt.

Fundamental changes in the world energy balance are inevitable within the next 60-80 years. Today, humanity has entered into the so-called "transitional" period, which means the gradual movement from the consumption of energy resources to the consumption of alternative sources (solar, wind, tidal, geothermal, water) of energy. This is a very difficult and expensive task. There are incredible difficulties in replacing petroleum products with other energy sources in several industries (for example transport), that new energy sources should have the same power and effect as, for example, petrol or aviation kerosene. But sooner or later this must happen because, in addition to the limited supply of organic fuel resources, the widespread use of this raw material has already brought severe environmental consequences.

The Problem of Renewable Energy Resources

Coal. The share of proven coal reserves in the entire geological reserves is quite small (about 7%, or 700 billion tons), but given the current level of consumption and future trends, these reserves alone will be available for mankind for more than 200 years.

The sharp rise in oil and gas prices in recent years has significantly increased the role and importance of coal. The volume of coal stockpiles has also increased. All this indicates that the share of coal in organic fuels and energy resources will increase significantly.

Oil. According to modern estimations, there are 1.65 trillion barrels of proven oil reserves in the world as of 2016 (About 170 billion tons) (Wordometers, 2016). About 40% of this amount belongs to the traditional oil category. Annual world oil production exceeds 35 billion Barrels (More than 5 billion tonnes) (Wordometers, 2016). Almost 2/3 of the stock comes to the Middle East region. The main consumers are the European states, North America and Japan, although the share of China, India and Southeast Asia is also rapidly growing.

It is expected that the rather high volume of oil production will be maintained until the 2040s. Then, a decrease in values is expected, on the one hand, due to the depletion of stocks and, consequently, a sharp rise in prices, and on the other hand, due to the development and introduction of alternative energy sources.

Natural gas. Natural gas is one of the most convenient and high-calorie fuel-energy resources. Its special popularity and growing role are due to its high thermal capacity, as well as the relative easiness of extraction and transportation, as well as ecological cleanliness. It is therefore not surprising that the imbalance between liquid and gaseous fuels shifts in favor of the latter, especially when the world supply of natural gas to mankind will last for a longer time than oil supply. However, the share of oil is expected to increase again shortly, as the supply of oil that is technically feasible (but not yet in operation today) is significantly higher than the total supply of natural gas. In addition, transporting gas across the ocean (i.e. liquefying gas and converting it to a gaseous state on the ground) increases the cost several times and is highly unprofitable. As of January 1, 2020, there were an estimated 7,257 trillion cubic feet Tcf (more than 150 trillion Cubic meters) of total world-proved reserves of natural gas (US Energy Information Administration, 2021).

Nuclear fuel. The main source of nuclear energy is natural uranium ore. However, it is important to consider the costs associated with its extraction and processing, as uranium mining is unprofitable due to the high cost. Many experts thought that nuclear power was the energy of the future and that its share would grow steadily, but this position has been changed after the 1986 Chernobyl disaster and the 2011 Fukushima earthquake in Japan, when many countries are intending to close the existed atomic power stations and accordingly are refusing the construction of new nuclear power stations. According to current trends, the share of nuclear energy will decrease in the future and the importance of non-traditional, renewable energy sources will increase at its expense. As of 2017, identified uranium reserves recoverable at US$130/kg were 6.14 million tons (compared to 5.72 million tons in 2015). At the rate of consumption in 2017, these reserves are sufficient for slightly over 130 years of supply (Chitadze, 2017).

Renewable energy resources - The high capital capacity is a limiting factor for the consumption of hydropower resources. Only very expensive large hydropower plants, with their large volume of dams (which in turn cause flooding of huge areas and serious environmental problems), can produce the same amount of electricity as large thermal or nuclear power plants. The construction of small-capacity hydropower plants is less profitable. It should also be noted that most of such hydropower plants depend on the regime of the rivers and may stop altogether during low levels of water.

Some countries use geothermal resources as their energy source. In countries such as Iceland and New Zealand, where there is a high volume of geysers and groundwater, geothermal power plants may be able to satisfy the entire electricity demand. This direction is promising in several other regions, but overall, worldwide, geothermal resources are unlikely to play an important role in solving energy problems.

Some researchers were optimistic about the inexhaustible sources of energy (solar, wind, tidal energy) and considered them to be the future of world energy and the solution to energy problems. Clearly, with technical progress and scientific advances, the role of non-traditional energy sources will increase significantly and to some extent replace even traditional fuel and energy resources, but their consideration as a solution for mankind's energy problems is exaggerated and far from reality.

Another possible resource for solving energy problems is considered biomass, having been obtained from the recycling of agricultural, industrial, and household organic waste, from which energy can be received in the form of biogas. However, this resource also only can be considered as an additional source. As for growing different crops as an energy source (e.g., extracting alcohol from sugar cane and using it as fuel), this is probably a less promising way, due to its yield and high cost.

All this shows that renewable and inexhaustible sources of energy can only play a supporting role in solving energy problems. So far they can not compete (except for hydro, and in some regions geothermal resources) with non-renewable fuel and energy resources. The latter will likely remain a major resource in the first half of the 21st century.

Possible Consequences of the Natural Resources Consumption

In the 19th century, the spread of engines that convert thermal energy into mechanical and electrical energy led to an unprecedented increase in the consumption of fossil resources. In our era, the rapid development of nuclear energy, rocket technology, and radio electronics has turned radioactive, metallic, and non-metallic elements (uranium, germanium, silicon, and others) into the most important types of natural resources. Technical progress sometimes even cuts off one or another element of nature from the composition of natural resources, for example, small deposits, water streams, small beds of land, etc. In modern technical conditions, they have lost their original meaning.

In the process of production of material wealth, a man actively affects a single element of nature and changes the whole complex of natural resources. One part of the natural resources are inexhaustible because, in the wake of their correct exploitation, self-renewal is also taking place. There are also natural resources whose supply is gradually decreasing and will eventually run out. Therefore, the rational use of natural resources is the primary concern of mankind. The modern scientific-technological revolution further expands the field of the use of natural resources and allows us to identify new types of natural resources.

As it is known, most probably, the high standard of living depends entirely on the availability of natural resources. Apart from abiotic and biotic raw materials, humanity uses water, soil, air, biodiversity, and land as habitats and for recreational purposes; and for energy, it is used wind power, solar power, and tidal flows. These resources also serve as emission sinks, waste dumps, and as indispensable production factors for farming and forestry.

But unfortunately – and inevitably – resource use across the entire supply chain generates environmental pollution; plus worldwide resource use is growing steadily.

The way humanity uses resources often provokes irreversible environmental changes. The extraction and processing of non-regenerative raw materials is often an energy-intensive activity associated with large-scale disturbance to ecosystems and water balance, resulting in air, soil, and water pollution. Even the extraction and production of renewable resources are often associated with the intensive use of energy, materials, chemicals, and, in some cases, water; and all these result in pollution. New land is often converted to arable land, and in some cases, entire ecosystems are destroyed in the process (Umwelt and Bundesamt, 2019).

In short, the extraction and processing of raw materials always have an impact on the environment, leading to soil degradation, water scarcity, loss of biodiversity, disruption of ecosystem functions, and exacerbation of global warming. And that is not all. The use of products made from raw materials almost always results in greenhouse gas emissions, pollution, ecosystem damage, and/or loss of biodiversity. Products need energy and water, as well as land for delivery, marketing, and use. Improper use of the product provokes harmful emissions that can enter our water, soil, and air. The very elements of infrastructure that society takes for granted, such as our homes, not to mention the myriad daily activities, often require extensive use of resources and cause new land to be paved, damaging ecosystems and ruining the beauty of nature.

And even at the end of the supply chain, harm to the environment is inevitable. For example, recycling requires energy, using waste to generate energy generates greenhouse gases and other pollutants, and new land is constantly filled with waste dumps.

Thus, the use of resources already somewhat exceeds the regenerative capacity of the Earth because non-renewable natural resources are finite and their quality is often mediocre. Growing pressure on natural resources as a result of sustained population growth around the world could provoke competition from other potential uses.

Resource Use Also Has Social Implications

Countless social consequences of resource use are linked to issues such as the distribution of raw materials, free access to clean water, and world food security.

The consumption of raw materials per capita in the industrialized world is estimated to be four times higher than in less developed countries. However, while the lion's share of value added from resource use is generated in industrialized countries, less developed countries often bear the brunt of the environmental and social impacts of raw material production.

Production of Raw Materials

Residents of the regions in question report abuses such as severe human rights violations or irreversible environmental damage from the extraction of raw materials, often causing health problems as a result of air and drinking water pollution. Other consequences include the expulsion of the local population from their land and forcing them to settle elsewhere, not to mention growing poverty. In areas that suffer from these impacts, sustainability efforts by companies involved in mining and similar operations are few and far between. Moreover, in some states, profits from the production of raw materials are used to finance armed conflicts. According to UN statistics, natural resources play a key role in 40 percent of all intrastate conflicts (Umwelt and Bundesamt, 2019).

From a product lifecycle perspective, we here in Germany are at least partly responsible for these environmental and social outcomes due to our growing reliance on imported raw materials and products made from them. In some cases, this also applies to sustainable raw materials, such as animal feed and energy crops, which we use in such abundance and for which large tracts of fertile land are used. The use of fertilizers and pesticides is not controlled to some extent and no protective measures are taken. This may adversely affect the health of the local population. Displacement and forced relocation of local populations, as well as land grabbing, can undermine the food supply of local people, and unsustainable production methods often lead to soil degradation and water scarcity, as well as destroying much-needed fertile land.

Disposal of End-of-Life Products

The disposal of end-of-life products can also have social consequences. Improper and illegal disposal of the waste being removed can lead to toxic emissions and serious illnesses – not to mention the fact that such work is often done by children.

Towards Transparency and Sustainable Development

In different regions of the world, people are pursuing various strategies aimed at more efficient use of resources and rational use of resources to keep the negative social and environmental consequences of resource use within reasonable limits. Also, the key to such efforts is waste management programs and product liability laws.

To achieve these goals, countries participate in the transfer of knowledge and technology aimed at promoting the efficient use of resources and rational use of resources.

However, sustainable and efficient use of resources is often only achievable if sustainable development standards are defined and followed. Certification measures are a key tool in this regard and to increase transparency in the production of raw materials. A good example of this is the Extractive Industries Transparency Initiative (EITI).

To achieve these goals, UBA is working to strengthen sustainability standards as well as certification systems for land use and the production and use of abiotic and biotic raw materials (Umwelt and Bundesamt, 2019).

CONCLUSION

Natural resources — natural elements constituting the natural environment, which is used to meet the material needs of human society at a given stage of the development of productive forces. The main types of natural resources are mineral, climatic, water, soil, plant, and fauna resources. From the second half of the 20th century, atomic planetary-space resources were also added to them: nuclear, geothermal, sea circulation, and other energy sources. The history of mankind testifies that in the wake of the development of productive forces, there was an inexorable increase in the types of natural resources. Not only the quantity increase but also the share of each individual's consumption and the utility of society.

In the 19th century, the spread of engines that convert thermal energy into mechanical and electrical energy led to an unprecedented increase in the consumption of fossil resources. In our era, the rapid development of nuclear energy, rocket technology, and radio electronics has turned radioactive, metallic, and non-metallic elements (uranium, germanium, silicon, and others) into the most important types of natural resources. Technical progress sometimes even cuts off one or another element of nature from the composition of natural resources, for example, small deposits, water streams, small beds of land, etc. In modern technical conditions, they have lost their original meaning.

In the process of production of material wealth, a man actively affects a single element of nature and changes the whole complex of natural resources. One part of the natural resources are inexhaustible because, in the wake of their correct exploitation, self-renewal is also taking place. There are also natural resources whose supply is gradually decreasing and will eventually run out. Therefore, the rational use of natural resources is the primary concern of mankind. The modern scientific-technological revolution further expands the field of the use of natural resources and allows us to identify new types of natural resources.

At the same time, it exists a necessity to discuss several aspects related to the determination of the concepts of the rational and irrational consumption of natural resources.

Rational use of natural resources - To consider all aspects of this concept, you must first try to explain it. What is environmental management and what is it? The rational use of natural resources is a production activity aimed at satisfying human needs through the full exploitation of extracted resources: the recovery of renewable natural resources and the use of production waste are ensured, which, in turn, contributes to the preservation of the environment. In other words, rational management of nature is waste-free, ecologically clean production, the goal of which is to meet the needs of society. The main goal of the rational management of nature is the scientifically justified effective use of natural resources, which contributes to the preservation of the natural environment as much as possible and minimizes the possibility of self-regeneration of biogeocenoses. Therefore, the rational use of natural resources should be very useful both for the person himself and for all the living beings around him. First, it saves the environment from unprocessed excess production and release of harmful substances in it, which is important for the health of any living organism; secondly, it allows saving and saving resources; Thirdly, it provides people with a means of livelihood and, fourthly, it encourages the development of science and the emergence of new technologies. So, the rational use of natural resources allows you to protect nature from the negative effects of production factors. how does this happen To protect the environment from the harmful effects of the enterprise, it is necessary to optimize the production and find the type of resources that were used as much as possible by people and caused relatively little damage to nature. Relatively clean nuclear power plants can become one example of environmental management. Unlike thermal power plants, nuclear power plants do not emit harmful substances into the atmosphere. The second advantage of nuclear power plants can be the absence of oxygen consumption, while thermal power plants consume about 8 million tons of oxygen per year to oxidize fuel. In addition, coal-fired power plants release more radioactive substances into the environment than nuclear power plants. Another advantage of nuclear power plants is the diversion of energy for heating and hot water supply in cities, which also helps to reduce unproductive heat losses. In addition, wave power plants are another example. This type of power plant can become a wave absorber that protects ports, shores, and harbors from destruction. In addition, wave power plants also save resources and are more profitable compared to wind power plants. They also protect the environment from harmful emissions. Another type of ecological power plant is the sun. Their main advantage, first of all, lies in the availability and exhaustion of the energy source in the conditions of the constant increase in the prices of traditional energy sources. In addition, with the current level of consumption, a special advantage is absolute safety for the surrounding world. Also, waste-free production can be the multiple use of water from rivers, lakes, wells, and other sources in the technological process, because the water used is cleaned and again participates in the production process without harming the environment. Irrational use of natural resources is a production system in which readily available natural resources are developed on a large scale, while their rapid depletion occurs due to incomplete processing. Thus, a large amount of waste is distributed and environmental pollution occurs. This kind of management of nature is characteristic of the rapid development of the economy due to the lack of sufficiently developed scientific and technical potential, and although initially, such activities may bring good results, later they still lead to harmful consequences for the ecological environment. An example of irrational use of natural resources is the virgin land development campaign in the USSR in 1955-1965. The reasons for the failure of this company were many factors: The reasons for the failure of this company were many factors: the development of virgin lands began without infrastructure and preparation - there were no roads, no trails, and no qualified personnel. The natural conditions of the steppe were not taken into account: sandstorms and dry winds were not taken into account, and there were no-tillage methods and grain varieties adapted to this type of climate. It should be noted that plowing the land was done at a forced pace and with huge costs. Thanks to such a colossal concentration of funds and people, as well as natural factors, the new lands, gave a very high yield in the first years, and from the mid-1950s, from half to a third of the grain produced in the USSR. However, they were not able to achieve stability: during harsh years on virgin lands, they could hardly collect the seed fund. In addition, in 1962-1963 due to ecological imbalance and soil erosion. A shower of dust appeared. Either way, the development of virgin land has reached a crisis stage, and the efficiency of cultivation has decreased by 65% (Tltaudit, 2023). All these data only indicate that the soil was cultivated in a wide way, but, despite this, it did not produce results. On the contrary, the structure of the soil deteriorated, the level of harvest decreased significantly, and the funds did not justify their investment. All this, without a doubt, testifies to the inefficient use of resources in an attempt to quickly and immediately solve the problems of agriculture, without a solid foundation, neither science, nor high-quality technology, nor an appropriate level of infrastructure, so that the result is completely different.

Previously, we compared the two concepts of rational and irrational use of natural resources, and, having given them examples, we can connect their meanings, and compare and reveal the fundamental differences between them. These differences can essentially be defined as two ways of development: intensive and extensive. The first way fully corresponds to the rational use of natural resources. It indicates the efficient use of resources, which makes a significant contribution both to production in general and to high-quality and waste-free technologies, thus making production environmentally friendly and not harmful to nature. In addition, the intensive way often fully meets the cultural and material needs of the community. The second way, on the contrary, is used for the irrational use of natural resources. Its main characteristics are the disproportionate ratio between the resources spent and the result, the orientation towards spatial (quantitative) and not high-tech (qualitative) value, and also, often, the insecurity of social needs. Finally, a vast way in nature causes colossal damage through actions that are not based on any scientific development or technology, releasing chemically harmful and dangerous substances and other production wastes into the environment. Sometimes this damage can reach an ecological disaster and become the cause of negative global processes and phenomena around the world.

REFERENCES

AtlasW. (2013). https://www.worldatlas.com/aatlas/infopage/oceans.htm

Bliadze, M., Chanturia, G., & Kereselidze, D. (2021). *World Geography*. Bakur Sulakauri Edition.

Chitadze, N. (2004). *Caspian oil projects and transit role of Georgia*. Tbilisi State University.

Chitadze, N. (2017). *World Geography. Political, Economic and Demographic Dimensions*. Scholars Press.

Christie, A., & Brathwaite, R. (2011). *Mineral Commodity Report 14 — Gold.* Institute of Geological and Nuclear Sciences Ltd.

Dictionary, F. (2022). *Recreation*. Retrieved from: https://www.thefreedictionary.com/recreation

Elizbarashvili, N., & Davitashvili, Z. (2012). *Global Geography*. Tbilisi State University.

Energy, G. (2010). *World Natural Resources*. Retrieved from: http://greenenergyhomedesign.tk/tag/green-energy/page/237/

FAO. (2022). *Annual Report*. Retrieved from: https://www.fao.org/common-pages/search/en/?q=Annual%20Report%202022

Global Change. (2021). *Current Lectures on Land Degradation*. Retrieved from: http://www.globalchange.umich.edu/globalchange2/current/lectures/land_deg/land_deg.html

Guide, C. I. M. O. (2008). *Measurement of sunshine duration*. World Meteorological Organization.

Karumidze, A. (2021). *International Organizations* (7th ed.). Universali.

Leckie, G. (1993). Global oil, and gas fields, sizes tallied and analyzed. Oil and Gas Journal.

Maksakovsky, V. (2009). World Social and Economic Geography. Aspects Press.

Map of World. (2018). *Forests for the World Map*. Retrieved from: https://www.mapsofworld.com/world-natural-forest.htm

Marinebio. (2001). *Ocean Resources*. Retrieved from: https://marinebio.org/oceans/ocean-resources

Neidze, V. (2004). *World Economic and Social Geography*. Institute of Geography of Georgia.

Nellemann, C., & Corcoran, E. (2010). *Dead Planet, Living Planet - Biodiversity and Ecosystem Restoration for Sustainable Development: A Rapid Response Assessment. United Nations Environment Program*. GRID-Arendal.

Ocean resources Defense Council. (2014). *Oceans*. Retrieved from: https://www.nrdc.org/oceans/

Pinet, P. R. (1996). *Invitation to Oceanography* (3rd ed.). West Publishing Co.

Rechner. (2010). Letter to the Editor: Outdoor recreation stimulates the economy. *Washington Post.*

Reliefweb. (2005). *Senator Paul Simon Water for the Poor Act: 2005 Report to Congress*. Retrieved from: https://reliefweb.int/report/world/senator-paul-simon-water-poor-act-2005-report-congress

Tltaudit. (2023). *Examples of rational use of natural resources*. Retrieved from: https://tltaudit.ru/ka/cardiovascular-disease/racionalnoe-ispolzovanie-prirodnyh-resursov-primery-prirodopolzovanie-3---referat/

Umwelt and Bundesamt. (2019). *Resource use and its consequences*. Retrieved from: https://www.umweltbundesamt.de/en/topics/waste-resources/resource-use-its-consequences

UN. (2022). *International Day for Biological Diversity*. Retrieved from: https://www.un.org/en/observances/biological-diversity-day

UNEP. (2023). *Annual Report, 2022*. Retrieved from: https://www.unep.org/resources/annual-report-2022

United Nations. (2005). *Basic Facts about the United Nations*. UN Department of Public Information.

University of Oxford. (2022). *Science Encyclopedia*. Oxford University Press.

Water Footprint Network. (2014). *Water Footprint Network partners with WWF to create the "Living Planet Report 2014"*. Retrieved from: https://waterfootprint.org/en/about-us/news/news/living-planet-report/

World Ocean Review. (2010). *Marine Minerals*. Retrieved from: https://worldoceanreview.com/en/wor-1/energy/marine-minerals/

World Resources Institute. (2010). *Forests*. Sustaining Forests for People and Planet. Retrieved from: https://www.wri.org/our-work/topics/forests

Zgenti, K., & Kharitopnashvili, L. (2010). *World Economic and Social Geography*. Nekeri.

KEY TERMS AND DEFINITIONS

Coal: A combustible black or brownish-black sedimentary rock, formed as rock strata called coal seams.

Land Degradation: A process in which the value of the biophysical environment is affected by a combination of human-induced processes acting upon the land.

Mineral Wealth: Natural mineral substances contained in the depths in solid, liquid, or gaseous state (including ground water and therapeutic muds) fit for usage in material production, or hydrocarbons.

Natural Gas: A naturally occurring mixture of gaseous hydrocarbons consisting primarily of methane in addition to various smaller amounts of other higher alkanes. Usually low levels of trace gases like carbon dioxide, nitrogen, hydrogen sulfide, and helium are also present.

Non-Renewable Energy: Comes from sources that will run out or will not be replenished in our lifetimes—or even in many, many lifetimes. Most non-renewable energy sources are fossil fuels: coal, petroleum, and natural gas. Carbon is the main element in fossil fuels.

Oil: Any nonpolar chemical substance that is composed primarily of hydrocarbons and is both hydrophobic and lipophilic.

Raw Materials: Materials or substances used in the primary production or manufacturing of goods.

Renewable Energy: Energy that is collected from renewable resources that are naturally replenished on a human timescale. It includes sources such as sunlight, wind, the movement of water, and geothermal heat.

Chapter 4
An Essay on the Energy and Equity Relationship With a Panel Data Analysis (2011–2020)

Muhlis Selman Saglam
Canakkale Onsekiz Mart University, Turkey

ABSTRACT

This study investigated how income inequality is shaped according to energy-related variables. For this, income inequality is the independent variable; energy efficiency, energy import dependency, and energy efficiency are included in the model as independent variables. The data of the 17 founding countries of the OECD between 2011-2020 is used. According to the F Test results, the model has a one-way unit effect. According to the robust Hausman test, the random effect is valid. There are heteroscedasticity and autocorrelation in the context of deviations from the assumption. On the other hand, the data fit a normal distribution. Also, there is no multicollinearity and no inter-unit correlation. In light of all these, the model was estimated with Arellano, Froot, and Rogers Estimator. According to the model estimation results, the model is not statistically significant. This shows that the relationship between income distribution and energy, which Illich expressed qualitatively, is not quantitatively valid for the 17 OECD founding countries.

DOI: 10.4018/978-1-6684-6727-5.ch004

INTRODUCTION

The crises experienced on a global scale inevitably determine the economic literature during and after the crisis. Several important events in the 1970s also shaped mainstream economics literature in this direction. The Bretton Woods system, which expresses an essential stage in the acceleration of globalization through economic institutions (such as the World Bank and IMF), has been seen as the reason for the spread of the problem of inequality and poverty (Soba et al., 2018). Globalization also reveals accumulated inequality (Voutsa, Borovas, 2015). This inequality problem is among the reasons for the collapse of the Bretton Woods system in 1973. The year 1973 is also crucial because it was the year that the OPEC energy crisis emerged. Therefore, it would not be wrong to describe the 70s as the years when increasing inequality became a chronic problem, and the effects of the energy crisis were seen and discussed.

Energy and inequality are the main topics discussed in economics literature in the 1970s. Skorov (1973), Yan (1975), Fuss (1977), Berry et al. (1978), Greenman (1979), Atkinson (1970), Allingham (1972) put this issue at the center of their discussions. Griffin (1974), Gastwirth (1975), Shorrocks (1978)) are not only discussed in the discipline of economics, but also in the field of social sciences, which examines the social aspect of these debates. The publications of Ahluwalia (1976), Kafandaris (1978), Scardigli (1975), and Garnier and Hout (1976) are the examples.

Ivan Illich, from the discipline of sociology, is one of the names who worked on the issue of energy and inequality in the 70s. Illich has adopted a critical approach to fields such as education, energy, and institutions in these fields in a way that also touches economics. His contribution to the literature on the relationship between education and inequality, the "deschooling of society," has made it known to a broader audience. Illich has carried out studies dealing with the issue of inequality in the field of energy as well as in the field of education. According to Illich, there is a relationship between the amount of energy and the struggle of that country against inequality:

Often overlooked is the fact that equity and energy can only develop together up to a point... More energy abundance means less distributional control over that energy. (Illich, 1974).

Accordingly, increasing a country's energy abundance reduces that country's control over energy. The reduction of control will initially create an inequality in the energy field. However, this inequality will spread to different areas in the process.

In our study, the problem of inequality expresses the inequality of income distribution. The relationship of this inequality with energy-dependent variables will be modeled. In the 1970s, developments in the field of energy were experienced with globalization; however, in this process, there has been a period in which indicators of increasing social and economic inequality have begun to emerge. However, the number of studies dealing with the relationship between energy and inequality in a critical perspective is not sufficient for this period. Our study aims to contribute to the literature by testing the claim of Illich, who has made a name for himself with his critical approaches to this literature, and by examining the data of OECD countries where energy and inequality debates are at the center. The following sections will include a literature review, data set, variable selection, model selection, interpretation, and future research recommendations.

Literature Review

The debate on energy and inequality has an important place in the mainstream economics literature of the 70s, with the effect of the global energy crisis. White (1973) stated that the U.S. energy crisis was a potential economic crisis during the OPEC crisis. The symptoms of this crisis are increased energy consumption, decreased local reserves, and environmental degradation. Incze et al. (1973), Sagner (1974), Ullrich (1974), Shinnar (1975), and Dvoskin and Heady (1976) are some of the researchers who published on the effects of the energy crisis in this period.

Sagner (1974) analyzed the impact of the energy crisis on transportation in selected American cities and offered recommendations. Ullrich (1974) discussed whether the energy crisis experienced in the same year would lead to a food crisis in the future and how it should be tackled. Shinnar (1975), on the other hand, examined the energy crisis through its impact on the transportation sector. Especially after the crisis, claims that owning a personal car will become obsolete in the future and a carless society may emerge. Shinnar (1975) stated that these views do not have a strong basis. According to him, it is possible to save more energy with the improvements made in the design of personal vehicles compared to public transportation. Dvoskin and Heady (1976) used a cross-regional, large-scale, linear programming model to assess the economic and environmental impacts of the energy crisis on agricultural production in the United States. The study examines the behaviour of crop production in the context of energy minimization, energy scarcity, high energy prices, and high agricultural exports accompanied by high energy prices. The overall impact of an energy crisis on the farm environment is somewhat unclear.

On the other hand, reducing energy use will also reduce the amount of tillage. Since the OPEC Crisis emerged due to a sharp decrease in oil supply, the USA was also at the forefront of the countries most affected by this crisis, at the beginning of the oil-dependent countries at that time. Therefore, as seen in the examples given above, there have been many academic studies on the effects of the energy crisis on the USA in the 1970s.

Mayer (1978) discussed how the sharp decrease in oil-related energy supply affects demand. As the Arab oil embargo demonstrates, the energy crisis has spurred the private and public sectors to reduce domestic demand for natural gas and oil. In this study, carried out in the case of America, it was seen that the residential consumer was encouraged to reduce the energy demand at home through various mechanisms. As a result, the need for natural gas decreased significantly.

Although studies on the U.S. scale constitute the central axis in the literature on the oil crisis, it is possible to come across studies dealing with the effects of the crisis on different countries. Haberstroh (1978) examined the impact of the crisis on Hungary's economic reform process. The article analyzes the reaction of the Hungarian economy to the increase in world oil prices since 1973, dependent on foreign trade, insufficient energy resources, tight-knit to the USSR, and having the most liberal of Eastern European economic reforms. In this process, Hungary accelerated its domestic energy production and shifted its energy use structure from oil to coal. However, the energy crisis did not cause Hungary to abandon its economic reform. Schuh (1978) examined how the Brazilian economy reacted to the energy crisis. This article argues that Brazil responded to the energy crisis by returning to its policies of autarky through import substitution. Oil imports have been replaced by nuclear energy and domestic oil. In order to manage the balance of payments with fixed exchange rates, even the imports of goods that have been chiefly imported until now have been replaced by domestic production. Liew (1980) examined the effects of the increase in oil prices on the Korean economy. The first serious problem faced by the Korean economy in those years was stagflation caused by the rising energy cost of imported oil. The study sought to determine how much the oil crisis contributed to Korean inflation and unemployment. In these years, when the effects of the energy crisis were felt, Uri (1979) examined the impact of energy demand and consumption on Indian trade. Tsurumi (1978) discussed how the Japanese economy coped with the oil crisis. Targets for leading export industries and income distribution were supported through "managed competition" and pricing policies. Managed competition was a key policy in the solid economic growth of the 1960s. A notable policy change is the promotion of sales in Japan by sizeable foreign oil companies.

The energy crisis of the 1970s was primarily discussed in the economics literature as which policies countries overcame the crisis. At the same time, it is possible to see studies that interpret the effects of the crisis based on specific sectors of the countries in the literature. We can say that these contents constitute most of the studies carried out in the energy field, especially until the 1980s. However, it is possible to say that there is less focus on the effects of the energy crisis on the inequality problem. In particular, the potential social and economic impacts of increasing energy costs have not been adequately discussed. Yan (1975) stated in his study that since the 1973 energy crisis, the energy goal has generally been to achieve independence through conserving energy consumption and expanding local supply. However, the possible social and economic implications of this policy's significantly higher energy costs are not fully understood. This article examines one aspect of the repercussions of energy cost, namely the income distribution effect. The study estimated price changes caused by high energy costs and changes in consumption expenditures for different income groups.

The literature on the relationship between energy and inequality has generally been examined through energy efficiency. Respecting the link between energy efficiency and income inequality, some researchers have directly discussed possible causality. Berndt and Morrisson (1979) discussed the relationship between the energy crisis and the problem of inequality through income distribution. The article examines the income distribution and employment effects of increasing energy prices in U.S. manufacturing. Previous research has focused almost entirely on consumption effects. It is concluded that energy price increases will decrease considerably when only direct energy expenditures are taken into account but will decrease significantly less with the inclusion of indirect energy expenditures. Although the share of total labor cost is not affected by rising energy prices, rising energy prices in the total labor force increase the demand and total income of blue-collar workers and cause a decrease in the relative employment and total income of white-collar workers.

On the socioeconomic effects of the energy crisis during this period, Frankena (1978) argued that declining net energy would affect major changes in the regional distributions of economic activity and population in industrialized societies. It is argued that economic theory generally is static and growth-dependent, thus incapable of addressing the human ecological processes involved in this transition. In this case, the approach of disciplines such as sociology to the issue is critical.

Illich (1974), as a sociologist, continued his approach that centered the problem of inequality on energy. However, being a researcher in the discipline of sociology allowed him to approach the issue of socioeconomic inequality from a broader perspective. According to him, the energy input of a country will increase inequality, inefficiency, and personal weakness after a certain point. Illich claims that the concept of the energy crisis has been widely adopted by researchers unaware of this reality. Examining this through the USA and Mexico, Illich argues that neither the shortage of oil nor the ineffective use of available energy is the cause of the danger of social collapse in both countries. This threat comes from industries imposing on society the amount of energy degrades and impoverishes people.

As can be seen, Illich, as a sociologist, includes the dynamics of the society in his interpretation of inequality. In doing so, it also receives support from economic concepts and institutions. However, he did not test his claims with quantitative data and models. In this case, it will not be possible to say the same for the discipline of economics, no matter how much their discourse resonates and is accepted in its field. The originality of our work emerges at this point. In our study, some of the claims of Illich, who examined the energy crisis in line with its relationship with the problem of inequality in his period, and touched upon extremely remarkable points that also had socioeconomic aspects, were tested. For this, the claims in question were modeled on data. The data of the 17 founding countries of the OECD between 2011-2020 were used. Since the econometric testing of the claims of Illich, a sociologist, was carried out for the first time in our study, we think the study will make an essential contribution to the literature.

The fact that the model we have established in our study is based on Illich's claims does not mean there is no relationship with today's literature. The relationship between the energy crisis and the problem of inequality is as important today as it was in the past. Energy poverty is seen as one of the main phenomena of the energy industry in the coming years. In addition, Illich expressed his claims in a language that covers the period and the future. In addition, when we look at the current literature on the energy crisis, there are hardly any studies using panel data analysis on the relationship between inequality and energy. Sinha et al. (2022) discussed the relationship between energy and inequality through unequal access to energy. Energy poverty is a critical problem in the world. According to the article, the interaction between the labor force and the energy market can pave the way for solving the problem. From this point of view, this study examined the factors that create inequality in access to energy. The inequality of access to energy is one of the points that Illich also mentioned. Hence, Sinha et al. (2022) are one of the studies that chose a research topic that is relatively close to our study.

On the other hand, Hassan et al. (2022) drew attention to the issue of energy poverty. Bardazzi et al. (2021) similarly emphasized the point of energy poverty. However, they also included the income inequality we used in our study. Energy poverty is associated mainly with the concept of income poverty. Recent studies in the relevant literature explore whether energy poverty is related to a state of social vulnerability based on structural inequalities. However, it is crucial to how valid this relationship is for income inequalities.

Lee et al. (2022) started with a research question similar to our study by examining income inequality by associating it with the energy issue. While the literature examines income inequality and the various factors that affect the impact of energy security on the economy, information on the link between energy security and income inequality is quite limited. Based on six indicators that capture four aspects of energy security, including usability, accessibility, scalability, and acceptability, this research applied the dynamic panel threshold approach to investigate whether 68 countries affected income distribution through global panel data between 2001 and 2018. According to the study, while energy security tends to deteriorate income distribution in the early stages of economic development, it improves income distribution after reaching a certain level of development. Adom et al. (2021) conducted a study in this direction, as the empirical evidence supporting the claim that energy efficiency improvements drive economic growth is still mixed. In a world of widening inequality, how income inequality dynamics affect growth and the energy efficiency link is critical, but empirical research exploring the role of income inequality is lacking. This study examines the regulatory role of income inequality in the connection between economic growth and energy efficiency. The study is based on an unbalanced annual panel dataset for 51 African countries from 1991 to 2017. The point where our research differs from these studies is to increase the number of changes in the energy title. Lee et al. (2022) use energy security as a variable, while Adom et al. (2021) prefer energy efficiency. In our study, energy efficiency, energy import dependency, and energy efficiency are included under the title of energy.

Turning to renewable energy sources to reduce energy and income inequalities is one of the topics researched in the literature. Priesmann et al. (2022) investigated the effects of renewable energy support taxes on income inequality and energy poverty. Representative pooled cross-sectional data covering more than 40,000 German households between 2003 and 2018 were used to estimate the price and income elasticity of energy consumption and to simulate the effects on income inequality and energy poverty. The results show that renewable energy support taxes increase income inequality and energy poverty. Uzar (2020) investigated whether income inequality is a driving force for renewable energy consumption. Many studies have investigated the impact of income inequality on environmental problems. The question of whether income inequality affects renewable energy consumption has become important with the increasing awareness of renewable energy. Uzar (2020) carried out the first known study on this relationship. In the study, the effect of income inequality on renewable energy consumption in 43 developed and developing countries for the years 2000-2015 was examined theoretically and empirically. The results show that a decreased income inequality will increase renewable energy consumption. In other words, policymakers have the opportunity to reduce income inequality and environmental degradation simultaneously.

Our study will contribute to the literature examining the relationship between energy and inequality in terms of empirically testing Illich's claim, which brings his own critical approach to this field, for OECD countries. The trajectory of the energy and inequality literature indicates an increasing trend to focus on renewable energy sources. As stated above, our study dealt with the relationship between the unique variables in the energy title and income inequality. The model can be expanded in future studies by following the literature course and including renewable energy sources data.

Data Set, Model, and Interpretation

Our study investigated how income inequality is shaped according to energy-related variables. For this, income inequality is the dependent variable; energy efficiency, energy import dependency, and energy productivity are included in the model as independent variables. The classical representation of the regression model is as shown in equation (1) below:

$$IoID_{it} = \beta_0 + \beta_1(enereff)_{it} + \beta_2(enerimdep)_{it} + \beta_3(enerprod)_{it} + u_{it} \quad (1)$$

Data were obtained from the Eurostat website. For the model, the classical two-dimensional panel data consisting of unit and time dimensions was used using the data of the 17 founding countries of the OECD between 2011-2020. Income distribution inequality is the ratio of the total income of the 20% of the population with the highest income (the highest 20%) to the total income of the 20% of the population with the lowest income (the lowest 20%). Income should be understood as equivalent disposable income (Eurostat, 2022). Energy efficiency refers to millions of tons of oil equivalents (Eurostat, 2022). Energy import dependency is shown as a percentage (Eurostat, 2022). Finally, energy productivity is the gross domestic product (GDP) for a given calendar year divided by the gross usable energy. It measures the productivity of energy consumption and provides a picture of the degree to which energy use diverges from GDP growth (Eurostat, 2022).

While creating the model, logarithms of *enereff, enerilmep,* and *enerprod* variables were taken due to the scale difference. In the model estimated by the fixed effects in-group estimator, the existence of a possible unit effect in the model was observed as a prior observation. As a priori information, it was seen that there was no time effect in the model. According to the F Test results, while there is a unit effect in the model, there is no time effect. Therefore, we can say that the one-way unit effect is valid in the model. According to the resistant Hausman test statistic, random effects are valid in the model.

It was observed that heteroscedasticity and autocorrelation were present in the tests performed to test the presence of deviations from the assumption. On the other hand, the data fit a normal distribution. Also, there is no multicollinearity and no inter-unit correlation. In light of all this, the model can be estimated with the Arellano, Froot, and Rogers Estimator. It is an estimator developed by Arellano

(1987), Froot (1989) and Rogers (1993) which provide robust parameter estimates in the case of autocorrelation and heteroscedasticity. The model estimation result is shared in ***Table 1*** below. According to the model estimation results, the model is not statistically significant. Assuming a time effect, the same model and all variables are significant. However, it would not be correct to interpret this. On the other hand, the model and variables are not statistically significant in estimating resistant regression under the presence of unit effect. This shows that the relationship between income distribution and energy, which Illich expressed qualitatively, is not quantitatively valid for the 17 OECD founding countries. Although Illich's claim made an impact on the relationship between energy and inequality in the 70s, during the consequences of the energy crisis were experienced, his claim was not supported by quantitative data. Although there are studies that quantitatively explain the relationship between energy and economic inequality, this claim is not confirmed in the model we have built on OECD countries.

Table 1. Arellano, Froot, and Rogers Random Effects Estimator

loID	Coef.	Robust Std. Err.	Z	P>lzl
lenereff	.2250596	.1759222	1.28	0.201
lenerimdep	.1709622	.1746633	0.98	0.328
lenerprod	-.3212121	.565673	-0.57	0.570
cons	4.103139	1.380739	2.97	0.003

CONCLUSION

In the 70s, the mainstream economics literature was surrounded by an energy crisis, the effects of this crisis, and solutions and policy proposals for it. This situation attracted attention to the environment in social sciences, especially economics. The socioeconomic impacts of the crisis were handled with different approaches. Especially in sociology, the social effects of the crisis were investigated on an inequality basis. Illich is a researcher who examines the economic impacts of various institutions (such as energy, and education) through inequality and draws attention to this approach. Therefore, at a time when the issue of energy and equality came to the fore again after the crisis, he argued with some qualitative claims that the reason for the social collapse was not the oil or other energy crises but rather the increasing industrialization and therefore the imposition of the amount of energy on the society. However, this approach to the relationship between energy and inequality did not end in the literature of the 70s. Especially today, the literature has begun to evolve into the relationship between renewable energy sources and income distribution inequality with the increasing awareness of renewable energy sources.

In our study, starting from the energy and equality debate context, we examined the relationship between income distribution inequality and three main energy-related variables. We expected to obtain a statistically meaningful model in line with Illich's claims. It was seen by tests that there is a unit effect in the model but no time effect. Contrary to our expectations, the model estimated by the resistive estimators in the presence of unit effect did not turn out to be significant. Similarly, it was seen that the variables other than the constant parameter were not statistically significant. On the contrary, when the time effect was assumed, the model and the variables were statistically significant. Still, this model could not be interpreted because there was no time effect.

As a result, contrary to our expectations, the model established based on some of Illich's qualitative claims did not reveal a statistically meaningful relationship. However, as mentioned in the literature review section, data on renewable energy has an increasing trend in the energy and equity literature. In future studies, models including this variable may give different results.

REFERENCES

Adom, P. K., Agradi, M., & Vezzulli, A. (2021). Energy efficiency- economic growth nexus: What is the role of income inequality? *Journal of Cleaner Production, 310*, 1–20. doi:10.1016/j.jclepro.2021.127382

Ahluwalia, M. S. (1976). Inequality, poverty and development. *Journal of Development Economics, 3*(4), 307–342. doi:10.1016/0304-3878(76)90027-4 PMID:12339111

Allingham, M. G. (1972). The measurement of inequality. *Journal of Economic Theory, 5*(1), 163–169. doi:10.1016/0022-0531(72)90124-X

Arellano, M. (1987). Computing robust standart errors for within group estimators. *Oxford Bulletin of Economics and Statistics, 49*(4), 431–434. doi:10.1111/j.1468-0084.1987.mp49004006.x

Atkinson, A. B. (1970). On the measurement of inequality. *Journal of Economic Theory, 2*(3), 244–263. doi:10.1016/0022-0531(70)90039-6

Berndt, E. R., & Morrisson, C. J. (1979). Income redistribution and employment effects of rising energy prices. *Resources and Energy, 2*(2-3), 131–150. doi:10.1016/0165-0572(79)90026-4

Berry, R. S., Salamon, P., & Heal, G. (1978). On a relation between economic and thermodynamic optima. *Resources and Energy*, *1*(2), 125–137. doi:10.1016/0165-0572(78)90002-6

Dvoskin, D., & Heady, E. O. (1976). Farming practices, environmental quality, and the energy crises. *Agriculture and Environment*, *3*(1), 1–13. doi:10.1016/0304-1131(76)90002-3

Eurostat. (2022a, October 27). *Energy efficiency*. https://ec.europa.eu/eurostat/databrowser/view/nrg_ind_eff/default/table?lang=en

Eurostat. (2022b, October 27). *Energy imports dependency*, https://ec.europa.eu/eurostat/databrowser/view/nrg_ind_id/default/table?lang=en

Eurostat. (2022c, October 27). *Energy productivity*. https://ec.europa.eu/eurostat/databrowser/view/t2020_rd310/default/table?lang=en

Eurostat. (2022d, October 27). *Inequality of income distribution*. ec.europa.eu/eurostat/databrowser/view/tespm151/default/table?lang=en

Frankena, F. (1978). Regional socioeconomic impacts of declining net energy. *Urban Ecology*, *3*(2), 101–110. doi:10.1016/0304-4009(78)90001-3

Froot, K. A. (1989). Consistent covariance matrix estimation with cross-sectional dependence and heteroscedasticity in financial data. *Journal of Financial and Quantitative Analysis*, *24*(3), 333–355. doi:10.2307/2330815

Fuss, M. A. (1977). The demand for the energy in Canadian manufacturing: An example of the estimation of production structures with many inputs. *Journal of Econometrics*, *5*(1), 89–116. doi:10.1016/0304-4076(77)90036-7

Garnier, M., & Hout, M. (1976). Inequality of educational opportunity in France and the United States. *Social Science Research*, *5*(3), 225–246. doi:10.1016/0049-089X(76)90001-6

Gastwirth, J. L. (1975). The estimation of a family of measures of economic inequality. *Journal of Econometrics*, *3*(1), 61–70. doi:10.1016/0304-4076(75)90067-6

Greenman, J. V. (1979). An energy analysis of the investment problem. *Energy Economics*, *1*(4), 219–223. doi:10.1016/0140-9883(79)90004-5

Griffin, K. (1974). The international transmission of inequality. *World Development*, *2*(3), 3–15. doi:10.1016/0305-750X(74)90091-6

Haberstroh, J. R. (1978). The case of Hungary: Liberal socialism under stress. *Journal of Comparative Economics*, *2*(2), 111–125. doi:10.1016/0147-5967(78)90043-4

Hassan, S. T., Batool, B., Zhu, B., & Khan, I. (2022). Environmental complexity of globalization, education, and income inequalities: New insights of energy poverty. *Journal of Cleaner Production*, *340*, 1–12. doi:10.1016/j.jclepro.2022.130735

Illich, I. (1974). *Energy and equity*. Marion Boyars.

Incze, J. S., & Merriam, J. Jr. (1973). The energy crisis. *Lancet*, *302*(7834), 918. doi:10.1016/S0140-6736(73)92057-6 PMID:4126964

Kafandaris, S. (1978). Restructuring images of inequality. *Habitat International*, *3*(1), 83–87. doi:10.1016/0197-3975(78)90040-1

Lee, C. C., Xing, W., & Lee, C. C. (2022). The impact of energy security on income inequality: The key role of economic development. *Energy*, *248*, 1–14. doi:10.1016/j.energy.2022.123564

Liew, C. K. (1980). The impact of higher energy prices on growth and inflation in an industrializing economy: The Korean experience. *Journal of Policy Modeling*, *2*(3), 389–408. doi:10.1016/0161-8938(80)90030-7

Mayer, L. S. (1978). Estimating the effects of the onset of the energy crisis on residential energy demand. *Resources and Energy*, *1*(1), 57–92. doi:10.1016/0165-0572(78)90014-2

Priesmann, J., Spiegelburg, S., Madlener, R., & Praktiknjo, A. (2022). Does renewable electricity hurt the poor? Exploring levy programs to reduce income inequality and energy poverty across German households. *Energy Research & Social Science*, *93*, 1–20. doi:10.1016/j.erss.2022.102812

Rogers, W. H. (1993). Regression standart errors in clustered samples. *Stata Technical Bulletin*, *13*, 19–23.

Sagner, J. S. (1974). The impact of the energy crisis on American cities based on dispersion of employment, utilization of transit, and carpooling. *Transportation Research*, *8*(4-5), 307–316. doi:10.1016/0041-1647(74)90049-5

Scardigli, V. (1975). Lifestyles, inequalities and the process of social change in France. *Futures*, *7*(3), 197–208. doi:10.1016/0016-3287(75)90064-6

Schuh, G. E. (1978). The case of Brazil: Import substitution revisited. *Journal of Comparative Economics*, *2*(2), 97–110. doi:10.1016/0147-5967(78)90042-2

Shinnar, R. (1975). The effect of the energy crisis on the private car in the U.S. *Transportation Research*, *9*(2), 87–95. doi:10.1016/0041-1647(75)90043-X

Shorrocks, A. (1978). Income inequality and income mobility. *Journal of Economic Theory*, *19*(2), 376–393. doi:10.1016/0022-0531(78)90101-1

Sinha, A., Lorente, D. B., Zafar, M. W., & Saleem, M. M. (2022). Analyzing global inequality in access to energy: Developing policy framework by inequality decomposition. *Journal of Environmental Management*, *304*, 1–17. doi:10.1016/j.jenvman.2021.114299 PMID:34923413

Skorov, G. (1973). The green revolution and social progress. *World Development*, *1*(11), 13–21. doi:10.1016/0305-750X(73)90252-0

Soba, M., Kumcu, S, & Akıncı, Ö. (2018). Bretton Woods ikizler: aşırı yoksulluk ve derinleşen eşitsizlik. *Süleyman Demirel Üniversitesi İktisadi ve İdari Bilimler Fakültesi Dergisi*, *23*, 669-687.

Ullrich, H. D. (1974). Energy crisis today, food crisis tomorrow? *Journal of Nutrition Education*, *6*(1), 4. doi:10.1016/S0022-3182(74)80171-8

Uri, N. D. (1979). Energy demand and interfuel substitution in India. *European Economic Review*, *12*(2), 181–190. doi:10.1016/0014-2921(79)90021-7

Uzar, U. (2020). Is income inequality a driver for renewable energy consumption? *Journal of Cleaner Production*, *255*, 1–11. doi:10.1016/j.jclepro.2020.120287

Voutsa, M. E., & Borovas, G. (2015). The operation of the Bretton Woods institutions in the modern globalized environment: Problems and challenges. *Procedia Economics and Finance*, *33*, 168–180. doi:10.1016/S2212-5671(15)01702-5

White, D. C. (1973). The U.S. energy crisis: A scientist's view. *Energy Policy*, *1*(2), 130–135. doi:10.1016/0301-4215(73)90035-9

Yan, C. (1975). Income distribution effect on higher energy costs. *The Journal of Behavioral Economics*, *4*(2), 45–53. doi:10.1016/0090-5720(75)90062-5

APPENDIX

The 17 OECD founding countries whose data are used in the study are listed below in alphabetical order:

Austria, Belgium, Denmark, France, Germany, Greece, Iceland, Ireland, Italy, Luxembourg, Netherlands, Norway, Portugal, Spain, Sweden, Turkey, United Kingdom

Chapter 5
Efficiency of the Turkish Labor Market:
Future Employment Prospects

Can Bekaroğlu
Canakkale Onsekiz Mart University, Turkey

ABSTRACT

This chapter examines the employment efficiency in the Turkish labor market between 1988 and 2021 and measures the efficient level of Turkish employment by 2030, using the data obtained from TUİK (Turkish Statistical Institute), employing an alternative methodology named "Regressive Envelopment Analysis" or simply REA. The study uses a two-stage analysis where the first stage measures how efficiently the labor force is created from the pool of economically active population, while the second stage measures how efficiently the labor force is put to employment. The overall (1988-2021) employment efficiency average is found to be about 94%, and the bulk of the efficiency loss is the result of the first stage, which is fundamentally shaped by the structural unemployment. It is suggested that long-term unemployment rates, even at the efficient employment levels, will steadily rise from an average of 8% in 2000 to 12% in 2030 for both sexes, but especially for females, nearing 14%.

INTRODUCTION

The Turkish labor market, like most developing countries, has many inherent, intricate, and everlasting structural problems that tend to fester over time rather than self-heal unless drastic measures and structural reforms are put into effect. Turkish employment levels have hovered below and around 50% (TUİK, 2022a) for the last

DOI: 10.4018/978-1-6684-6727-5.ch005

decade, far below the OECD or Eurozone average of 69%, let alone Germany at 77% (OECD, 2022a). Likewise, Turkish labor force participation rates are well below the 80% OECD average, standing at merely 61-62% (OECD, 2022b).

This huge, almost 20% gap between the Turkish labor market and the OECD average, on the surface, comes from female labor force participation (FLFP) and employment figures (OECD, 2022c), which Turkey has been desperately trying to increase for the last two decades to little avail. FLFP in Turkey has dipped from a peak of 36% in 1989 all the way down to 23% in 2004 (TUİK, 2022b) only to rebound to 34% in 2019 (TUİK, 2022c), in contrast to the OECD average, which has been continuously increasing from 58% in 1990 to 63% in 2015. Not only is there a huge gap between total and female employment and participation rates, but it has also been increasing since the 1990s. Of course, this has both Turkish (domestic) and OECD (foreign) aspects to it, but we will only tackle the domestic part in this book.

Even though the FLFP and employment rates had been declining until 2005, as previously mentioned, according to the figures we have compiled from Karabıyık (2012), the results are actually both more promising and more humbling than once thought. Non-agricultural female employment has more than doubled, constantly increasing from 1.37 million (7.6%) in 1989 to 3.7 million (7.6%) in 2010. In other words, Turkish female employment, apart from urbanization, has been on a constant and rapid upswing since the late 80s, as clearly seen and experienced in urban life.

The caveat here is that urbanization in Turkey started much later than in Western nations (Işık, 2005) and experienced a rushed, haphazard, and unbalanced development, with highly limited urban employment opportunities for women who lost their agricultural jobs in the countryside. This is particularly true for countries like Turkey with very low (foreign and domestic) investment and funding opportunities (OECD, 2021) in order to boost employment, especially in the service sector (Karanassou & Salvador, 2008). So, given the financial opportunities and the limited time frame, it would be erroneous to call the relatively low level of female employment a failure. It certainly is far from the desired outcome, but it would only be beneficial to understand the true nature of the issue before an actual solution can be found.

The contributions of the study will be to introduce a new and simple regression-based efficiency analysis technique, explore the root causes of Turkish unemployment, by using a two-stage efficiency analysis, and estimate future (un)employment figures based on historical efficiency trends. This chapter is, therefore, composed of five parts. In Section 1, we touch upon the employment struggles in Turkey, particularly with regard to female employment. In Section 2, we define efficiency and the related concepts, with a brief look at the seminal steps in efficiency measurement in the literature as well as the relevant studies in Turkey, before explaining the methodology used in this study. Section 3 will tackle the efficiency measurement of the Turkish labor market in two stages, while Section 4 will make predictions regarding future (un) employment based on the analysis in Section 3. Finally, we will conclude the chapter in Section 5 with a summary and solutions to increase employment in the future.

BACKGROUND

1. **Basic Definitions Regarding Labor Markets:** *Employment* is usually defined as the inclusion of labor (people) in the production process, often in return for a certain financial compensation. (Ülgener, 1980:73). *Unemployment* is defined as a situation in which people who are able and willing to work are unable to find work after a period of job searching (Ünsal, 2000: 14) or simply the loss of work in which willing and able individuals are unable to participate in production (Ekin, 2003). In order to be classified as unemployed, an individual must be actively looking for a job and accept a legitimate job offer at the current market wage. (Reynolds, 1974:125). Alternative definitions and measurements exist and vary among countries. The American Bureau of Labor Statistics, for instance, lists six different measures of unemployment (U.S. Bureau of Labor Statistics, 2022a). There are five main types of conventional unemployment. *Frictional unemployment* is the result of temporary and economically expected factors such as relocation, quitting, resigning, and job search (Talas, 1997:138). *Cyclical unemployment* occurs because of fluctuations in the economy (business cycles) and is usually easier to manage compared to frictional unemployment (Lilien, 1982:777). *Seasonal unemployment* occurs due to the discretionary (Biçerli, 2009:463) or non-discretionary (Mourdoukoutas, 1988:315) ups and downs of certain job activities, such as increasing construction work or agricultural production in summer. *Structural unemployment*, which is the most troublesome and undesirable type of unemployment, is the result of the disparity or mismatch between the skill sets that the markets demand and the labor force supplies (Gilpatrick, 1966:205). Biçerli (2009:452) finds that particularly regular high school and college graduates suffer from structural unemployment, suggesting that it is the type of education they received, rather than the pure quantity of it that matters. Finally, *institutional unemployment*, which is often counted as a type of structural unemployment, as we will do in this chapter, emerges as a result of state policies and legal frameworks such as minimum wage laws and high rates of unionization (Hayes, 2002). *Civillian Noninstitutional Population* (CNP) is the entire population of the country, excluding active-duty members of the armed forces, people confined to or living in institutions or facilities such as prisons, jails, correctional facilities, residential care facilities, or detention centers, as well as citizens of foreign

countries residing within the domestic country's borders but not on the premises of an embassy. (U.S. Bureau of Labor Statistics, 2022b). The *labor force* (LF), which is a subset of the CNP, is the total number of all employed and unemployed people. *Labor force participation rate* (LFPR) is defined as the ratio of LF to CNP, while *employment rate* (ER) is the ratio of the employed (E) to CNP. Similarly, the *unemployment rate* (UR) is the ratio of the unemployed (U) to the labor force (Borjas, 2013:23-24).

2. **Capital Accumulation and the (Female) Employment:** As we mentioned in the introduction, capital accumulation is positively correlated with employment and negatively correlated with unemployment levels (Karanassou & Salvador, 2008), while financial funding and investment levels in Turkey are comparably very low (OECD, 2021), only worsened by market failures and shallow markets (Aslan, 2004), all of which result in low labor force participation and employment rates. Even more surprisingly, it is the relative, rather than absolute, investment and capital accumulation levels compared to other countries that are instrumental in determining the employment opportunities for middle- and high- income countries (Malley & Moutos, 2001), which means that it is the global financial system that determines and creates a zero-sum merchantilist result (Dell, 1987). For example, an ILO report (Elder & Smith, 2010) on gender differences in labor force and employment rates in 189 countries finds only a 1.5% increase in average female employment over the last 20 years. In short, the stark difference between developed and developing nations will never really shrink in the long run, despite the possibility of some countries switching leagues. As illustrated in the comparative analysis of Verick (2014), female employment tends to decrease for various reasons, such as urbanization and increasing levels of education among the youth, as low-income countries climb the development ladder (Elder & Smith, 2010). To illustrate, one can look at the poorest countries, such as Nepal and Cambodia, with the highest FLFP rates, approaching 80%, much higher than highly egalitarian and developed European countries. Similarly, a much more conservative and poorer Turkey in 1923 had a much higher labor force participation rate of 73.5% than modern Turkey at around 50% (Biçerli, 2018:71). Interestingly, both Malaysia and Turkey are relatively developed countries with similar levels of development. However, Malaysia, which is rich in financial flows and trade per capita, has a 50% FLFP rate (Verick, 2014), while Turkey, with much lower levels of financial flows and trade per capita, has a mere 30% FLFP rate (TUİK, 2022b).
3. **The Increasingly Egalitarian Approach and the U-Shaped Female Employment** As shown by Korkmaz & Korkut (2012) in their Turkey-wide analysis, despite an increasingly more liberal and egalitarian environment for women, Turkish FLFP rates had been dropping until 2005 across all age groups

and education levels, only to be partially reversed by aggressive and passionate state policies to increase FLFP, including but not limited to so-called positive discrimination policies (Norris, 2001). Such policies do indeed somewhat increase female employment, but only at the expense of male employment, while total employment does not change (OECD, 2022c). Substituting male security personnel with females for low-risk positions, for instance, may artificially boost female employment, but this is only a short-sighted and haphazard solution that also decreases trust in the economic system that is undermined by such sexist policies that are not based on merit. Such low levels of female employment cannot be attributed to cultural norms or social coercion because: a) fertility rates (or children per woman) are at historically low levels, well below 2.1 (TUİK, 2021); b) time allocated to or required for housework is also at historically low levels (Fred, 2022a); c) the number of female university students has surpassed that of their male counterparts and now exceeds 50% (YÖK, 2022); d) the education system, the press, the government, and the private sector all strongly support and advocate for higher rates of female employment. Haller & Höllinger (1994) also showed that the role of the cultural norms of a society had a much smaller effect on female employment, compared to the highly exaggerated claims by feminist circles. Contrary to expectations, relatively conservative Hungary, for instance, has a much higher FLFP rate than the much more liberal Netherlands. Likewise, Malaysia, an Islamic country based on Sharia Law, has a much greater level of FLFP than secular Turkey due to financial flows and much greater trade per capita. Although the complexity of the question prevents a definitive answer on the issue, the most popular and reasonable argument addressing the decreasing or non-increasing levels of FLFP in developing countries is the Feminization U Hypothesis (Sinha, 1967), which suggests that low-income countries enjoy rising productivity as their income rises, in addition to the immigration of rural populations to urban areas (urbanization), leading to decreasing low-productivity agricultural activity, which destroys much of the female workforce. However, a subsequent upswing in the increasingly female-intensive services sector is expected to result in a rebound in female employment levels once the country reaches and exceeds the middle-income zone. Even though this theory was initially accepted as a stylized fact (Goldin, 1995), researchers such as Gaddis & Klasen (2014) have shown that the relationship between female employment and development levels is not automatic or robust, lacking definitive empirical support. In brief, there is no definitive and automatic relationship between development and female employment levels in a country. However, female employment tends to be higher either in low-income and low-productivity countries or in high-income countries with high levels of investment and financial flows, in contrast to middle-income countries with low levels of investment and financial flows, where female employment tends to be low (Elder & Smith, 2010). In other words, female employment follows money and tends to be extremely dependent on financial flows (Christiansen, 2016). In essence, with rapidly rising female education levels and college graduates, the Turkish female labor supply has sufficiently matured and society has been adequately transformed, while the economic bottleneck remains the lack of a specific demand for a female labor force that targets a very limited set of skills and jobs, as well as the increasingly tight monetary, financial, and fiscal measures the country has been enduring, putting a tight grip on female and total employment levels (Aydoğan, 2017). A genuine and substantial increase in demand for female labor can only be a result of increased economic production, which requires a better structured, more robust, and developed financial system (Eroğlu & Bursal, 2020), as well as an exit strategy from the uncertainty and fragilities of energy dependence (Saritaş et al., 2018).

4. **Turkish Economic Capacity and Female Intensive Growth** As we mentioned in the introduction, non-agricultural female employment has been constantly increasing from 1.37 million (7.6%) in 1989 to 3.7 million (13.7%) in 2010 (Karabıyık, 2012), which heralds both good and bad news. It is good in the sense that the type of female employment that really matters and that people actually wish to pursue, rather than being compelled to, has been rising. However, this implies that female employment was overestimated in the 1990s and the Turkish economic capacity for job creation appears to be rather limited (Uslu, 2020; Timur & Doğan Çalışkan, 2015). This is particularly a result of the financial strain in the economy, associated with low (nominal) income in the international markets, which leads to job-saving measures, not necessarily by cutting jobs but through informal labor markets and/or increased work hours due to relatively high job and wage flexibilities (Onaran, 2002; Onaran, 2007; İlkkaracan & Selim, 2003; İlkkaracan & Yörükoğlu, 2004; Demir & Erdem, 2010). Long story short, female employment has been on a constant upswing, apart from the rushed urbanization consequences, though further robust increases in employment require increased labor-intensive production, and capital accumulation as a result of a strong financial system.

So far, we have seen that the relatively low female employment numbers in Turkey have little to do with culture or discrimination against women but rather are a product of economic woes, financial strains, and, as we will see later in the chapter, structural unemployment. If this is all true, then there should not be much of a difference between the expected and actual levels of female employment. One way to measure this is to use the correlation between total and female-only employment figures. A simple OLS regression analysis using the cross-sectional data compiled from the ILO across 180 countries (International Labour Organization, 2019a; 2019b; 2019c), where female employment is the dependent variable and total employment is the independent variable, yields the results in Table 1. The regression is estimated once with the raw data (estimate 1) and again with adjusted data (estimate 2) to correct for a few oil-rich small Gulf states with unusually high male ratios due to very large numbers of foreign male workers. Both P values are close to zero, both intercepts are negative, and the coefficients are 1.18 and 1.26, respectively, suggesting that 60-63% of new employment is going to women (see Graph 1).

Table 1. OLS regression of total employment on female employment

	Estimate 1	Estimate 2
R-Squared	0,7746	0,8369
Observations	180	180
Intercept	-20,6358	-25,0247
Coefficients	1,1827	1,2616
P-Values	0,0000	0,0000

Figure 1. Female employment estimate based on total employment rates

Basically, the analysis predicts that male employment always dominates in countries with low overall employment levels. If, for instance, total employment is 40%, we expect 55% male employment and 25% female employment. However, further increases in employment disproportionately increase female employment. If the employment rate rose to 50%, male employment would merely rise to 62% while female employment would skyrocket to 38%. Similarly, a 60% total employment rate will result in 69% male employment and 51% female employment, along the lines of Malaysia. But how does it compare with Turkey, with 30% female employment? Analysis suggests that Turkey is supposed to have a 34% female employment rate, with 4–5% falling short of the estimated values, most likely due to financial strains and a real depreciation in the Turkish Lira (TCMB, 2022).

Reasons for Turkish Unemployment in the Literature

There are many other studies that investigate the reasons for unemployment in Turkey, which has been going through a substantial structural transformation, especially in the post-2000 period. Ay (2012) provides a comprehensive review of the employment woes in Turkey prior to 2012, suggesting three main conclusions that can be drawn from the literature. First of all, urbanization is still an ongoing process in Turkey, with the rural population rapidly declining, necessitating the creation of massive amounts of new employment in order to address the newly urbanized populations even though Turkey could not reach its industrialization targets (Ercan, 1998; Tatlıdil & Xanthacou, 2002; Dayıoğlu & Ercan, 2010). Furthermore, this hasty and haphazard urbanization has resulted in an increase in the uneducated urban population who have discovered that their once useful skill-set is now obsolete. In addition to the aforementioned studies, Ercan (2007a) and Tunal (2003) highlight the growing structural unemployment as a result of rapid urbanization and the emergence of skills mismatch. Finally, studies such as Boztepe (2007) and Parlaktuna (2010) find that there is a huge gap between Turkey and the OECD countries in terms of female as well as youth employment (Gürsel & Atasoy, 2000; Ercan, 2007b; Ercan, 2010a; Ercan, 2010b), which also shows that this is a structural issue rather than a cultural one.

Authors such as Onaran (2002), İlkkaracan & Selim (2003), İlkkaracan & Yörükoğlu (2004); Onaran & Aydıner-Avşar (2006), and Boztepe (2007), unsurprisingly, find a negative correlation between labor demand and labor costs, which consist of two main parts, namely compensation and taxes. Naturally, they conclude that a reduction in either part of the cost will increase total employment. The rapid economic growth in the real sector between 2000 and 2020 also had a very limited impact on the labor market (Uslu, 2020; Timur & Doğan Çalışkan, 2015), which was more than reversed due to the extreme financial strains in the recent years with (un)employment rates at historically dismal levels (TUİK, 2022b). Given the recent examples such as the Netherlands, reduction in weekly work hours or work sharing through part time work etc. have been proposed as a solution to boost employment (Karabulut, 2007).

Like we mentioned in the previous part, Turkey appears to show relatively high job and wage flexibilities compared to most OECD countries, which is a result of an abundant labor supply and relatively expensive non-labor production factors due to the highly devalued local currency. In essence, firms try to compensate for the razor thin profit margins with relatively abundant labor, either by paying lower wages than officially sanctioned wages through the informal sector or by overworking their employees to increase the return on their labor costs (Onaran, 2002; Onaran, 2007; İlkkaracan & Selim, 2003; İlkkaracan & Yörükoğlu, 2004; Demir & Erdem, 2010).

Figure 2. 2000-2021 unemployment rates: 5-year moving averages

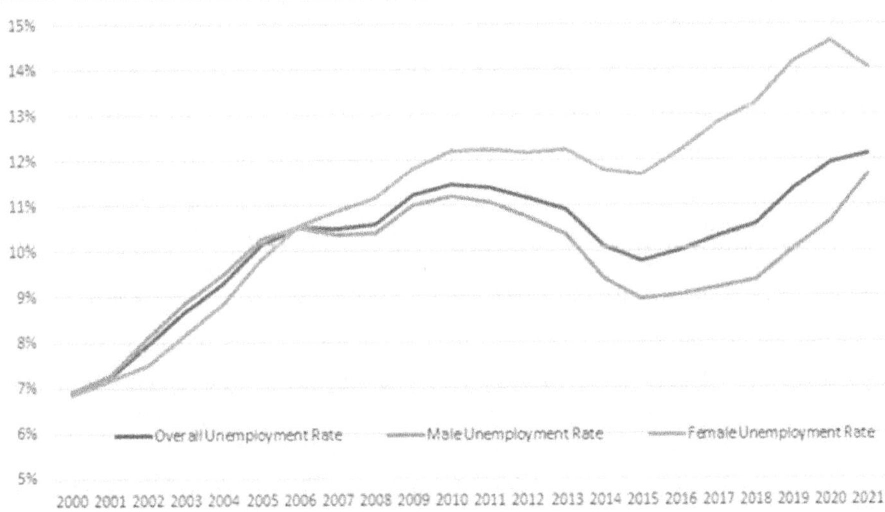

Post-2000 Unemployment Trends

As seen in Graph 2, which was created using the data complied from TUİK (2022a; 2022b), Turkish unemployment rates have been on an increasing trend and are expected to further increase. The 2000–2021 period could be broken down into three sub-periods in terms of unemployment: 1) Post-2001 Crisis of increasing unemployment; 2) Post-2010 Period of decreasing unemployment due to increased foreign direct investment; 3) Post-Gezi Park Period of increasing unemployment due to rapidly decreasing foreign investment and currency shocks. Despite the idiosyncratic characteristics of each period, increasing unemployment appears to be the dominant overarching trend here.

A closer look at the aforementioned TUİK data will reveal the obvious reason, "the elephant in the room", so to speak, behind the rapidly increasing unemployment trends for both genders. The FLFP rate dropped all the way down to 23% in 2004, only to rise back up to 34% in 2019, which resulted in the female unemployment rate rising from around 10% to 15% without a corresponding increase in the female employment rate. Let's look at it this way: Between 2000 and 2018, labor force participation (LFP) in raw numbers increased by merely 29% for men but a record-breaking 69% for women. Similarly, employment figures only rose by 25% for men but by 56% for women (more than double that of men). In essence, the state- and country-wide positive discrimination policies surpassed expectations, and female employment progressed by leaps and bounds. Despite all this, the historically rapid increase in the female labor force due to the increasing share of female university students overshadowed the increase in female employment, leading to a record jump in unemployment for both genders.

A 2019 report on the overall situation of female employment, across multiple years and all regions in Turkey (Sosyo Politik Saha Araştırmaları Merkezi, 2019) finds that the female employment rate is highest in the agricultural sector (44-46%), followed by the services sector (29-33%) and the industrial sector (18-19%). Interestingly, it turns out that the much more conservative countryside provides much higher and more egalitarian employment opportunities for women compared to the much more modern and liberal urban setting. Thus, increasing urbanization leads to a decline in the equality of employment. Except for Mardin, Batman, Şırnak, and Siirt, which have an average female employment rate of 20%, it would be incorrect to assume that the employment gap is the result of gender discrimination or a cultural phenomenon, because even the provinces with the highest female (and agricultural) employment rates, such as Kastamonu, Çankırı, and Sinop, cannot exceed 38.5%. Similarly, the female employment differences across provinces are minimal, with hardly any difference between relatively conservative Istanbul (32%) and highly liberal Izmir (32.5%), for instance, which demonstrates that the disparity between genders is mostly based not on culture but on the economy.

MEASURING EFFICIENCY

Basic Concepts

Three related concepts, namely efficiency, effectiveness, and productivity, are often confused both in everyday life and in academia, so it would be beneficial to clearly define the concepts we will be dealing with in the rest of the chapter. Though these are commonly known microeconomic terms, we will stick with Ray & Chen (2015) as the main source for definitions.

a. *Productivity* = the ratio of output per input
b. Efficiency has two types, depending on orientation.
    i. *Output Oriented Efficiency* = Actual Output / Maximum (Best) Output
    ii. *Input Oriented Efficiency* = Minimum (Best) Input / Actual Input
c. *Effectiveness* = Actual output / Target Output
d. *Decision Making Unit (DMU):* The unit, firm, party evaluated

Basically, productivity is a relationship between input and output, while efficiency is a relationship between all outputs or all inputs separately. Effectiveness is very similar to output-oriented efficiency, though the actual output is compared to the "target" output rather than the "best or maximum" output. These are all related terms, however, as when efficiency increases, effectiveness also most definitely increases. The relationship between productivity and efficiency is a bit trickier and requires some caveats. Given the input level, increasing productivity implies both increasing (output-oriented) efficiency and effectiveness. However, if input is not fixed, increasing productivity might be accompanied by decreasing or increasing efficiency and/or effectiveness. Likewise, an effective unit might be inefficient.

Let's do a simple example and define the production function as $y = f(x) = x^{0.5}$, which comprises the efficient frontier and the actual x-y bundles below. Let's also target 90% efficiency for effectiveness.

In table 2, there are 8 DMUs with inputs ranging from 1 to 6. The efficient production levels, as well as effective (defined as 90% of efficient) and actual production levels, are also given. Productivity, efficiency, and effectiveness levels are then calculated. The production frontier here clearly demonstrates decreasing returns to scale (hence $f(x) = x^{0.5}$) and productivity will surely decrease even if efficiency may increase or stay the same. From DMU 1 to 2, or from DMU 3 to 4, actual output and productivity increases along with an increase in efficiency and effectiveness. However, from DMU 1 to any other DMU, both efficiency and effectiveness increase despite a decline in productivity!

Table 2. Example for productivity, efficiency, and effectiveness

DMU	x (Input)	y (Efficient)	y (Effective)	y (Actual)	Productivity	Efficiency	Effectiveness
1	1	1,00	0,90	0,80	0,80	0,80	0,89
2	1	1,00	0,90	0,90	0,90	0,90	1,00
3	2	1,41	1,27	1,15	0,58	0,81	0,90
4	2	1,41	1,27	1,25	0,63	0,88	0,98
5	3	1,73	1,56	1,40	0,47	0,81	0,90
6	4	2,00	1,80	1,65	0,41	0,83	0,92
7	5	2,24	2,01	1,90	0,38	0,85	0,94
8	6	2,45	2,20	2,15	0,36	0,88	0,98

Methodologies to Measure Efficiency

There are fundamentally two separate schools of efficiency measurement in academic circles. The first one is *Data Envelopment Analysis* (DEA), while the other one is composed of various types of regression-based techniques. Another distinction in efficiency measurement techniques comes down to whether the functional form is known or needed. Parametric techniques such as *Corrected Ordinary Least Squares* (COLS) or *Stochastic Frontier Analysis* (SFA) do require a functional form, while DEA or *Concave Nonparametric Least Squares* (CNLS), which are computer programming approaches, do not require any functional form. Another important distinction comes from the interpretation of the deviations. On one hand, the *Ordinary Least Squares* approach (OLS) associates all deviations with random noise, while deterministic frontier approaches such as DEA and COLS associate all deviations with inefficiency. SFA and Stochastic Nonparametric Envelopment of Data (StoNed), on the other hand, divide deviations into two distinct parts: inefficiency and random noise.

The simplest and probably oldest technique to estimate efficiency is the OLS estimation technique, dating back to Legendre and Gauss (Kuosmanen, 2008:4) and Cobb-Douglas (1928). OLS assumes that all firms are rational and efficient, so all deviations are attributed to random noise. Of course, interpretations regarding whether firms are above or below the expected regression line and their negative or positive deviations can be used to interpret inefficiency and rank DMUs accordingly. The Corrected Ordinary Least Squares (COLS) approach (Winsten, 1957; Aigner & Chu, 1968) attributes all deviations to inefficiency, effectively shifting the frontier upwards to obtain all non-positive deviations, which not only shifts the intercept but also changes the coefficients (Kuosmanen, 2008:6). The main issue with respect to COLS, apart from the common characteristics of the parametric techniques, is that it is not very robust to outliers. Furthermore, there is no definitive link between the microeconomic fundamentals of the nonparametric techniques like DEA and the statistical approach like COLS, which simply pushes the frontier based on relatively strong assumptions. Another very similar approach, Corrected Absolute Mean Deviation (CAMD), is based on Mean Absolute Deviation (MAD) rather than Ordinary Least Squares (OLS), and the estimated regression passes through the median, rather than the mean (Ray et al., 2015:13).

Both OLS and COLS require a functional form, which poses a problem if the functional form is unknown or incorrect, in which case one can resort to nonparametric regression techniques, as comprehensively reviewed by Yatchew (1998; 2003). The oldest such technique is Concave Nonparametric Least Squares (CNLS), dating back to Hildreth (1954), which is subject to continuity, monotonicity, and concavity constraints, and shown to be a consistent technique by Hansen & Pledger (1976). Despite its benefits (a consistent statistical approach that does not require a functional form), CNLS is a compute-intensive and generally difficult to solve technique as it selects the best fit function from an infinite number of functions. CNLS also requires concavity and sorting of inputs in ascending order, which is difficult when the input is a vector (Kuosmanen, 2008:7). Corrected Concave Nonparametric Least Squares (C2NLS) is a nonparametric variant of the COLS model in which nonparametric least squares subject to monotonicity and concavity constraints replace the first-stage parametric OLS regression, which is also shown to be consistent, and the main advantage to alternative methods such as DEA is robustness to small-sample error (Kuossmanen & Johnson, 2010:153-154, 159).

DEA is an axiomatic approach to efficiency analysis that uses mathematical programming techniques and does not rely on any particular functional form to construct the efficient production frontier (Ray et al., 2015:118). DEA is actually a family of models that differ in their specification, returns to scale, and efficiency measure orientation (Kuosmanen, 2008:8). The technique was first introduced by Charnes et al. (1978) and further developed by Banker et al. (1984) based on Farrell's simple measure of firm efficiency that accounted for multiple inputs (Farrell, 1957). Under the usual assumptions of monotonicity, concavity, free-disposability, and feasibility, the main advantages of DEA are its nonparametric treatment of the production frontier, its ability to simultaneously address multiple inputs and outputs, and its conceptual simplicity despite its computationally heavy nature. Its main shortcoming is that it attributes all deviations from the frontier to inefficiency (Ray et al., 2015:118). Many authors, such as Schmidt (1985), also criticize DEA for its non-statistical nature, though other authors addressed these concerns by showing that the DEA estimators are statistically consistent (Banker, 1993), and DEA can be written as a form of nonparametric least squares analysis (Kuossmanen & Johnson, 2010:152).

Stochastic Frontier Analysis (SFA), which was first proposed by Aigner et al. (1977) and Meeusen et al. (1977), utilizes parametric regression techniques, which require specific functional forms not only for the frontier but also for the inefficiency distribution. The probabilistic modeling of SFA is able to decompose deviations from the frontier into two parts, namely, the inefficiency and the noise term for the unobserved heterogeneity of DMUs and environmental effects, etc. DEA and SFA are the two most commonly used techniques, each with their own strengths. SFA has the advantage of modeling inefficiency and noise, whereas DEA does not need a functional form, can impose axiomatic properties, and can sustain multiple outputs (Ray et al., 2015:118). For further information, readers can take a look at Kumbhakar and Lovell (2000), which provide a detailed analysis and explanation of SFA.

Despite all these advances in measuring efficiency, the gap between the two schools of thought, namely parametric and nonparametric approaches, remained a frustrating issue; that is, until a new approach named Stochastic Nonparametric Envelopment of Data (StoNEd) emerged under a unified framework of production analysis, combining SFA and CNLS, following the recent works on CNLS by Kuosmanen & Johnson (2010) and Kuosmanen & Kortelainen (2012). StoNED is basically a paradigm shift that provides a more flexible platform for efficiency analysis and allows us to combine the existing tools and techniques in refreshing ways across the DEA-SFA spectrum (Ray et al., 2015:118) in spite of its compute-heavy, linear programming nature and often complicated model structures.

Studies in Turkey

There are about a dozen studies in the literature analyzing partly or exclusively the efficiency of employment in Turkey. DEA, in various forms, is used in almost all of these studies for efficiency analysis. About half of these articles study employment efficiency at the provincial level, while the other half look into the issue at the regional level, with two exceptions: one studies country-level data, and the other studies the efficiency of the public sector.

The oldest study at the provincial level is a study by Algan & Akdoğan (2001) that examines the socio-economic efficiency levels of the 18 provinces across the Black Sea Region using the simple CCR DEA technique. Aydemir (2002) measures the country-wide technical and scale efficiency levels of each province by focusing on resource efficiency and value added in terms of regional competitiveness. Atan et al. (2004) use statistical analysis in addition to DEA to measure the efficiency of provinces as well as their development levels as they appear as a reference in the analysis. Kıran (2008) measures the efficiency of priority provinces for development between the years 1995 and 2000 by using the super SBM (slack-based model) DEA. Finally, Cam & Atan (2018) measure the technical and technological efficiency as well as factor productivity of provinces between the years of 2013 and 2016, using the ratio of the actively employed to the non-institutionalized population as inputs and average labor force participation, average earnings, and employment rate as outputs.

The first study that measures employment efficiency at the regional level is the doctoral dissertation by Öncel (2010), who analyzes resource efficiency with an input-oriented DEA technique and uses 26 Level-2 Nomenclature of Territorial Units for Statistics (NUTS-2), concluding that less developed regions are also less efficient in resource use. Şengül et al. (2013) use a DEA approach in addition to Tobit analysis to measure employment efficiency during the 2007-2008 period, and public investments per capita were used as the main input. Köne (2014), who incorporates environmental factors into her analysis to investigate economic regional efficiency, discovers that regional economic efficiency is highly correlated with national income per capita. Bakırcı et al. (2014) study the effects and efficiency of regional development policies on 12 NUTS-1 regions, suggesting annual efficiency fluctuations and a higher overall efficiency level for the western regions compared to the east.

To our knowledge, the only study in the Turkish literature regarding employment efficiency at a country level is by Noyan (2012), who utilizes the relatively new Fuzzy-DEA technique (Sengupta, 1992) to measure the efficiency of 41 mostly OECD countries, including Turkey, between 2008 and 2010, suggesting a relatively low level of employment efficiency for Turkey, which did not appear to improve over the years. Finally, Oğuz & Oğuz (2016) investigate the efficiency of public policies regarding entrepreneurship, competition, and employment with alternative techniques and suggest that employment-focused policies hamper economic growth and development and that governments should focus on production, which will gradually and naturally improve employment. Interestingly, what is missing in the literature is a two-stage efficiency analysis to decompose unemployment into structural and frictional parts and forecast future employment frontiers.

Methodology and Data

Despite the novel approach of StoNED under a more flexible and unified framework of production analysis, the compute-heavy linear programming nature of nonparametric analyses and the often overly complicated structure of models are still an issue. In many cases where the problem is relatively simple, quick, intuitive, and easy to compute, efficiency analysis techniques similar to COLS but better integrated with DEA principles remain uninvestigated. In this chapter, we will try to provide a simpler yet novel approach to measuring efficiency by combining the main DEA principles with a COLS-style regression-based parametric output-oriented efficiency analysis, which we call "*Regression Envelopment Analyses*" (REA).

The data is entirely compiled from TUİK (2022b; 2022c). There are three main variables: the number of people in the non-institutionalized civillian (active) population, the number of people in the labor force, and the number of employed people. The analysis consists of two stages (see Table 3), where the active population (x) is the input of the first stage and the entire mechanism. The labor force (y) is the intermediary variable, serving as the output of the first stage and the input of the second stage. Finally, employment (z) is the output of the second stage and the entire mechanism (X => Y => Z).

Table 3. Two-stage efficiency analysis framework

Stage	Input	Output
First Stage	# of people in Active population (X)	# of people in Labor force (Y)
Second Stage	# of people in Labor force (Y)	# of people in Employment (Z)

The aforementioned mechanism results in three different measures of output-oriented efficiency, namely the first-stage efficiency (α), the second-stage efficiency (β), and the overall efficiency (γ), which mathematically can be defined as $\boldsymbol{\gamma = \alpha\beta}$. The first-stage and the overall efficiencies (α and γ) can be directly calculated from the data, but the second-stage efficiency (β) may need to be indirectly derived using the formula $\boldsymbol{\beta = \gamma/\alpha}$. So, once we calculate the α and γ values, it is only trivial to derive β but how do we calculate α and γ?

a) $\alpha = y / y^*$, given x; α = actual y / efficient level of y, given actual level of x
b) $\beta = z / z^*$, given y; β = actual z / efficient level of z*, given actual level of y
c) $\gamma = z / z^*$, given y*; γ = actual z / efficient level of z*, given efficient level of y

The efficient results can then be written as

d) $y^* = y / \alpha$, given x; where y* is the efficient level of labor force (y)
e) $z^* = z / \beta$, given y; where z* is the 1st stage efficient level of employment (z)
f) $z^{**} = z / \gamma$, given y*; where z** is the 2nd stage efficient level of employment (z)

Please note that calculation of each individual coefficient may not be possible even though the derivation of efficiency coefficient is shown in the model specifications above, making the separate calculation of γ superfluous. In the case of multiple inputs, γ can only be indirectly calculated as formulated above though this is not the case in our example with a 1 input-1output combination in each stage.

A two-stage analysis of employment efficiency is especially important as it helps decompose the source and probable causes of inefficiency. If, for instance, the bulk of the inefficiency occurs at the first stage, it signals a problem in the active population's participation in the labor force, most likely due to a long-term structural issue that requires or is caused by a social transformation. On the other hand, the second-stage inefficiencies tend to signal rather short-term frictional, seasonal, or cyclical fluctuations that can be directly addressed with macroeconomic, or in some cases, microeconomic policies. Omitting one of these stages or simply combining the entire mechanism into one stage will undoubtedly lead to an underestimation of the inefficiency figures, yielding ambiguous results for the type of unemployment.

Model Assumptions to Construct the Production Frontier

The following are the assumptions and constraints to construct the efficient production frontier:

a. All actually observed input-output bundles are feasible and the efficient output for any given input is greater than or equal to the maximum output among all input-output bundles, where the evaluated input is greater than or equal to the comparison group inputs. That is, for a given (X_j, Y_j) bundle, $Y_J^* \geq \max(Y_i)$ for all $X_j \geq X_i$.

b. Production function is monotonistically increasing. That is, $Y_{J+1}^* \geq Y_J^*$ given that $X_{J+1} \geq X_J$ for two input-output bundles, (X_j, Y_j) and (X_{j+1}, Y_{j+1}).

c. The productivity decreases as the number of inputs increases. That is, for two consecutive bundles (X_j, Y_j) and (X_{j+1}, Y_{j+1}), $Y_J^{**}/X_J \geq \max(Y_J^*/X_J, Y_{J+1}^*/X_{J+1})$ given that $X_{J+1} \geq X_J$.

d. Regression estimates based on conservatively efficient input-output bundles (X_j, Y_j^{**}) are also feasible, and the final efficient output is depicted by the maximum of the two feasible outputs, Y_J^{**} or Y_j^r. That is, for a given (X_j, Y_j) bundle, $Y_J^{***} = \max(Y_J^{**}, Y_j^r)$, where $Y_j^r = \beta_0 + \beta_j^* X_j$.

e. Inputs are freely disposable. That is, if input-output bundle (X_j, Y_j) is feasible and $X_{J+1} \geq X_J$, then (X_{j+1}, Y_j) is also feasible.

f. Ouputs are freely disposable. That is, if input-output bundle (X_j, Y_j) is feasible and $Y_{J+1} \leq Y_J$, then (X_j, Y_{j+1}) is also feasible.

Please note that the notations above regarding the model assumptions are arranged in accordance with the first stage, where X is the input, and Y is the output. The same assumptions also apply at the second stage, with a slight change in notations where Y becomes Z and X becomes Y.

In the analysis, input-output bundles need to be arranged in ascending order. In order to achieve this, a composite input is created from the input vector in the case of multiple inputs and estimated by using OLS regression based on actual data. The order of the composite input and the OLS estimated values are the same since the estimated efficient frontier strictly increases with the input. However, the order of the efficient outputs and the estimated inputs may differ since actual and efficient values are different. Thus, the data may need to be bootstrapped by reordering and using the efficient output estimates until the order of the input and output vectors concur.

The method does not assume a concave production function but merely a decreasing overall productivity level, which is a much weaker assumption than concavity, and REA tends to be a more conservative approach than COLS or DEA-VRS. However, the curse of dimensionality does not affect REA and is much more robust to outliers than COLS, which simply raises the OLS intercept and may draw a rather unrealistic frontier. Therefore, the REA efficiency score may be greater or lower than its DEA or COLS counterparts, though with a tendency to converge with DEA when the number of inputs is small, and with COLS in the absence of outliers. The main advantages and disadvantages of the REA technique are listed below.

The main advantages of REA;

a) REA is simple to calculate and implement,
b) is robost to outliers and more realistic and/or conservative in approach,
c) does not require linear programming and is not compute heavy,
d) marries the main DEA (microeconomic) principles with regression analysis,
e) does not suffer from the curse of dimensionality,
f) can be configured with changing patterns of (constant or decreasing) returns,
g) already has super-efficiency embedded within.
h) yields similar results with much more complicated methods.
i) REA is less sensitive to the number of observations than parametric methods such as DEA.

The main disadvantages of REA;

a) REA is ultimately a regression analysis and unable acommodate more than one output,
b) is a deterministic frontier analysis and does not take stochastic effects into account,

c) production function is not concave, so marginal returns are not monotonistic.
d) when inputs are a vector, the order of the inputs must be estimated and bootstrapped.
e) REA, much like DEA, tends to overestimate the efficiency of the DMUs.

EFFICIENCY OF THE TURKISH LABOR MARKET

Constructing the Frontiers: 1st Stage – Active Population to Labor Force

In order to measure the efficiencies and determine future trends, we first need to construct the production frontiers according to the model specifications explained in the previous part. Graph 3 constructs the efficient labor force (LF) frontier for the 1988–2030 period and compares it with the actual LF through 2022. Please see Graphs 3b and 3c in the appendix for gender-specific (male and female) employment frontiers. In Graph 3, one will notice that the overlap between the actual LF and the efficient frontier is the greatest in the early and late periods, with a big gap in between. Similar patterns emerge in the other graphs, albeit less apparent for male LF in Graph 3b and overly apparent for female LF in Graph 3c. This implies that the LF levels were relatively efficient till the mid-90s and increasingly inefficient through 2006, after which they started to rapidly gain efficiency, reaching a high point in 2018, only to be halted by the financial crisis of 2019 and the subsequent pandemic. A closer examination of Graphs 3b and 3c reveals that the significant ups and downs in female LF, rather than the cyclical changes that largely shape male LF, are the main contributors to this fluctuation in efficiency. Starting in 1988, the female LF stagnated at around 6-6.5 million, while the efficient figures continued to grow. Only after 2006, did the female LF start to climb, reaching a theoretical maximum in 2018, which is a record year in terms of LF participation for both genders (TUİK, 2022c).

Actual and efficient figures of male, female, and combined LF for select years can be found in Table 4, which also incorporates the efficiency rates (actual and efficient LFs). LF efficiency averages around 94.9%, with a relatively high male LF efficiency of 98.1% and a relatively low female efficiency of 86.2%. Please see Table 10 in the appendix for all years between 1988 and 2030.

Figure 3. Overall efficient labor force frontier for the 1988-2030 period

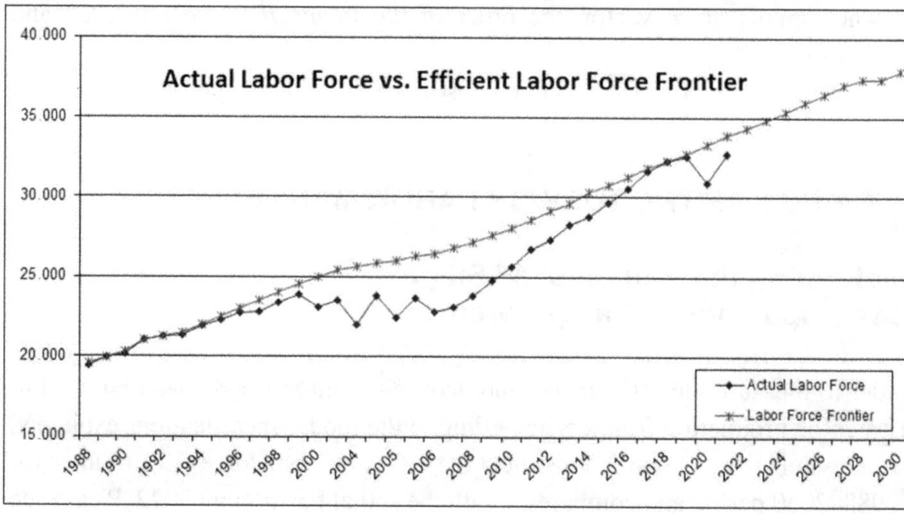

Table 4. Comparison of actual and efficient labor force levels for select years

	Actual Labor Force			Efficient Labor Force			Efficiency Rate		
	Total	Male	Female	Total	Male	Female	Total	Male	Female
1988	19.390	13.536	5.854	19.600	13.536	6.170	0,989	1,000	0,949
1990	20.150	13.990	6.160	20.288	14.087	6.300	0,993	0,993	0,978
1995	22.285	15.858	6.427	22.515	15.858	7.203	0,990	1,000	0,892
2000	23.078	16.890	6.188	24.967	17.180	8.029	0,924	0,983	0,771
2005	22.454	16.704	5.750	26.013	17.629	8.487	0,863	0,948	0,678
2010	25.641	18.257	7.383	28.050	18.892	9.189	0,914	0,966	0,803
2015	29.678	20.453	9.225	30.784	20.759	10.062	0,964	0,985	0,917
2018	32.274	21.801	10.473	32.274	21.801	10.531	1,000	1,000	0,994
2019	32.549	21.863	10.686	32.708	21.863	10.686	0,995	1,000	1,000
2020	30.873	21.105	9.768	33.298	21.952	10.772	0,927	0,961	0,907
2021	32.716	22.156	10.560	33.897	22.294	10.952	0,965	0,994	0,964
2025				35.921	23.386	11.599	Average		
2030				37.926	24.933	12.491	0,949	0,981	0,862

* All numbers are in millions.

Constructing the Frontiers: 2nd Stage – Labor Force to Employment

Graph 4 constructs the 1988–2030 period's efficient employment frontier and compares it with the actual employment levels until 2022. Please see Graphs 4b and 4c in the appendix for gender-specific (male and female) employment frontiers. One will notice that all graphs, regardless of gender, follow very similar cyclical patterns and fluctuations. Employment efficiency tends to increase till 1996, followed by a downward spiral until 2010 and an upward swing till 2017, after which drastic fluctuations ensue. There are two prominent points here: a) Employment efficiency levels (97.6%) are much higher than LF efficiency levels (94.9%), especially for women (96.5% vs. 86.2%) rather than men (98.1% vs. 97.9%); b) While female employment efficiency (96.5%) is again lower than that of males (97.9%), the difference is much smaller this time (1.4%) for the 1988-2021 period average. This means once women reach the labor market, their employment efficiency, or potential employability, is more or less the same as men's. Therefore, the bulk of the problem does not stem from the inner workings of the market but rather from the inability to create appropriate (female) labor that the market seeks, or simply put, a mismatch of skills.

Figure 4. Overall efficient employment frontier for the 1988-2030 period

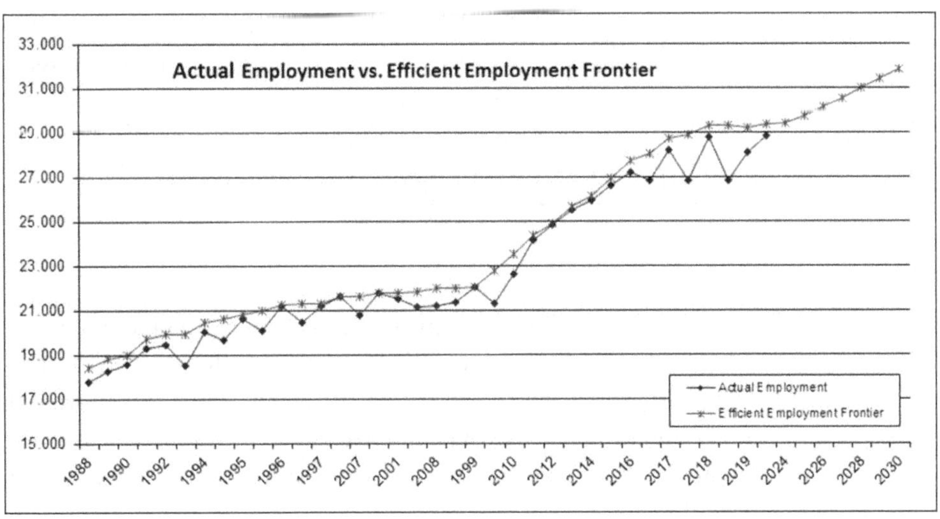

Apart from cyclical fluctuations, male employment efficiency has not changed much over the years, whereas female employment efficiency has improved from a low of 93.5% in 2004-2005 to a high of 98.9% in 2012 or 98.1% in 2018, which compares favorably to men's 99% efficiency. On one hand, this is good news because women's employment efficiency has caught up with men's; on the other hand, it is bad news because there is hardly any more room to improve. In other words, the capacity of the Turkish economy to create more female employment without perfectly substituting male employment appears to have come to a dead end.

Table 5. Comparison of actual and efficient employment levels for select years

	Actual Employment			Efficient Employment			Efficiency Rate		
	Total	**Male**	**Female**	**Total**	**Male**	**Female**	**Total**	**Male**	**Female**
1988	17.754	12.520	5.234	18.372	12.841	5.544	0,966	0,975	0,944
1990	18.538	12.901	5.637	18.992	13.218	5.794	0,976	0,976	0,973
1995	20.586	14.628	5.958	20.839	14.832	6.045	0,988	0,986	0,986
2000	21.581	15.780	5.801	21.581	15.780	5.820	1,000	1,000	0,997
2005	20.066	14.958	5.108	20.997	15.606	5.462	0,956	0,958	0,935
2010	22.594	16.170	6.425	23.474	16.759	6.752	0,962	0,965	0,952
2015	26.620	18.562	8.058	26.945	18.706	8.208	0,988	0,992	0,982
2018	28.737	19.719	9.018	29.301	19.939	9.194	0,981	0,989	0,981
2019	28.080	19.156	8.924	29.195	19.887	9.363	0,962	0,963	0,953
2020	26.812	18.506	8.306	28.030	19.303	8.671	0,957	0,959	0,958
2021	28.797	19.792	9.005	29.345	20.069	9.263	0,981	0,986	0,972
2025				29.696	20.212	9.571	Average		
2030				31.819	21.187	10.617	0,976	0,979	0,965

* All numbers are in millions.

Actual and efficient figures of male, female, and combined employment for select years can be found in Table 5, which also incorporates the efficiency rates of actual and efficient employment. Employment efficiency averages at around 97.6%, with a slightly higher male employment efficiency at 97.9% and a slightly lower female efficiency at 96.5%. For all years between 1988 and 2030, please see Table 11 in the appendix.

Overall Employment (In)efficiency Results

Constructing the two-stage production frontiers yields us two different efficiency results: one for both stages combined ($\gamma = \alpha\beta$), another only for the second stage (β only), from which, the first-stage efficiency can be easily derived ($\alpha = \gamma / \beta$). Graph 5 shows efficiency (left side, $\boldsymbol{\gamma}$, α, β) and inefficiency (right side, $1- \boldsymbol{\gamma}$, $1- \alpha$, $1- \beta$) results for both stages as well as the overall results. Please see Graphs 5b and 5c in the appendix for gender-specific (male and female) efficiency results. Combined with the numerical data from Table 6-A, we see that the employment efficiency has come down from 96.8% in the 1988-2000 period to 91.5% in the 2000-2021 period. A similar trend inflicted both men, whose efficiency has dropped from 98.3% to 95.7% (Table 6-B), and women, whose efficiency similarly fell from 89.1% to a mere 82.4 (Table 6-C).

Figure 5. Overall employment efficiency and inefficiency results for the 1988-2021 period

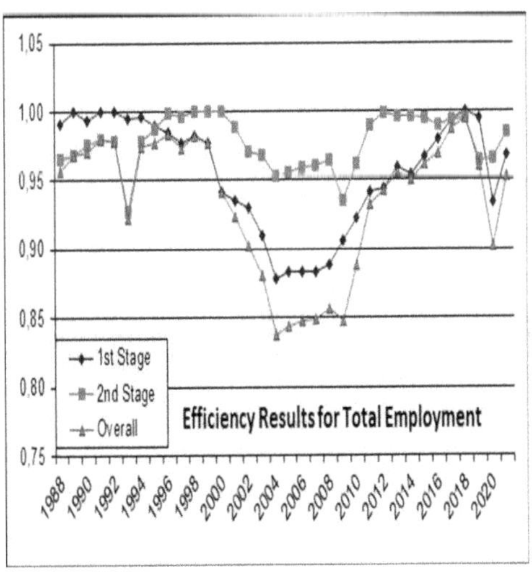

The bulk of the inefficiency (66.7% on average) comes from the first stage, whose share has increased from 41.3% in the 1988-2000 period to 73.3% in the 2000-2021 period. This implies that the inefficiency in employment has been increasingly geared towards the first stage of the labor market, which signals a structural, rather than a frictional or cyclical, issue. Gender breakdown reveals that this is true for both genders, as the share of first-stage employment inefficiency has increased from merely 7.3% to 53.9% for men (Table 6-B, Graph 5b in the appendix), and from 71.9% to 80.7% for women (Table 6-C, Graph 5c in appendix). It turns out that structural employment issues have always been the main issue for women, while they have recently started to be the primary issue for men as well.

Table 6. Stage 1, Stage 2, and combined employment efficiency levels

Years	Efficiency1	Efficiency2	Efficiency	Share (1)	Share (2)
			A. Overall		
1988-2021	0,955	0,978	0,935	0,667	0,333
1988-2000	0,987	0,981	0,968	0,413	0,587
2000-2021	0,936	0,977	0,915	0,733	0,267
Change	-5,1%	-0,5%	-5,3%	0,320	-0,320
			B. Men		
1988-2021	0,984	0,981	0,965	0,448	0,552
1988-2000	0,999	0,984	0,983	0,073	0,927
2000-2021	0,976	0,980	0,957	0,539	0,461
Change	-2,3%	-0,4%	-2,6%	0,466	-0,466
			C. Women		
1988-2021	0,880	0,965	0,850	0,776	0,224
1988-2000	0,920	0,969	0,891	0,719	0,281
2000-2021	0,853	0,965	0,824	0,807	0,193
Change	-6,7%	-0,4%	-6,7%	0,088	-0,088

PREDICTING THE FUTURE

Overall Employment Trend

Graph 6 shows the overall employment trends by comparing the actual and efficient employment frontiers. The first-degree efficient frontier (*) only assumes full efficiency at the second stage of the labor market, where the labor force as given is efficiently allocated to employment, which is maximized given the capacity constraints of the market. The second-degree efficient frontier (**), on the other hand, assumes efficiency in both stages of the market, LF participation and employment, therefore shooting up to a greater level of the frontier. Naturally, the second degree efficiency will always be less than or equal to the first-degree efficiency figures due to a higher level of frontier. The gap between the two efficiency frontiers is due to the first-stage efficiency (regarding the LF participation), while the gap between the first degree efficiency frontier and the actual data regards the second-stage employment efficiency.

As Table 7 shows, the average first-stage efficiency for the 1988-2021 period was 95.5%, substantially lower than the second-stage efficiency of 97.8%, so the bulk (66.7%) of the inefficiency came from the first-stage (see Table 6-A). However, the share of the first-stage in terms of inefficiency drastically fluctuated, ranging from 100% in 2000 to 0% in 2018, and was heavily correlated with LF participation and employment rates. For instance, 2018 had the highest LF participation (53.2%) and employment (49.9%) rates in the dataset (TUİK, 2022b; 2022c). However, beginning in 1999, there was a sudden decline in employment and efficiency levels that lasted over a decade of stagnation in the labor market, which only began to recoup its losses around 2010, primarily due to increased efficiency in the first stage. This *honeymoon period* lasted until 2018, with record levels of LF participation and employment, only to go downhill because of the subsequent financial crisis, followed by the 2020-2021 COVID-19 pandemic's toll on employment. As of 2021, even though employment figures partially bounced back due to increasing efficiency from 90.2% to 95.3%, they are still far from the 2018 level of 99.5%

Figure 6. Overall employment trends for the 1988-2030 period

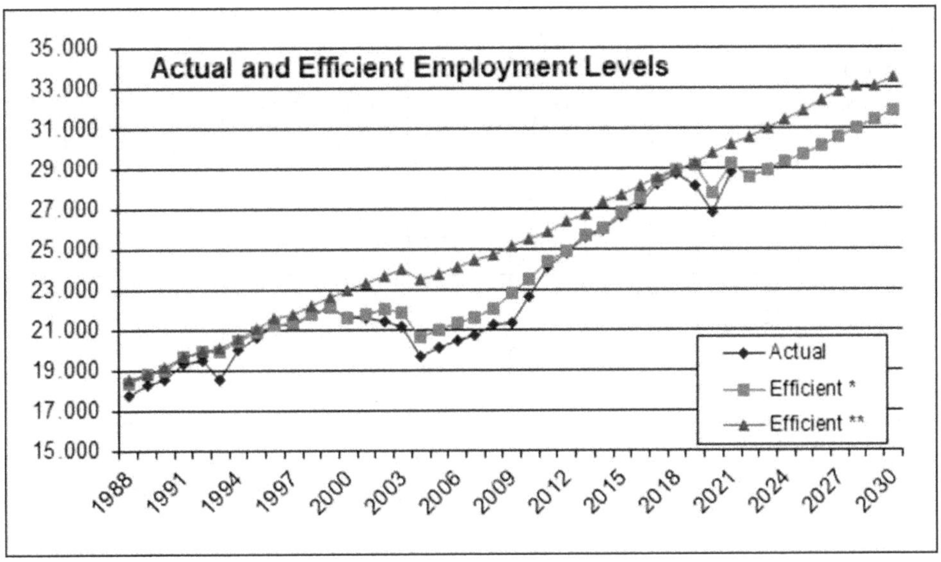

Table 7. Comparison of two-stage employment efficiency rates for select years

	Overall Employment			Male Employment			Female Employment		
	1st Stage	2nd Stage	Total	1st Stage	2nd Stage	Total	1st Stage	2nd Stage	Total
1988	0,991	0,966	0,957	1,000	0,975	0,975	0,955	0,944	0,902
1990	0,994	0,976	0,970	0,994	0,976	0,970	0,978	0,973	0,951
1995	0,990	0,988	0,978	1,000	0,986	0,986	0,915	0,986	0,901
2000	0,941	1,000	0,941	0,995	1,000	0,995	0,801	0,997	0,799
2005	0,883	0,956	0,844	0,961	0,958	0,921	0,716	0,935	0,670
2010	0,923	0,962	0,888	0,969	0,965	0,935	0,826	0,952	0,786
2015	0,967	0,994	0,962	0,987	0,999	0,985	0,925	0,982	0,909
2018	1,000	0,995	0,995	1,000	1,000	1,000	0,995	0,981	0,976
2019	0,996	0,965	0,960	1,000	0,970	0,970	1,000	0,953	0,953
2020	0,933	0,966	0,902	0,965	0,968	0,933	0,916	0,962	0,881
2021	0,968	0,985	0,953	0,994	0,990	0,984	0,968	0,972	0,941
Average	0,955	0,978	0,935	0,984	0,981	0,965	0,880	0,965	0,850

Table 8. Comparison of actual and two-stage efficient employment levels for select years

	Overall Employment			Male Employment			Female Employment		
	Actual	Efficient*	Efficient**	Actual	Efficient*	Efficient**	Actual	Efficient*	Efficient**
1988	17.754	18.372	18.543	12.520	12.841	12.841	5.234	5.544	5.804
1990	18.538	18.992	19.105	12.901	13.218	13.298	5.637	5.794	5.925
1995	20.586	20.839	21.055	14.628	14.832	14.832	5.958	6.045	6.610
2000	21.581	21.581	22.924	15.780	15.780	15.865	5.801	5.820	7.263
2005	20.066	20.997	23.778	14.958	15.606	16.238	5.108	5.462	7.624
2010	22.594	23.474	25.440	16.170	16.759	17.286	6.425	6.752	8.179
2015	26.620	26.770	27.672	18.562	18.582	18.836	8.058	8.208	8.870
2018	28.737	28.889	28.889	19.719	19.719	19.719	9.018	9.194	9.241
2019	28.080	29.113	29.243	19.156	19.752	19.752	8.924	9.363	9.363
2020	26.812	27.745	29.725	18.506	19.123	19.826	8.306	8.637	9.430
2021	28.797	29.249	30.214	19.792	19.995	20.110	9.005	9.263	9.573
2025		29.696	31.865		20.126	21.016		9.571	10.084
2030		31.819	33.502		21.187	22.300		10.617	10.790

* All numbers are in millions.

Actual and efficient figures of male, female, and combined employment for select years can be found in Table 8, from which the efficiency figures in Table 7 were derived. Table 8 also includes efficient levels of employment estimates until 2030. The employment level is estimated to reach 33.5 million by 2030 (up by 4.7 million from 2021 figures) assuming full efficiency in both stages of the labor market. However, it also signals a ceiling for the employment figures, especially for female employment. For all years between 1988 and 2030, please see Tables 12 and 13 in the appendix.

Employment Trend by Gender

Male employment, as shown in Graph 7 and enumerated in Tables 7 and 8, starts at 12,520 in 1988 and reaches 19,792 in 2021. The efficient numbers are not so different either, starting at 12,841 in 1988 and reaching 20,110 in 2021. The worst performing year for male employment efficiency was 2004 at 91%, when the economy as well as employment hit the bottom, while 2018 was the best performing year with a 100% efficiency for male employment, a record LF participation rate of 72.7%, and an employment rate of 65.7%, exactly 3% higher than in 2004. The overall male labor market efficiency is pretty evenly split between the first and second stages; 98.4% and 98.1% efficient, respectively, resulting in a 96.5% overall efficiency. However, as we have shown before, the share of first-stage inefficiency has significantly increased since 2000. Tables 7 and 8 also tell us that male labor efficiency is already fairly high and that a mere 318,000 additional jobs would render male employment efficient in 2021. By 2030, the Turkish economy is estimated to create only 1,395,000 additional (for a total of 21,187,000) jobs for males, even if the economy were fully efficient.

Female employment, as shown in Graph 8 and enumerated in Tables 7 and 8, starts at 5,234 in 1988 and exceeds 9 million in 2018, stagnating until 2021. The efficient numbers differ significantly, starting at 5,804 in 1988 and reaching 9,573 in 2021. The worst performing year for female employment efficiency was the 2004-2005 period, at around 67%, when economic stagnation, in addition to economic transformation fueled by rapid urbanization, caused a huge gap between efficient and actual female employment levels. Similar to the trends, 2018 was the best-performing year, with a 97.6% efficiency rate for female employment, a record LF participation rate of 34.2%, and an employment rate of 29.4%, almost 9% higher than in 2004. The overall female labor market efficiency is highly biased towards the first-stage (88%), in contrast to the second stage (96.5%), producing an average overall efficiency of 85%. However, female employment efficiency has substantially increased since 2005, reaching as high as 97.6% in 2018, only to drop to 94.1% after the COVID-19 pandemic. Tables 7 and 8 also tell us that female labor efficiency is relatively high given the capacity of the economy and that 568,000 additional jobs would render female employment efficient in 2021. By 2030, the Turkish economy is estimated to create a maximum of 1,785,000 additional (for a total of 10,790,000) jobs for females, assuming a fully efficient economy.

Figure 7. Male employment trends for the 1988-2030 period

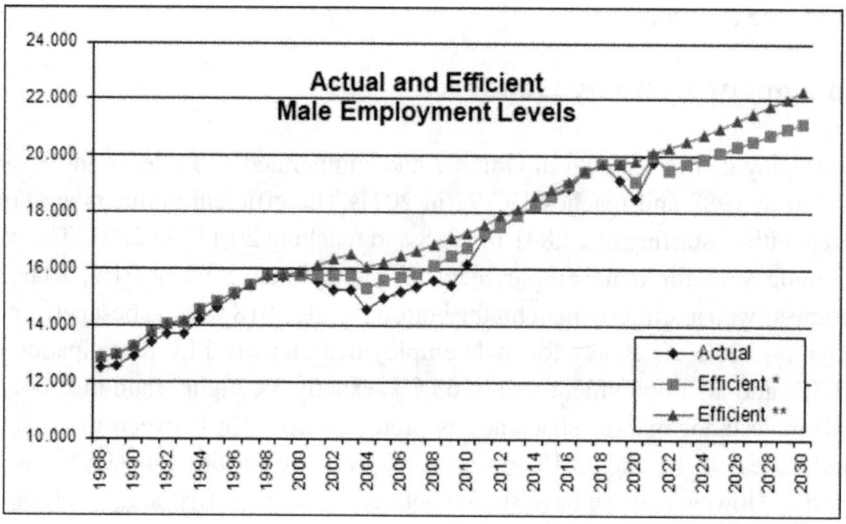

Figure 8. Female employment trends for the 1988-2030 period

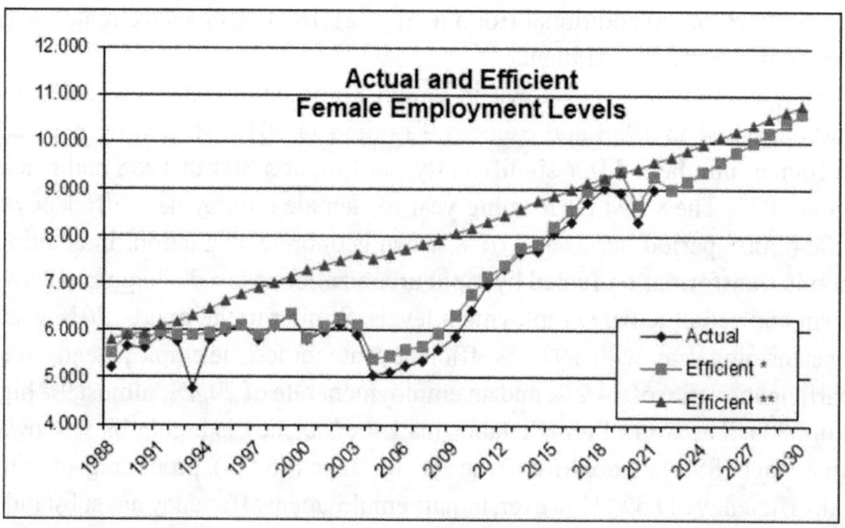

Unemployment Trend

As we previously discussed (Background, Part 6), unemployment has been consistently increasing for both genders, despite the fact that employment increased at record high levels from 2000 to 2018, an average of 33% against a population increase of 31% over a period of 18 years, and the economy was operating efficiently, at least given current economic capacity, with both overall and female employment rates at historical highs. The reason is obvious; the increase in the LF between 2000 and 2018 is almost 40%, much higher than the aforementioned increase in employment rate, creating an increasingly bigger gap, which is the number of unemployed people who are looking for very specific, usually urban jobs. A gender breakdown of this phenomenon reveals a mere 29% increase in male LF between 2000 and 2018, despite a 31% population increase, implying a decreasing LF participation rate for males aligned with developed countries. So it is not the male LF that is causing the issue here. On the other hand, female LF increased by a record 69% between 2000 and 2018, more than the 56% increase in female employment, which is already over double the 25% increase in male employment (TUİK, 2022b; 2022c). So, the gap between the LF and employment growth rates is the driving factor for the increasing unemployment in Turkey.

Graph 9 and Table 9 depict the unemployment trends between 2000 and 2030 (for detailed information regarding all years, please see Table 14 in the appendix). Actual unemployment has been fluctuating from a low 6.5% figure all the way to a high of 14% in 2009, and 13.7% in 2019, with an obviously rising trend that is true for both genders. Similarly, male unemployment fluctuated between 6.6% in 2000, 13.9% in 2009 and 12.4% in 2019, while female unemployment, which tends to be more volatile, increased from 6.3% in 2000 to 14.3% in 2009 and all the way up to 16.5% in 2019 (see Graphs 9b and 9c in the appendix). Unemployment has been rising and will most likely continue to rise even if both stages of the economy are fully efficient (efficient** in the graph). The efficient unemployment rate has increased from 5.4% in 1988 to 10.7% in 2020, and is expected to further increase to 11.7% by 2030. The trends are similar for both genders, albeit somewhat lower for males (5.1% to 10.6%) and higher for females (5.9% to 13.6%). Table 9 also reveals an interesting result for the employment market: the efficient employment level decreased from 54.9% in 1988 to 49.6% in 2000 and is expected to further decrease to 46% by 2030. The caveat here is that the efficient male employment rate will continue to decrease to 62% (from 77.1% in 1988 and 64% in 2020), while the efficient female employment rate will settle around 29.3% by 2030 (only slightly decreasing from 29.8% in 2020).

Figure 9. Overall employment trends for the 1988-2030 period

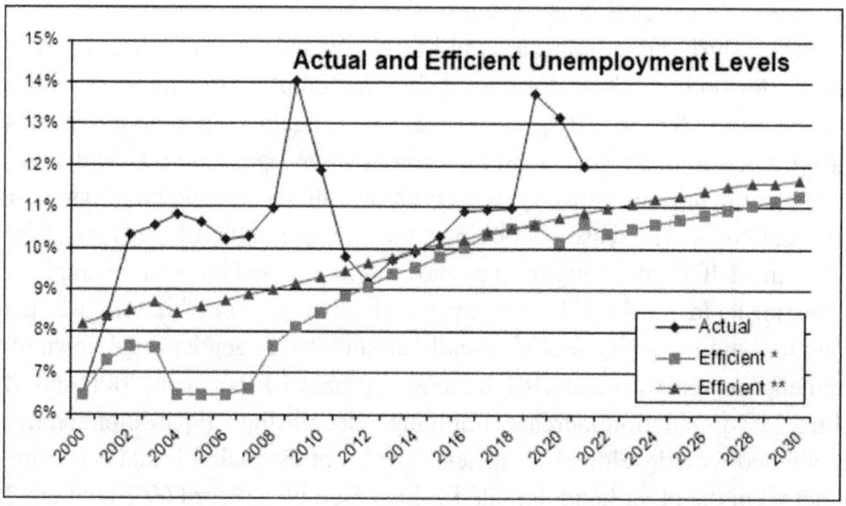

Table 9. Trends of employment and unemployment rates for select years

	Overall Unemployment			Male Unemployment			Female Unemployment		
	Unemployment		Employment	Unemployment		Employment	Unemployment		Employment
	Actual	Efficient**		Actual	Efficient**		Actual	Efficient**	
1988	0,084	0,054	0,549	0,075	0,051	0,771	0,106	0,059	0,340
1990	0,080	0,058	0,537	0,078	0,056	0,757	0,085	0,059	0,328
1995	0,076	0,065	0,511	0,078	0,065	0,727	0,073	0,082	0,318
2000	0,065	0,082	0,496	0,066	0,077	0,692	0,063	0,095	0,312
2005	0,106	0,086	0,486	0,105	0,079	0,669	0,112	0,102	0,309
2010	0,119	0,093	0,484	0,114	0,085	0,670	0,130	0,110	0,306
2015	0,103	0,101	0,478	0,092	0,093	0,659	0,127	0,119	0,303
2018	0,110	0,105	0,476	0,096	0,096	0,657	0,139	0,123	0,302
2019	0,137	0,106	0,476	0,124	0,097	0,650	0,165	0,124	0,301
2020	0,132	0,107	0,475	0,123	0,097	0,640	0,150	0,125	0,298
2021	0,120	0,109	0,474	0,107	0,098	0,638	0,147	0,126	0,298
2025		0,113	0,472		0,101	0,630		0,131	0,295
2030		0,117	0,460		0,106	0,620		0,136	0,293

CONCLUSION

In this study, we have found that unemployment has been on an increasing trend since the late 1980s, especially with the rapid urbanization and economic transformation from an agrarian to a more service-based economy. Female employment dipped to a record low level of 23% in 2004, urging the government to take bold steps to boost female employment, often at the expense of male employment, which has also been continuously decreasing from 77% in 1988 to a mere 63% in 2021. In short, there are fewer jobs today than ever before. So, it is no coincidence that decreasing employment levels are accompanied by increasing unemployment levels. However, it is surprising that labor force (LF) participation rates are on the rise, particularly due to government policies to increase female university graduation and LF participation rates in general, which leads to higher female unemployment rates due to excess female LF on one hand, and higher male unemployment rates due to women substituting for men partially as a result of positive discrimination policies on the other hand, increasing unemployment en masse.

When we look at the employment data, the Turkish economy is found to be surprisingly efficient, at least judging by recent years. Historically speaking, employment efficiency averages are 96.5% for men and 85% for women. However, the efficiency average for the last 5 years is 97.7% for men and 94.1% for women, significantly higher than the historical levels, implying a substantial increase, mostly due to the efficiency gains for female employment, which increased from 67% in 2004 to 97.6% in 2018. The efficiency losses in the Turkish labor market mostly stemmed from the first stage, where people decide to join the labor force. If, however, they did not have the necessary skills, implying a skills mismatch, they would not find a job and have to leave the market, keeping the labor force small. Before and after 2000, the share of first-stage inefficiency increased from 41% to 73%. This has always been the case for women (increasing from 72% to 81%) but has also started to significantly affect men (from merely 7% to 54%). In short, most of the employment woes in Turkey concern structural issues that are independent of economic fluctuations.

Efficiency rates basically give us a ceiling on what can be potentially done within the current capacity of the Turkish economy. With all that said, the Turkish labor market is relatively efficient, and even though increasing efficiency would definitely help, it would not create miracles to change the general scope of the labor market. The year 2021, for instance, was already 95.3% efficient, and only an additional 1,417,000 jobs would render it fully efficient, boosting the employment rate to 47.4% (up from 45.2%). By 2030, we expect efficient male employment rates to drop to 62% and efficient female employment rates to hover around 30%. The unemployment rate will likely increase by at least 2%, maintaining a 3% gender gap (about 11% for men and 14% for women) even with a fully efficient economy.

Unemployment results from the increasing gap between the labor force and employment. A labor force that grows faster than employment creates a potentially growing chronic unemployment problem with a relatively large and persistent difference. An effective solution would be either to a) get rid of the excess labor force by decreasing or limiting the labor supply or b) increase employment by creating additional jobs in the economy. As we mentioned above, the Turkish labor market appears relatively efficient, which means, with the given capacity and structure of the economy, as many jobs as possible have already been created, and employment cannot grow much faster than it already is. Thus, the only way to reduce unemployment in the short term is to reduce labor supply (growth) and accept a relatively low female employment level; otherwise, unemployment will inevitably continue its rising trend, as predicted in this study.

Obviously, limiting female labor supply to reduce unemployment is completely at odds with the dominant political view, so a more preferable and long-term solution would be to find ways to directly increase employment. The government has done more than its share to increase employment by rapidly increasing the number of government employees from 2 million in 2007 to 5 million in 2022, a significant portion of whom are women (Strateji ve Bütçe Başkanlığı, 2022). The problem here is that further public employment does not seem like a viable option for the government with its limited budget and resources, as well as the diminishing prospects for growth in education and healthcare. The only sustainable policy to boost employment is simply to increase production. Increasing labor-intensive production will certainly increase labor demand and overall employment, which will inevitably be female labor (and service) intensive, as we demonstrated at the beginning of the chapter. As a result, the following actions should be taken:

a) A labor-intensive production oriented economic model should be adopted,
b) A strong and robust financial system should be established,
c) Energy independence and current account surplus should be targeted,
d) Turkey should be transformed into a center of attraction for foreign capital,
e) Universities should be modernized and become professionally oriented,
f) Number of universities and their capacities should be reduced and flexible to circumstances,
g) Students should be more technically educated and professionally oriented.

What all developed nations have in common is a strong currency and high rates of capital accumulation, which facilitate funding for and decrease the cost of various projects and investments, most of which lead to more production and employment. The current account balance is an important factor, but it is far from being the only factor. Many countries, such as the United States or the United Kingdom, have constant current account deficits accompanied by appreciating currencies, while others, such as Iran or Russia, experience the opposite (CIA, 2022; OECD 2022d). Despite significant depreciation of the Turkish Lira in recent years, Turkey continues to run a record current account deficit (TUİK, 2022d). The underlying factor here is to create demand for the country's currency, and transform the economy accordingly. The dollar will always be valuable as long as people and/or countries demand it for one reason or another. Likewise, the Turkish Lira or Iranian Rial will always be undervalued because hardly anybody or any country demands it. Increased investment and capital accumulation will result in increased employment opportunities, particularly for women, if Turkey can achieve the status of being a financial center of attraction.

REFERENCES

Aigner, D., Lovell, C. A. K., & Schmidt, P. (1977). Formulation and estimation of stochastic frontier production function models. *Journal of Econometrics*, *6*(1), 21–37. doi:10.1016/0304-4076(77)90052-5

Aigner, D. J., & Chu, S. (1968). On Estimating the Industry Production Function. *The American Economic Review*, *58*, 826–839.

Algan, N., & Akdoğan, M. (2001). Veri Zarflama Analiz Tekniği Kullanılarak Karadeniz Bölgesi'nde Yer Alan İllerin Etkinlik Değerlendirilmesi. In Ulusal Bölge Bilimi, Bölge Planlama Kongresi Bildiriler Kitabı (pp. 241-251). Trabzon.

Aslan, M. (2005). Political Economic Analysis of Turkish Economy: Structural Problems and the Role of the State. *Eskişehir Osmangazi Üniversitesi Sosyal Bilimler Dergisi*, *6*(2), 15-27. Retrieved from https://dergipark.org.tr/tr/pub/ogusbd/issue/10987/131485

Atan, M., Özgür, E., & Güler, H. (2004). Çok Değişkenli İstatistiksel Analizler ve VZA ile İllerin Gelişmişlik Düzeylerinin Karşılaştırılması. *Gazi Üniversitesi İktisadi ve İdari Bilimler Fakültesi Dergisi*, *6*(2), 25-42. Retrieved from https://dergipark.org.tr/en/pub/gaziuiibfd/issue/28341/301243

Ay, S. (2012). Türkiye'de İşsizliğin Nedenleri: İstihdam Politikaları Üzerine Bir Değerlendirme. *Yönetim ve Ekonomi Dergisi*, *19*(2), 321–341.

Aydemir, Z. C. (2002). *Bölgesel Rekabet Edebilirlik Kapsamında İllerin Kaynak Kullanım Görece Verimlilikleri: Veri Zarflama Analizi Uygulaması* [Master's Thesis]. DPT Yayınları.

Aydın, N. (2012). Türkiye'de İstihdam Politikalarının Bulanık Veri Zarflama Analizi Yöntemiyle Etkinlik Analizi. *Finans Politik & Ekonomik Yorumlar Dergisi*, *49*, 15–28.

Aydoğan, E. (2017). Foreign Direct Investment in Turkey: A Comparison with the Central and Eastern European Countries. *Finans Politik ve Ekonomik Yorumlar Dergisi*, 626.

Bakırcı, F., Ekinci, E. D., & Şahinoğlu, T. (2014). Bölgesel Kalkınma Politikalarının Etkinliği: Türkiye Alt Bölgeler Bazında Bir Uygulama. *Atatürk Üniversitesi Sosyal Bilimler Enstitüsü Dergisi*, *18*(2), 281–298.

Banker, R. D. (1993). Maximum-likelihood, consistency and data envelopment analysis - A statistical foundation. *Management Science*, *39*(10), 1265–1273. doi:10.1287/mnsc.39.10.1265

Banker, R. D., & Cooper, C. A. (1984). WW. Some models for estimating technical and scale inefficiencies in data envelopment analysis. *Management Science*, *30*(9), 1078–1092. doi:10.1287/mnsc.30.9.1078

Biçerli, K. (2009). *Çalışma Ekonomisi* (5th ed.). Beta Yayınları.

Biçerli, K. (2018). *Çalışma Ekonomisi* (10th ed.). Beta Yayınları.

Borjas, G. J. (2013). Measuring Labor Supply. In *Labor Economics* (pp. 22–23). McGraw-Hill Education.

Boztepe, Y. (2007). *Türkiye'de İşsizlik Kavramı ve İşsizliğin Ortadan Kaldırılması ile İlgili Bir Model Oluşturulması* [Master's Thesis]. Yıldız Teknik Üniversitesi Sosyal Bilimler Enstitüsü, İstanbul.

Cam, E., & Atan, M. (2018). Türkiye'de İl Bazında İstihdam Politikalarının Etkinliği. *Bilgi Sosyal Bilimler Dergisi*, (1), 102–123.

Charnes, A., Cooper, W. W., & Rhodes, E. (1978). Measuring the efficiency of decision making units. *European Journal of Operational Research*, *2*(6), 429–444. doi:10.1016/0377-2217(78)90138-8

Christiansen, L. E., Lin, H., Pereira, J., Topalova, P. B., & Turk, R. (2016). Unlocking Female Employment Potential in Europe: Drivers and Benefits. *International Monetary Fund, Publication Services*. Retrieved June 5, 2022, from https://www.imf.org/external/pubs/ft/dp/2016/eur1601.pdf

CIA. (2022). *Country Comparisons - Current Account Balance*. The World Factbook. Retrieved July 31, 2022, from https://www.cia.gov/the-world-factbook/field/current-account-balance/country-comparison

Cobb, C. W., & Douglas, P. H. (1928). A Theory of Production. *The American Economic Review, 18*(Supplement), 139–165.

Dayıoğlu, M., & Ercan, H. (2010). *Developing More Effective Labour Market Policies and Institutions in Emerging Economies: Turkey. International Labour Organization Project, (76). Employment Analysis and Research Unit*. EMP/ANALYSIS.

Dell, E. (1987). Mercantilism. In *The Politics of Economic Interdependence* (pp. 168–185). Palgrave Macmillan. doi:10.1007/978-1-349-18874-1_11

Demir, F., & Erdem, N. (2010). Labor Market Performance after Structural Adjustment in Developing Countries: The Interesting But not so Unique Case of Turkey. In L. K. Valencia & B. J. Hahn (Eds.), *Employment and Labor Issues: Unemployment, Youth Employment and Child Labor* (pp. 1–55). Nova Science Publishers.

Ekin, N. (2003). Türkiye'de İşsizlik: İş Aramayan İşsizler – Kırsal Yoksullar – Kentsel Kayıtdışı Yapay İstihdamdakiler – Açık İşsizler. *Kamu-İş Dergisi*, (7), 2–21.

Elder, S., & Smith, A. (2010). *Women in labour markets: Measuring progress and identifying challenges*. International Labour Office.

Ercan, H. (1998). Non-Wage Labor Costs in the Turkish Manufacturing Industry: An International Comparison. In T. Bulutay (Ed.), *Impositions and Policies Related to the Turkish Labour Market* (pp. 113–130). Ankara State Institute of Statistics Publication.

Ercan, H. (2007a). Demographic Trends and Rural-to-Urban Migration in Turkey: Implications for the Labour Market and Education. In World Bank project on Investing in Turkey's Next Generation: Making Youth Transition from School to Work Contribute to Turkey's Development (pp. 1-7). World Bank Publications - Reports, The World Bank Group.

Ercan, H. (2007b). *Youth Employment in Turkey*. International Labour Organization Publications.

Ercan, H. (2010a). The Impact of the Global Financial Crisis on Employment in Turkey. In *Crisis and Turkey: Impact Analysis of Crisis Response Measures* (pp. 73–97). International Labour Office Publications.

Ercan, H. (2010b). *EEO Review: Youth Employment Measures 2010*. European Employment Observatory.

Eroğlu, İ., & Bursal, M. (2020). Türkiye'nin Cari Açık Dinamiklerinde Enerji Açığının Rolü ve Finansman Kalitesi. *MANAS Sosyal Araştırmalar Dergisi*, *9*(4), 2231–2247. doi:10.33206/mjss.689842

Farrell, M. J. (1957). The measurement of productive efficiency. *Journal of the Royal Statistical Society. Series A (General)*, *120*(3), 253–281. doi:10.2307/2343100

Fred. (2022a). *Average Hours of Work Per Week, Total, Household Survey for United States (2022)*. Retrieved April 22, 2019, from https://fred.stlouisfed.org/series/M08354USM310NNBR

Gaddis, I., & Klasen, S. (2014). Economic development, structural change, and women's labor force participation: A reexamination of the feminization U hypothesis. *Journal of Population Economics*, *27*(3), 639–681. doi:10.100700148-013-0488-2

Gilpatrick, E. (1966). On The Classification of Unemployment: A View of The Structural-Inadequate Demand Debate. *Industrial & Labor Relations Review*, *19*(2), 201–212. doi:10.1177/001979396601900203

Goldin, C. (1995). The U-Shaped Female Labor Force Function in Economic Development and Economic History. In *Schultz TP Investment in Women's Human Capital and Economic Development* (pp. 61–90). University of Chicago Press.

Gürsel, S., & Atasoy, V. (2000). Genç İşsiz, Kadın İşsiz. *Forum TOBB Dergisi*, *7*(6), 18–22.

Haller, M., & Hoellinger, F. (1994). Female Employment and the Change of Gender Roles: The Conflictual Relationship between Participation and Attitudes in International Comparison. *International Sociology*, *9*(1), 87–112. doi:10.1177/026858094009001006

Hanson, D. L., & Pledger, G. (1976). Consistency in concave regression. *Annals of Statistics*, *4*(6), 1038–1050. doi:10.1214/aos/1176343640

Hayes, A. (2002). What is Unemployment. *Investopia*. Retrieved June 6, 2001, from https://www.investopedia.com/terms/u/unemployment.asp

Hildreth, C. (1954). Point Estimates of Ordinates of Concave Functions. *Journal of the American Statistical Association, 49*(267), 598–619. doi:10.1080/01621459.1954.10483523

İlkkaracan, İ., & Selim, R. (2003). The Role of Unemployment in Wage Determination: Further Evidence on the Wage Curve from Turkey. *Applied Economics, 35*(14), 1589–1598. doi:10.1080/0003684022000040939

İlkkaracan, İ., & Yörükoğlu, A. (2004). Ücretlerin Belirlenme Süreçleri. In *Türkiye'de İşgücü Piyasasının Kurumsal Yapısı ve İşsizlik* (pp. 95–145). TÜSİAD.

International Labour Organization. (2019a). *Employment to population ratio, 15+, female (%) (modeled ILO estimate) - Country Ranking*. ILOSTAT database. Retrieved September 19, 2019, from https://www.indexmundi.com/facts/indicators/SL.EMP.TOTL.SP.FE.ZS/rankings

International Labour Organization. (2019b). *Employment to population ratio, 15+, male (%) (modeled ILO estimate) - Country Ranking*. ILOSTAT database. Retrieved September 19, 2019, from https://www.indexmundi.com/facts/indicators/SL.EMP.TOTL.SP.MA.ZS/rankings

International Labour Organization. (2019c). *Employment to population ratio, 15+, total (%) (modeled ILO estimate) - Country Ranking*. ILOSTAT database. Retrieved September 19, 2019, from https://www.indexmundi.com/facts/indicators/SL.EMP.TOTL.SP.ZS/rankings

Işık, Ş. (2005). Türkiye'de Kentleşme ve Kentleşme Modelleri. *Ege Coğrafya Dergisi, 14*(1-2), 57-71. Retrieved June 7, 2022, from https://dergipark.org.tr/tr/pub/ecd/issue/4878/66933

Karabıyık, İ. (2012). Türkiye'de Çalışma Hayatında Kadın İstihdamı. *Marmara Üniversitesi İktisadi ve İdari Bilimler Dergisi, 32*(1), 231–260.

Karabulut, A. (2007). *Türkiye'de İşsizliği Önlemede Aktif İstihdam Politikalarının Rolü ve Etkinliği* [Master's Thesis]. Çalışma ve Sosyal Güvenlik Bakanlığı Türkiye İş Kurumu Genel Müdürlüğü.

Karanassou, M., Sala, H., & Salvador, P. (2008). Capital Accumulation and Unemployment: New Insights on the Nordic Experience. *Cambridge Journal of Economics, 32*(6), 977–1001. doi:10.1093/cje/ben022

Kıran, B. (2008). *Kalkınmada Öncelikli İllerin Ekonomik Etkinlikleri-nin Veri Zarflama Analizi Yöntemi ile Değerlendirilmesi* [Unpublished Master's Thesis]. Çukurova Üniversitesi Sosyal Bilimler Enstitüsü, Adana.

Köne, A. Ç. (2014). Türkiye'de Bölgesel Eko-Verimliliğin Ölçülmesi: Veri Zarflama Analizi ile Bir Değerlendirme. *TİSK Akademi Dergisi*, *9*(16), 138–155.

Korkmaz, D., & Korkut, A. (2012). Türkiye'de Kadın İşgücüne Katılımın Belirleyicileri. *Süleyman Demirel Üniversitesi İktisadi ve İdari Bilimler Fakültesi Dergisi*, *17*(2), 41–65.

Kumbhakar, S., & Lovell, C. A. K. (2000). *Stochastic frontier analysis*. Cambridge University Press. doi:10.1017/CBO9781139174411

Kuosmanen, T. (2008). *Data Envelopment Analysis as Least-Squares Regression*. Working Paper. Retrieved July 30, 2022, from https://tinyurl.com/DEA-Least-Squares-Regression

Kuosmanen, T., & Johnson, A. (2010). Data Envelopment Analysis as Nonparametric Least-Squares Regression. *Operations Research*, *58*(1), 149–160. doi:10.1287/opre.1090.0722

Kuosmanen, T., & Kortelainen, M. (2012). Stochastic non-smooth envelopment of data: Semiparametric frontier estimation subject to shape constraints. *Journal of Productivity Analysis*, *38*(1), 11–28. doi:10.100711123-010-0201-3

Lilien, D. M. (1982). Sectoral Shifts and Cyclical Unemployment. *Journal of Political Economy*, *90*(4), 777–793. doi:10.1086/261088

Malley, J., & Moutos, T. (2001). Capital Accumulation and Unemployment: A Tale of Two "Continents". *The Scandinavian Journal of Economics*, *130*(1), 79–99. doi:10.1111/1467-9442.00231

Meeusen, W., & Vandenbroeck, J. (1977). Efficiency estimation from cobb-douglas production functions with composed error. *International Economic Review*, *18*(2), 435–445. doi:10.2307/2525757

Mourdoukoutas, P. (1988). Seasonal Employment, Seasonal Unemployment and Unemployment Compensation: The Case of the Tourist Industry of the Greek Island. *American Journal of Economics and Sociology*, *47*(3), 315–329. doi:10.1111/j.1536-7150.1988.tb02044.x

Norris, P. (2001). Breaking the Barriers: Positive Discrimination Policies for Women. In J. Klausen & C. S. Maier (Eds.), *Has Liberalism Failed Women?*

OECD. (2021). *FDI in Figures October 2021: Global FDI flows rebound to exceed pre-pandemic levels*. Foreign Direct Investment Statistics: Data, Analysis and Forecasts. Retrieved July 30, 2022, from https://www.oecd.org/investment/investment-policy/FDI-in-Figures-October-2021.xlsx

OECD. (2022a). Employment rate (indicator). https://data.oecd.org/emp/employment-rate.htm

OECD. (2022b). Labour force participation rate (indicator). https://data.oecd.org/emp/labour-force-participation-rate.htm

OECD. (2022c). *Labour force participation rate by sex, 15+,15-64 and 15-24 years old.* Retrieved July 30, 2022, from https://www.oecd.org/gender/data/labour-force-participation-by-sex-age.htm

OECD. (2022d). *Current Account Balance.* OECD Quarterly International Trade Statistics. Retrieved August 31, 2022, from https://data.oecd.org/trade/current-account-balance.htm

Oğuz, F., & Oğuz, S. (2016). Kamuda Girişimcilik, Rekabet ve İstihdam Politikalarının Etkinliği. *Maliye Dergisi. Temmuz-Aralık, 2016*(171), 34–54.

Onaran, Ö. (2002). Measuring Wage Flexibility: The Case of Turkey Before and After Structural Adjustment. *Applied Economics, 34*(6), 767–781. doi:10.1080/00036840110056934

Onaran, Ö. (2007). İşsizlik ve Alternatif Politikalar. In *Türkiye'de İşsizliğin Çözüm Yolları ve Sendikalar Çalıştayı* (p. 13). DİSK.

Onaran, Ö., & Aydıner-Avşar, N. (2006). The controversy over employment policy: Low labor costs and openness, or demand policy? A sectoral analysis for Turkey. *Department of Economics Working Paper Series,* (97), 24.

Öncel, A. (2010). *Türkiye'de Bölgelerarası Gelişmişlik Farklılıklarını Gidermeye Yönelik Politika Uygulamaları ve Etkin Kaynak Kullanımı Üzerine Bir Analiz* [Unpublished Doctoral Dissertation]. Sakarya Üniversitesi Sosyal Bilimler Enstitüsü, Adapazarı.

Parlaktuna, İ. (2010). Türkiye'de Cinsiyete Dayalı Mesleki Ayrımcılığın Analizi. *Ege Academic Review, 10*(4), 1217–1230. doi:10.21121/eab.2010419607

Ray, S., & Chen, L. (2015). Data Envelopment Analysis for Performance Evaluation: A Child's Guide. In S. Ray, S. Kumbhakar, & P. Dua (Eds.), *s), Benchmarking for performance evaluation: A production frontier approach.* Springer. doi:10.1007/978-81-322-2253-8_2

Ray, S. C., Kumbhakar, S. C., & Dua, P. (2015). *Benchmarking for performance evaluation: A production frontier approach.* Springer. doi:10.1007/978-81-322-2253-8

Reynolds, L. (1974). *Labor Economics and Labor Relations* (6th ed.). Prentice-Hall, Inc.

Sarıtaş, H., Genç, A., & Avcı, T. (2018). Türkiye'de Enerji İthalatı, Cari Açık ve Büyüme İlişkisi: VAR ve Granger Nedensellik Analizi. *Ekonomik ve Sosyal Araştırmalar Dergisi, 14*, 181–200.

Schmidt, P. (1985). Frontier production functions. *Econometric Reviews, 4*(2), 289–328. doi:10.1080/07474938608800089

Şengül, Ü., Eslemian, S., & Eren, M. (2013). Türkiye'de İstatistikî Bölge Birimleri Sınıflamasına göre Düzey 2 Bölgelerinin Ekonomik Etkinlikle-rinin VZA Yöntemi ile Belirlenmesi ve Tobit Model Uygulaması. *Yönetim Bilimleri Dergisi, 11*(21), 75–99.

Sengupta, J. K. (1992). A fuzzy systems approach in data envelopment analysis. *Computers & Mathematics with Applications (Oxford, England), 24*(8-9), 259–266. doi:10.1016/0898-1221(92)90203-T

Sinha, J. N. (1967). Dynamics of Female Participation in Economic Activity in a Developing Economy. In *United Nations, Department of Economic and Social Affairs. Proceedings of the World Population Conference*. UN Publications.

Sosyo Politik Saha Araştırmaları Merkezi. (2019). Kadının Genel İstihdamdaki Yeri Ve İstihdam Edildiği İş Kollarındaki Durumu Üzerine Çalışma. *Kadın Çalışmaları / Slider Category*. Retrieved February 13, 2022, from https://tinyurl.com/sahamerkezi-kadin-istihdami

Strateji ve Bütçe Başkanlığı. (2022). *Kamu sektörü istihdam sayıları (2007-2022)*. Retrieved August 31, 2022, from https://www.sbb.gov.tr/kamu-istihdami/

Talas, C. (1997). Toplumsal Ekonomi Çalışma Ekonomis (7th ed.). İmge Kitabevi.

Tatlıdil, E., & Xanthocou, Y. (2002). Türk İşgücünün Yapısı ve Avrupa Birliği İstihdam Politikaları. *Ege Academic Review, 2*(2), 1–14.

TCMB. (2002). *TÜFE Bazlı Reel Efektif Döviz Kuru (2003=100) (Temmuz 2022)*. Retrieved July 31, 2022, from https://tinyurl.com/TCMB-ReelKur-Temmuz2022

Timur, M. C., & Doğan Çalışkan, Z. (2015). İstihdam Yaratmayan Büyüme: Türkiye Analizi. Jobless Growth: Analysis of Turkey. *Ardahan Üniversitesi İİBF Dergisi, 1*, 231–247.

TUİK. (2021). *Doğum İstatistikleri, 2020. Sayı 37229*. Retrieved February 15, 2022, from https://data.tuik.gov.tr/Bulten/Index?p=Dogum-Istatistikleri-2020-37229

TUİK. (2022a). *İşgücü İstatistikleri, Haziran 2022*. Retrieved August 12, 2022, from https://data.tuik.gov.tr/Bulten/Index?p=Isgucu-Istatistikleri-Haziran-2022-45651&dil=1

TUİK. (2022b). *İşgücü İstatistikleri (1988-2013)*. Retrieved February 15, 2022, from https://biruni.tuik.gov.tr/isgucuapp/isgucu.zul

TUİK. (2022c). *İşgücü İstatistikleri (2014 ve Sonrası)*. Retrieved February 15, 2022, from https://biruni.tuik.gov.tr/medas/?kn=72&locale=tr

TUİK. (2022d). *Dış Ticaret İstatistikleri, Temmuz 2022*. Retrieved August 31, 2022, from https://biruni.tuik.gov.tr/medas/?kn=72&locale=tr

Tunalı, İ. (2003). Background Study on the Labour Market and Employment in Turkey. *European Training Foundation Final Report*, 125.

Ülgener, S. F. (1980). *Milli Gelir, İstihdam ve İktisadi Büyüm* (6th ed.). Der Yayınevi.

Ünsal, E. (2000). Makro İktisat (3rd ed.). İmaj Yayıncılık.

U.S. Bureau of Labor Statistics. (2022a). *Household Data. Table A-15. Alternative measures of labor underutilization*. Retrieved July 31, 2022, from https://www.bls.gov/news.release/empsit.t15.htm

U.S. Bureau of Labor Statistics. (2022b). *Concepts and Definitions. Labor Force Statistics from the Current Population Survey*. Retrieved July 31, 2022, from https://www.bls.gov/cps/definitions.htm

Uslu, H. (2020). İstihdam Yaratmayan Ekonomik Büyüme: Türkiye İçin Okun Yasası Çerçevesinde Ekonometrik Bir Analiz. *Optimum Ekonomi ve Yönetim Bilimleri Dergisi*, *7*(1), 101–126. doi:10.17541/optimum.578251

Verick, S. (2014). Female labor force participation in developing countries. *IZA World of Labor: Evidence-Based Policy Making*, *87*. Advance online publication. doi:10.15185/izawol.87

Winsten, C. B. (1957). Discussion on Mr. harrells paper. *J. Royal Statist. Soc. Sen A. Statist. Soc.*, *120*(3), 282–284.

Yatchew, A. (2003). *Semiparametric Regression for the Applied Econometrician*. Cambridge University Press. doi:10.1017/CBO9780511615887

Yatchew, A. J. (1998). Nonparametric Regression Techniques in Economics. *Journal of Economic Literature*, *36*, 669–721.

YÖK. (2022). *Öğrenci Sayıları Özet Tablosu. 2020-2021 Öğretim Yılı*. Yüksek Öğretim Bilgi Yönetim Sistemi. Retrieved Feburuary 2, 2022, from https://istatistik.yok.gov.tr/

APPENDIX 1

Figure 10. Male efficient labor force frontier for the 1988-2030 period

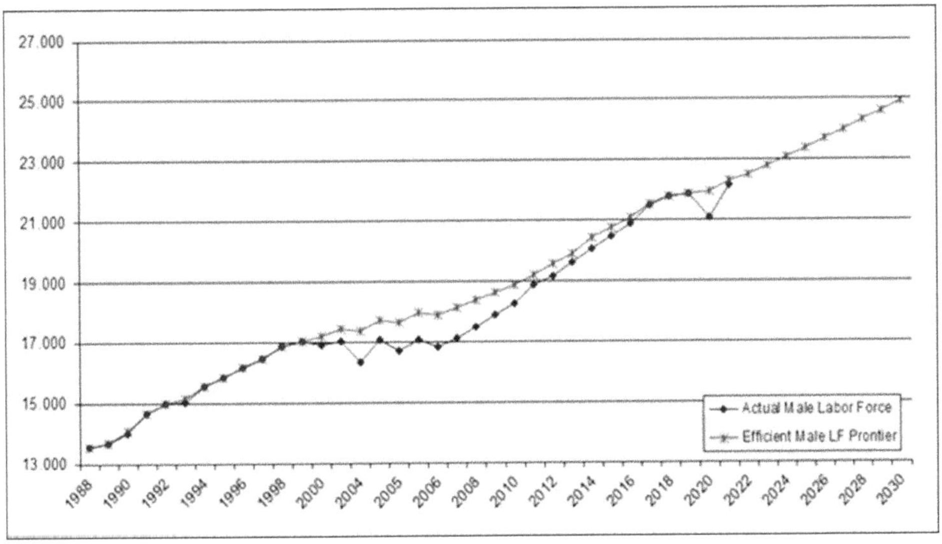

Figure 11. Female efficient labor force frontier for the 1988-2030 period

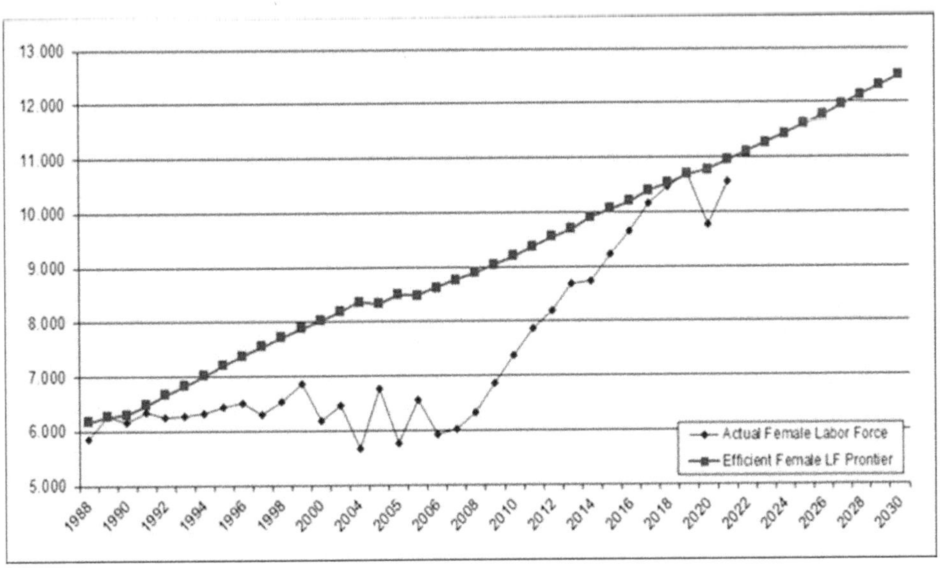

Figure 12. Male efficient employment frontier for the 1988-2030 period

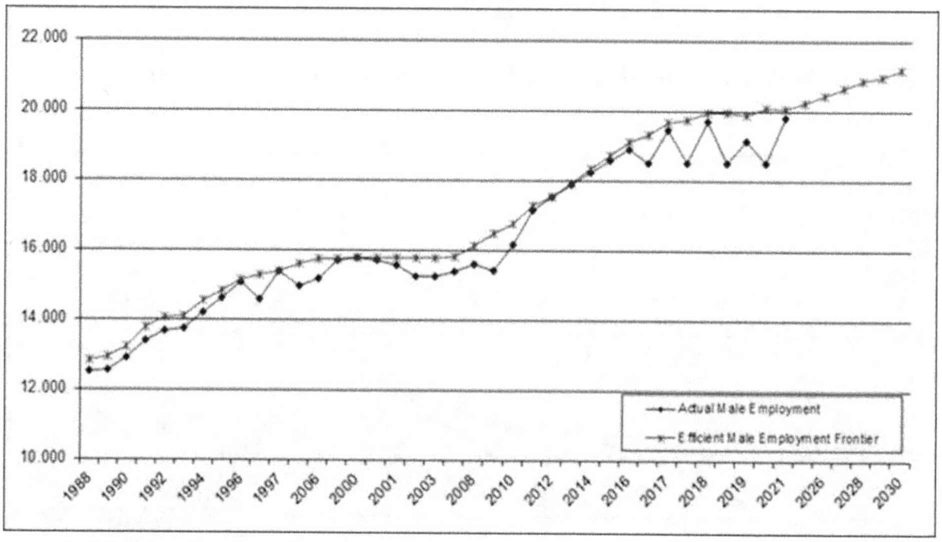

*Figure 13. Female efficient employment frontier for the 1988-2030 period**
** Please Note that the production frontier is not arranged chronologically but by employment levels.*

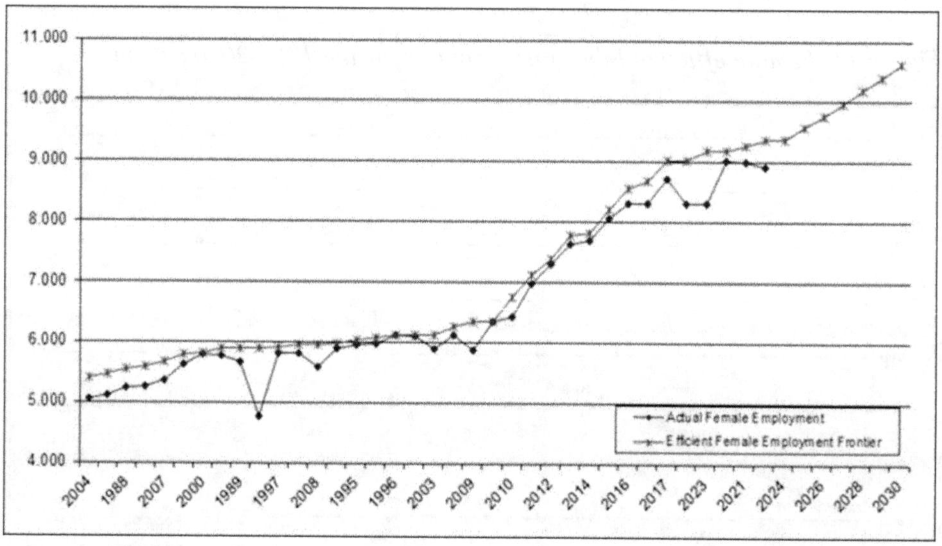

Figure 14. Male employment efficiency and inefficiency results for the 1988-2030 period

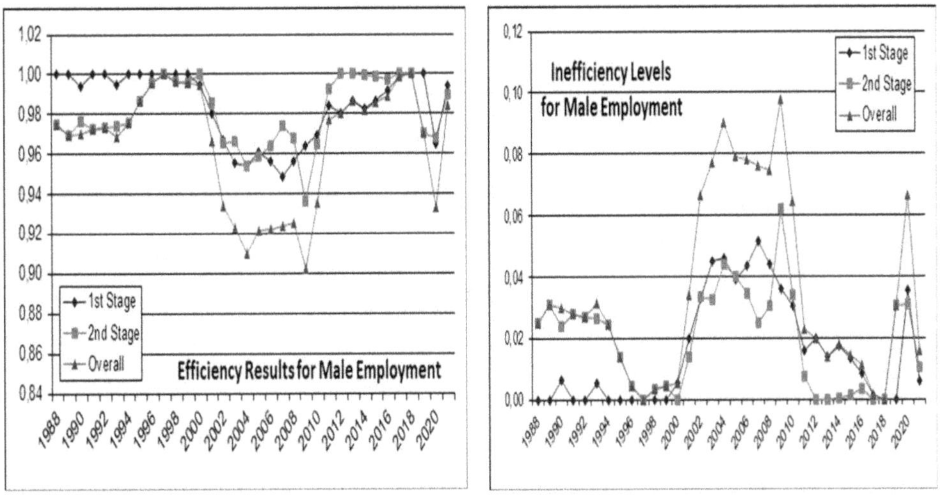

Figure 15. Female employment efficiency and inefficiency results for the 1988-2030 period

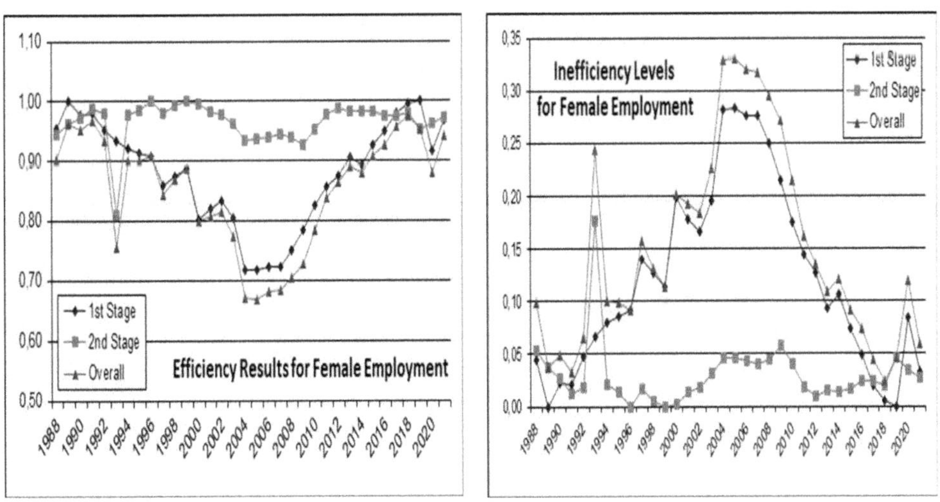

Figure 16. Male employment trends for the 1988-2030 period

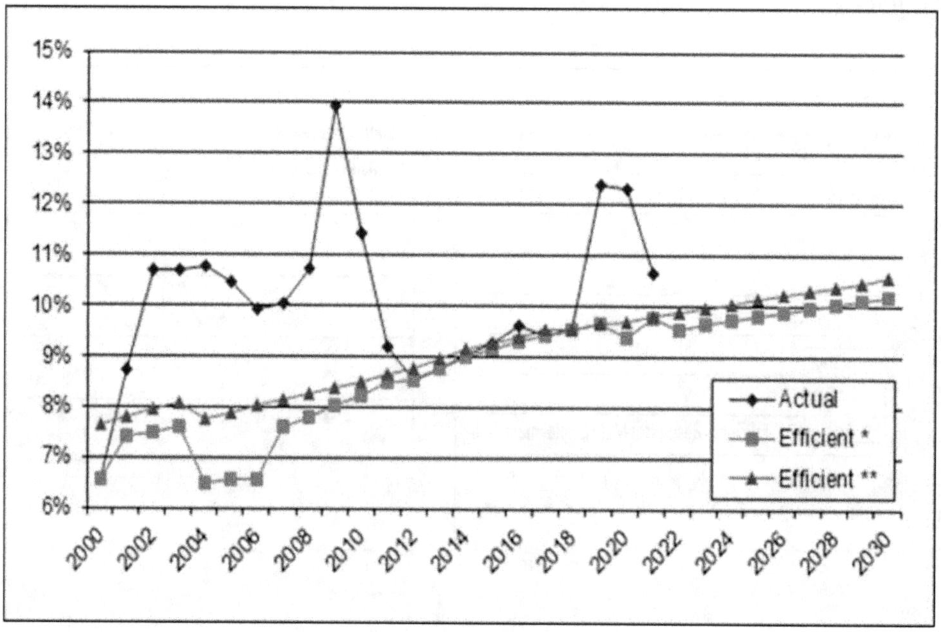

Figure 17. Female employment trends for the 1988-2030 period

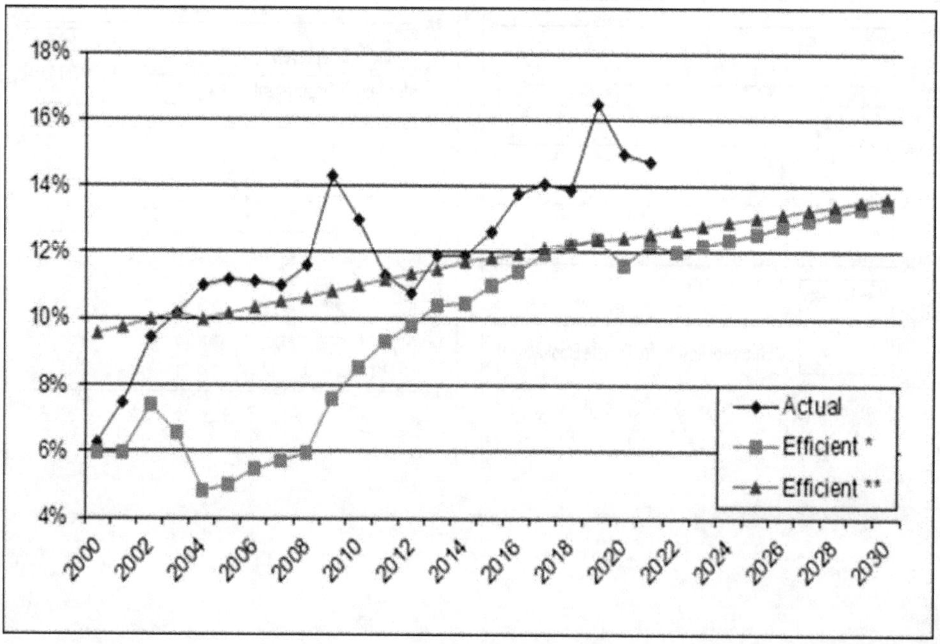

APPENDIX 2

Table 10. Comparison of actual and efficient labor force levels for all years

	Actual Labor Force			Efficient Labor Force			Efficiency Rate		
	Total	Male	Female	Total	Male	Female	Total	Male	Female
1988	19.390	13.536	5.854	19.600	13.536	6.170	0,989	1,000	0,949
1989	19.930	13.663	6.267	19.930	13.663	6.267	1,000	1,000	1,000
1990	20.150	13.990	6.160	20.288	14.087	6.300	0,993	0,993	0,978
1991	21.010	14.665	6.345	21.010	14.665	6.480	1,000	1,000	0,979
1992	21.264	15.002	6.262	21.264	15.002	6.662	1,000	1,000	0,940
1993	21.313	15.045	6.268	21.435	15.135	6.836	0,994	0,994	0,917
1994	21.876	15.552	6.324	21.961	15.552	7.017	0,996	1,000	0,901
1995	22.285	15.858	6.427	22.515	15.858	7.203	0,990	1,000	0,892
1996	22.696	16.182	6.514	23.035	16.182	7.378	0,985	1,000	0,883
1997	22.755	16.464	6.291	23.549	16.465	7.550	0,966	1,000	0,833
1998	23.385	16.848	6.537	24.034	16.848	7.714	0,973	1,000	0,847
1999	23.877	17.025	6.852	24.529	17.025	7.882	0,973	1,000	0,869
2000	23.078	16.890	6.188	24.967	17.180	8.029	0,924	0,983	0,771
2001	23.491	17.040	6.451	25.428	17.461	8.185	0,924	0,976	0,788
2002	23.818	17.058	6.760	25.858	17.721	8.331	0,921	0,963	0,811
2003	23.640	17.086	6.555	26.283	17.978	8.476	0,899	0,950	0,773
2004	22.016	16.348	5.669	25.616	17.379	8.358	0,859	0,941	0,678
2005	22.454	16.704	5.750	26.013	17.629	8.487	0,863	0,948	0,678
2006	22.751	16.836	5.916	26.410	17.879	8.618	0,861	0,942	0,686
2007	23.114	17.098	6.016	26.809	18.128	8.756	0,862	0,943	0,687
2008	23.805	17.476	6.329	27.188	18.368	8.885	0,876	0,951	0,712
2009	24.748	17.898	6.851	27.633	18.636	9.043	0,896	0,960	0,758
2010	25.641	18.257	7.383	28.050	18.892	9.189	0,914	0,966	0,803
2011	26.725	18.867	7.859	28.562	19.200	9.372	0,936	0,983	0,839
2012	27.339	19.147	8.192	29.119	19.581	9.544	0,939	0,978	0,858
2013	28.271	19.597	8.674	29.589	19.915	9.689	0,955	0,984	0,895
2014	28.786	20.057	8.729	30.323	20.449	9.911	0,949	0,981	0,881
2015	29.678	20.453	9.225	30.784	20.759	10.062	0,964	0,985	0,917
2016	30.536	20.899	9.637	31.245	21.093	10.202	0,977	0,991	0,945
2017	31.643	21.484	10.159	31.870	21.542	10.393	0,993	0,997	0,977

continues on following page

Table 10. Continued

	Actual Labor Force			Efficient Labor Force			Efficiency Rate		
	Total	**Male**	**Female**	**Total**	**Male**	**Female**	**Total**	**Male**	**Female**
2018	32.274	21.801	10.473	32.274	21.801	10.531	1,000	1,000	0,994
2019	32.549	21.863	10.686	32.708	21.863	10.686	0,995	1,000	1,000
2020	30.873	21.105	9.768	33.298	21.952	10.772	0,927	0,961	0,907
2021	32.716	22.156	10.560	33.897	22.294	10.952	0,965	0,994	0,964
2022				34.323	22.513	11.095			
2023				34.848	22.799	11.260			
2024				35.380	23.091	11.428			
2025				35.921	23.386	11.599			
2026				36.469	23.686	11.772			
2027				37.026	23.991	11.948			
2028				37.410	24.300	12.126			
2029				37.392	24.614	12.307			
2030				37.926	24.933	12.491			

Table 11. Comparison of actual and efficient employment levels for all years

	Actual Employment			Efficient Employment			Efficiency Rate		
	Total	**Male**	**Female**	**Total**	**Male**	**Female**	**Total**	**Male**	**Female**
1988	17.754	12.520	5.234	18.372	12.841	5.544	0,966	0,975	0,944
1989	18.222	12.548	5.674	18.813	12.946	5.895	0,969	0,969	0,963
1990	18.538	12.901	5.637	18.992	13.218	5.794	0,976	0,976	0,973
1991	19.288	13.395	5.893	19.694	13.778	5.968	0,979	0,972	0,987
1992	19.460	13.682	5.778	19.902	14.058	5.890	0,978	0,973	0,981
1993	18.499	13.722	4.777	19.942	14.093	5.896	0,928	0,974	0,810
1994	20.006	14.191	5.815	20.457	14.546	5.948	0,978	0,976	0,978
1995	20.586	14.628	5.958	20.839	14.832	6.045	0,988	0,986	0,986
1996	21.194	15.067	6.127	21.224	15.135	6.127	0,999	0,995	1,000
1997	21.203	15.399	5.804	21.279	15.399	5.917	0,996	1,000	0,981
1998	21.778	15.686	6.092	21.778	15.741	6.127	1,000	0,997	0,994
1999	22.048	15.713	6.335	22.048	15.780	6.335	1,000	0,996	1,000
2000	21.581	15.780	5.801	21.581	15.780	5.820	1,000	1,000	0,997
2001	21.524	15.555	5.969	21.778	15.780	6.068	0,988	0,986	0,984
2002	21.354	15.232	6.122	21.994	15.780	6.260	0,971	0,965	0,978

continues on following page

Table 11. Continued

	Actual Employment			Efficient Employment			Efficiency Rate		
	Total	Male	Female	Total	Male	Female	Total	Male	Female
2003	21.147	15.256	5.891	21.841	15.787	6.127	0,968	0,966	0,961
2004	19.632	14.585	5.047	20.588	15.291	5.398	0,954	0,954	0,935
2005	20.066	14.958	5.108	20.997	15.606	5.462	0,956	0,958	0,935
2006	20.423	15.165	5.258	21.275	15.730	5.593	0,960	0,964	0,940
2007	20.738	15.382	5.356	21.581	15.797	5.672	0,961	0,974	0,944
2008	21.194	15.598	5.595	21.982	16.111	5.953	0,964	0,968	0,940
2009	21.277	15.406	5.871	22.745	16.461	6.334	0,935	0,936	0,927
2010	22.594	16.170	6.425	23.474	16.759	6.752	0,962	0,965	0,952
2011	24.110	17.137	6.973	24.359	17.265	7.128	0,990	0,993	0,978
2012	24.821	17.512	7.309	24.860	17.512	7.392	0,998	1,000	0,989
2013	25.524	17.883	7.641	25.667	17.924	7.773	0,994	0,998	0,983
2014	25.933	18.244	7.689	26.135	18.344	7.816	0,992	0,995	0,984
2015	26.620	18.562	8.058	26.945	18.706	8.208	0,988	0,992	0,982
2016	27.206	18.893	8.313	27.724	19.114	8.555	0,981	0,988	0,972
2017	28.188	19.460	8.728	28.729	19.649	9.018	0,981	0,990	0,968
2018	28.737	19.719	9.018	29.301	19.939	9.194	0,981	0,989	0,981
2019	28.080	19.156	8.924	29.195	19.887	9.363	0,962	0,963	0,953
2020	26.812	18.506	8.306	28.030	19.303	8.671	0,957	0,959	0,958
2021	28.797	19.792	9.005	29.345	20.069	9.263	0,981	0,986	0,972
2022				28.875	19.738	9.018			
2023				29.309	19.959	9.184			
2024				29.392	20.072	9.375			
2025				29.696	20.212	9.571			
2026				30.106	20.438	9.770			
2027				30.523	20.638	9.975			
2028				30.975	20.848	10.184			
2029				31.380	20.970	10.398			
2030				31.819	21.187	10.617			

Table 12. Comparison of two-stage employment efficiency rates for all years

	Overall Employment			Male Employment			Female Employment		
	1st Stage	2nd Stage	Total	1st Stage	2nd Stage	Total	1st Stage	2nd Stage	Total
1988	0,991	0,966	0,957	1,000	0,975	0,975	0,955	0,944	0,902
1989	1,000	0,969	0,969	1,000	0,969	0,969	1,000	0,963	0,963
1990	0,994	0,976	0,970	0,994	0,976	0,970	0,978	0,973	0,951
1991	1,000	0,979	0,979	1,000	0,972	0,972	0,979	0,987	0,967
1992	1,000	0,978	0,978	1,000	0,973	0,973	0,953	0,981	0,935
1993	0,995	0,928	0,923	0,995	0,974	0,969	0,933	0,810	0,756
1994	0,996	0,978	0,974	1,000	0,976	0,976	0,920	0,978	0,900
1995	0,990	0,988	0,978	1,000	0,986	0,986	0,915	0,986	0,901
1996	0,985	0,999	0,984	1,000	0,995	0,995	0,908	1,000	0,908
1997	0,977	0,996	0,974	1,000	1,000	1,000	0,860	0,981	0,843
1998	0,983	1,000	0,983	1,000	0,997	0,997	0,874	0,994	0,869
1999	0,977	1,000	0,977	1,000	0,996	0,996	0,886	1,000	0,886
2000	0,941	1,000	0,941	0,995	1,000	0,995	0,801	0,997	0,799
2001	0,935	0,988	0,924	0,980	0,986	0,966	0,822	0,984	0,808
2002	0,930	0,971	0,903	0,967	0,965	0,934	0,834	0,978	0,816
2003	0,910	0,968	0,881	0,955	0,966	0,923	0,805	0,961	0,774
2004	0,878	0,954	0,837	0,954	0,954	0,910	0,718	0,935	0,671
2005	0,883	0,956	0,844	0,961	0,958	0,921	0,716	0,935	0,670
2006	0,883	0,960	0,847	0,956	0,964	0,922	0,724	0,940	0,680
2007	0,883	0,961	0,849	0,949	0,974	0,924	0,724	0,944	0,683
2008	0,889	0,964	0,857	0,956	0,968	0,926	0,750	0,940	0,705
2009	0,906	0,935	0,848	0,964	0,936	0,902	0,785	0,927	0,728
2010	0,923	0,962	0,888	0,969	0,965	0,935	0,826	0,952	0,786
2011	0,942	0,990	0,932	0,984	0,993	0,977	0,856	0,978	0,838
2012	0,945	0,998	0,943	0,980	1,000	0,980	0,874	0,989	0,864
2013	0,960	0,996	0,956	0,986	1,000	0,986	0,906	0,983	0,891
2014	0,954	0,996	0,950	0,982	1,000	0,982	0,893	0,984	0,879
2015	0,967	0,994	0,962	0,987	0,999	0,985	0,925	0,982	0,909
2016	0,979	0,990	0,970	0,992	0,997	0,989	0,950	0,974	0,926
2017	0,994	0,993	0,987	0,999	1,000	0,999	0,980	0,976	0,956
2018	1,000	0,995	0,995	1,000	1,000	1,000	0,995	0,981	0,976
2019	0,996	0,965	0,960	1,000	0,970	0,970	1,000	0,953	0,953

continues on following page

Table 12. Continued

	Overall Employment			Male Employment			Female Employment		
	1st Stage	2nd Stage	Total	1st Stage	2nd Stage	Total	1st Stage	2nd Stage	Total
2020	0,933	0,966	0,902	0,965	0,968	0,933	0,916	0,962	0,881
2021	0,968	0,985	0,953	0,994	0,990	0,984	0,968	0,972	0,941
Average	0,955	0,978	0,935	0,984	0,981	0,965	0,880	0,965	0,850

Table 13. Comparison of actual and two-stage efficient employment levels for all years

	Overall Employment			Male Employment			Female Employment		
	Actual	Efficient*	Efficient**	Actual	Efficient*	Efficient**	Actual	Efficient*	Efficient**
1988	17.754	18.372	18.543	12.520	12.841	12.841	5.234	5.544	5.804
1989	18.222	18.813	18.813	12.548	12.946	12.946	5.674	5.895	5.895
1990	18.538	18.992	19.105	12.901	13.218	13.298	5.637	5.794	5.925
1991	19.288	19.694	19.694	13.395	13.778	13.778	5.893	5.968	6.095
1992	19.460	19.902	19.902	13.682	14.058	14.058	5.778	5.890	6.183
1993	18.499	19.942	20.045	13.722	14.093	14.168	4.777	5.896	6.320
1994	20.006	20.457	20.537	14.191	14.546	14.546	5.815	5.948	6.463
1995	20.586	20.839	21.055	14.628	14.832	14.832	5.958	6.045	6.610
1996	21.194	21.224	21.541	15.067	15.135	15.135	6.127	6.127	6.748
1997	21.203	21.279	21.778	15.399	15.399	15.399	5.804	5.917	6.884
1998	21.778	21.778	22.163	15.686	15.741	15.741	6.092	6.127	7.014
1999	22.048	22.048	22.566	15.713	15.780	15.780	6.335	6.335	7.146
2000	21.581	21.581	22.924	15.780	15.780	15.865	5.801	5.820	7.263
2001	21.524	21.778	23.301	15.555	15.780	16.098	5.969	6.068	7.386
2002	21.354	21.994	23.652	15.232	15.780	16.314	6.122	6.260	7.502
2003	21.147	21.841	23.998	15.256	15.787	16.527	5.891	6.127	7.616
2004	19.632	20.588	23.454	14.585	15.291	16.030	5.047	5.398	7.522
2005	20.066	20.997	23.778	14.958	15.606	16.238	5.108	5.462	7.624
2006	20.423	21.275	24.102	15.165	15.730	16.446	5.258	5.593	7.729
2007	20.738	21.581	24.428	15.382	15.797	16.652	5.356	5.672	7.837
2008	21.194	21.982	24.737	15.598	16.111	16.851	5.595	5.953	7.939
2009	21.277	22.745	25.101	15.406	16.461	17.074	5.871	6.334	8.065
2010	22.594	23.474	25.440	16.170	16.759	17.286	6.425	6.752	8.179

continues on following page

Table 13. Continued

	Overall Employment			Male Employment			Female Employment		
	Actual	Efficient*	Efficient**	Actual	Efficient*	Efficient**	Actual	Efficient*	Efficient**
2011	24.110	24.359	25.858	17.137	17.265	17.542	6.973	7.128	8.324
2012	24.821	24.860	26.313	17.512	17.512	17.868	7.309	7.392	8.460
2013	25.524	25.621	26.697	17.883	17.883	18.135	7.641	7.773	8.575
2014	25.933	26.041	27.296	18.244	18.253	18.578	7.689	7.816	8.750
2015	26.620	26.770	27.672	18.562	18.582	18.836	8.058	8.208	8.870
2016	27.206	27.470	28.049	18.893	18.952	19.113	8.313	8.534	8.980
2017	28.188	28.373	28.558	19.460	19.460	19.485	8.728	8.946	9.131
2018	28.737	28.889	28.889	19.719	19.719	19.719	9.018	9.194	9.241
2019	28.080	29.113	29.243	19.156	19.752	19.752	8.924	9.363	9.363
2020	26.812	27.745	29.725	18.506	19.123	19.826	8.306	8.637	9.430
2021	28.797	29.249	30.214	19.792	19.995	20.110	9.005	9.263	9.573
2022		28.505	30.561		19.520	20.291		8.998	9.686
2023		28.895	30.989		19.719	20.529		9.184	9.817
2024		29.292	31.424		19.921	20.771		9.375	9.949
2025		29.696	31.865		20.126	21.016		9.571	10.084
2026		30.106	32.313		20.334	21.265		9.770	10.221
2027		30.523	32.768		20.543	21.518		9.975	10.360
2028		30.948	33.081		20.756	21.775		10.184	10.501
2029		31.380	33.067		20.970	22.035		10.398	10.644
2030		31.819	33.502		21.187	22.300		10.617	10.790

Table 14. Trends of employment and unemployment rates for all years

	Overall Unemployment			Male Unemployment			Female Unemployment		
	Unemployment		Employment	Unemployment		Employment	Unemployment		Employment
	Actual	Efficient**		Actual	Efficient**		Actual	Efficient**	
1988	0,084	0,054	0,549	0,075	0,051	0,771	0,106	0,059	0,340
1989	0,086	0,056	0,548	0,082	0,052	0,763	0,095	0,059	0,340
1990	0,080	0,058	0,537	0,078	0,056	0,757	0,085	0,059	0,328
1991	0,082	0,063	0,534	0,087	0,060	0,754	0,071	0,059	0,328
1992	0,085	0,064	0,524	0,088	0,063	0,746	0,077	0,072	0,323
1993	0,132	0,065	0,515	0,088	0,064	0,735	0,238	0,075	0,321

continues on following page

Table 14. Continued

	Overall Unemployment			Male Unemployment			Female Unemployment		
	Unemployment		Employment	Unemployment		Employment	Unemployment		Employment
	Actual	Efficient**		Actual	Efficient**		Actual	Efficient**	
1994	0,085	0,065	0,513	0,088	0,065	0,734	0,080	0,079	0,320
1995	0,076	0,065	0,511	0,078	0,065	0,727	0,073	0,082	0,318
1996	0,066	0,065	0,510	0,069	0,065	0,723	0,059	0,085	0,317
1997	0,068	0,075	0,503	0,065	0,065	0,718	0,077	0,088	0,315
1998	0,069	0,078	0,500	0,069	0,066	0,717	0,068	0,091	0,314
1999	0,077	0,080	0,498	0,077	0,073	0,703	0,075	0,093	0,313
2000	0,065	0,082	0,496	0,066	0,077	0,692	0,063	0,095	0,312
2001	0,084	0,084	0,494	0,087	0,078	0,688	0,075	0,098	0,311
2002	0,103	0,085	0,497	0,107	0,079	0,702	0,094	0,100	0,309
2003	0,105	0,087	0,500	0,107	0,081	0,694	0,101	0,101	0,315
2004	0,108	0,084	0,485	0,108	0,078	0,677	0,110	0,100	0,305
2005	0,106	0,086	0,486	0,105	0,079	0,669	0,112	0,102	0,309
2006	0,102	0,087	0,490	0,099	0,080	0,683	0,111	0,103	0,308
2007	0,103	0,089	0,489	0,100	0,081	0,679	0,110	0,105	0,308
2008	0,110	0,090	0,487	0,107	0,083	0,676	0,116	0,106	0,307
2009	0,140	0,092	0,486	0,139	0,084	0,673	0,143	0,108	0,306
2010	0,119	0,093	0,484	0,114	0,085	0,670	0,130	0,110	0,306
2011	0,098	0,095	0,482	0,092	0,086	0,666	0,113	0,112	0,305
2012	0,092	0,096	0,481	0,085	0,087	0,663	0,108	0,114	0,305
2013	0,097	0,098	0,480	0,087	0,089	0,662	0,119	0,115	0,304
2014	0,099	0,100	0,479	0,090	0,091	0,660	0,119	0,117	0,303
2015	0,103	0,101	0,478	0,092	0,093	0,659	0,127	0,119	0,303
2016	0,109	0,102	0,478	0,096	0,094	0,658	0,137	0,120	0,302
2017	0,109	0,104	0,477	0,094	0,095	0,657	0,141	0,121	0,302
2018	0,110	0,105	0,476	0,096	0,096	0,657	0,139	0,123	0,302
2019	0,137	0,106	0,476	0,124	0,097	0,650	0,165	0,124	0,301
2020	0,132	0,107	0,475	0,123	0,097	0,640	0,150	0,125	0,298
2021	0,120	0,109	0,474	0,107	0,098	0,638	0,147	0,126	0,298
2022		0,110	0,474		0,099	0,636		0,127	0,297
2023		0,111	0,473		0,100	0,634		0,128	0,297
2024		0,112	0,473		0,100	0,632		0,129	0,296
2025		0,113	0,472		0,101	0,630		0,131	0,295
2026		0,114	0,471		0,102	0,628		0,132	0,295
2027		0,115	0,471		0,103	0,626		0,133	0,294

continues on following page

Table 12. Continued

	Overall Unemployment			Male Unemployment			Female Unemployment		
	Unemployment		Employment	Unemployment		Employment	Unemployment		Employment
	Actual	Efficient**		Actual	Efficient**		Actual	Efficient**	
2028		0,116	0,468		0,104	0,624		0,134	0,294
2029		0,116	0,461		0,105	0,622		0,135	0,293
2030		0,117	0,460		0,106	0,620		0,136	0,293

Chapter 6
Is the Tourism-Induced Environmental Kuznets Curve Hypothesis Valid in the Most Visited Countries?
A Fresh Insight from Dynamic Common Correlated Effects (DCCE) Approach

Tunahan Hacıimamoğlu
Recep Tayyip Erdogan University, Turkey

ABSTRACT

This study aims to test the tourism-induced EKC hypothesis, which claims the existence of an inverted U-shaped relationship between environmental pollution and tourism activities, for the period 1995-2018 in the most-visited countries in terms of the number of international tourists (Spain, United States, China, Italy, Turkey, Mexico, Thailand, Germany, and England). In this study, in which ecological footprint, which is a comprehensive and alternative variable, was used as an indicator of environmental pollution, the relations between the variables were analyzed using second-generation test techniques. Long-term relationships were examined by the LM and DH test, and the coefficients were estimated with the common correlated effects mean group (CCE–MG) and dynamic common correlated effects (DCCE) estimators. As a result of the empirical analysis, it was found that there is a U-shaped relationship between environmental pollution and tourism activity, and the validity of the tourism-induced EKC hypothesis could not be confirmed.

DOI: 10.4018/978-1-6684-6727-5.ch006

INTRODUCTION

International tourism, one of the most important components of the service industry, has gained remarkable speed in the last few decades. While the number of international tourists worldwide was 166 million in 1970, it reached 1,460 million in 2019, with a 780% increase (UNWTO, 2020). However, the COVID–19 pandemic, which started in 2019 in Wuhan city of China, caused one of the most severe economic crises since the Great Depression in 1929. Like many other industries, tourism was also negatively affected by this pandemic. Due to the measures taken such as domestic and international travel bans, lockdowns, and limitations on international transportation, the number of international tourists decreased by approximately 1 billion (72%) in 2020 (UNWTO, 2022).

Together with the effects of globalization, the tourism industry playing a role as a source of revenue and employment for many countries is the driving force for economic development. The global revenue from the tourism industry in 2019 was $1.8 trillion and the number of individuals employed in these industries reached 333 million. Moreover, at the global level, the tourism industry constitutes 10.3% of global production, 28% of global service export, and 6.8% in total (WTTC, 2022). In summary, it can be seen that the tourism industry remarkably corroborates economic growth, employment, and exports on a global scale.

Besides its contributions to economic growth, the tourism industry also has environmental effects. This is because, as well as many economic activities, it is a fact that tourism activities might increase the pressure on the environment (Bojanic and Warnick, 2020). For this reason, countries make effort to eliminate the possible negative effects of tourism activities on the environment, protect nature and natural sources, and ensure the development of sustainable tourism. Moreover, increasing tourism activities might cause important environmental damages that cannot be reversed or might be expensive to reverse such as air pollution, loss of natural habitat, and soil erosion. In other words, an increase in tourism activities and the developments such as the construction of hotels, touristic facilities, and infrastructures might cause additional increases in energy consumption, decreases in natural and agricultural lands, and an increase in environmental pollution risk (El–Menyari, 2021).

Recent studies investigate the tourism–induced EKC hypothesis over various environmental pollution indicators. Introduced by Pigram (1980) and then developed by Tisdell (1987), the tourism–induced EKC hypothesis assumes that there is an inverse U-shaped relationship between environmental pollution and tourism activities. Accordingly, it is expected that, in the first phase where environmental awareness is low and environment-friendly production is ignored, the development of tourism activities would increase the environmental pollution to a specific turning point. In the second phase, in which environmental awareness increases and environment–friendly production technology is adopted, it is projected that environmental pollution would decrease with the development of tourism activities (Katırcıoğlu et al., 2018).

In the literature, environmental pollution is represented by indicators including carbon dioxide (CO_2) emissions, sulfur hexafluoride (SF_6), methane (CH_4), nitrogen peroxide (N_2O), perfluorocarbon (PFC), and hydrofluoric carbon (HFC), as well as ecological footprint (ECF). However, it is clear that CO_2 emissions have been more frequently used as an environmental pollution indicator in those studies (Baloch et al., 2019; Uddin et al., 2017). On the other hand, the point criticized is that environmental pollution, as a multidimensional concept, was represented using CO_2 emissions, which is a limited indicator measuring only the air pollution dimension (Solarin, 2019; Yilanci and Pata, 2020). For this reason, rather than CO_2 emissions, recent studies have started using ECF which is a more comprehensive and alternative indicator incorporating the water, air, and soil pollution dimensions. Introduced by Rees (1992) and developed by Wackernagel and Rees (1996), ECF measures the pressure of humankind on nature (demand for natural sources) and consists of six components: carbon footprint, cropland footprint, grazing land footprint, fishing grounds footprint, forest land footprint, and built-up land footprint (Global Footprint Network, 2022).

The present study aims to analyze the presence of the tourism–induced EKC hypothesis, which claims that there is an inverse U–shaped relationship between environmental pollution and tourism activities, in the nine countries with the highest international tourist arrivals, using annual data from 1995 to 2018. Contributions of this study to the literature can be explained as follows. First of all, the present study is one of the rare studies testing the tourism–induced EKC hypothesis by making use of ECF. Second, rather than conventional estimation methods, the latest and dynamic test methods considering the cross–sectional dependence were utilized in empirical analyses. Otherwise, conventional tests ignoring cross-sectional dependence can yield biased and inconsistent results. Third, there is still no consensus on the results reported in studies examining the relationship between tourism and the environment. Finally, for the country selection, the countries that attract the most tourists were preferred rather than any country or group of countries. Therefore, it is expected that more accurate results will be achieved in the evaluation of the environment and tourism relationship and in determining the necessary policies.

BACKGROUND

Since the 1990s, the Environmental Kuznets Curve (EKC) hypothesis has been comprehensively examined for developed and developing country groups, as well as individual countries. Besides that, it can also be seen that various variables including direct foreign investments, globalization, trade openness, technological advancement, energy consumption, and oil prices have been included in traditional EKC hypothesis analyses. However, although the environmental effects of those variables were intensively examined within the scope of the EKC hypothesis, the number of studies examining the tourism–induced EKC hypothesis, which assumes an inverse U–shaped relationship between environmental pollution and tourism activities, is still insufficient (Ozturk et al., 2016). Moreover, most of those studies measured environmental pollution by using CO_2 emissions, and ECF was used in very few of them (Katırcıoğlu et al., 2018).

In the literature review section, the pioneering and current studies examining the effects of tourism activities on CO_2 emissions and ECF are presented. The studies examining the effects of tourism activities on CO_2 emissions can be explained as follows:

Lee and Brahmasrene (2013) analyzed the effects of tourism on economic growth and CO_2 emissions in EU member countries for the period 1988–2009. The results of the fixed effect model showed that tourism affected CO_2 negatively. For the period 1995–2005, Tiwari et al. (2013) examined the relationships among tourism, energy consumption, and climate change in 25 OECD countries. They used Panel Vector Autoregressive (PVAR) model and no significant relationship was found between tourism and CO_2 emissions. Katırcıoğlu et al. (2014) explored the long-term relationships between tourism, energy consumption, and CO_2 emissions in Cyprus for the period 1970–2009. The results achieved from ARDL (Autoregressive Distributed Lag) model showed that tourism increased CO_2 emissions. Using the Dynamic Ordinary Least Squares (DOLS) estimator, Katırcıoğlu (2014) investigated the relationship between tourism and CO_2 emissions in Singapore for the period 1971–2010. The estimation results confirmed the validity of the tourism-induced EKC hypothesis. Solarin (2014) investigated the determinants of CO_2 emissions for Malaysia during the period 1972–2010. The results obtained from ARDL and DOLS tests showed a positive relationship between tourism and CO_2 emissions. In their study covering the period 1990–2010 and using the ARDL approach, Ben–Jebli et al. (2015) analyzed the relationships among economic growth, renewable energy consumption, CO_2 emissions, and tourism in Tunisia. The results they achieved proved that tourism increased CO_2 emissions. Paramati et al. (2017) tested the effects of tourism on economic growth and CO_2 emissions in the period 1995–2012 for 26 developed and 18 developing countries. Test results corroborated the tourism-induced EKC hypothesis. Using the data of Greece for the period 1970–2014, Işık et al. (2017) investigated the relationships among economic growth, international trade, financial development, touristic expenditures, and CO_2 emissions. The results obtained from the ARDL test showed that tourism positively affected CO_2 emissions. In their study on 12 Asia–Pacific countries, Shakouri et al. (2017) analyzed the effects of tourism and economic growth on CO_2 emissions for the period 1995–2013 by using PVAR and the Generalized Method of Moments (GMM). Analysis results revealed that the tourism–induced EKC hypothesis was valid. Using Pooled Mean Group (PMG) estimation method, Sherafatian–Jahromi et al. (2017) examined the relationship between tourism and CO_2 emissions in Malaysia, Indonesia, Thailand, Singapore, and the Philippines for the period 1979–2010. PMG estimation results showed that the tourism–induced EKC hypothesis was valid. Danish and Wang (2018) investigated the relationship among tourism, economic growth, and CO_2 emissions in BRICS (Brazil, Russia, India, China, and South Africa) economies by using the data of the period between 1995–2014. Dynamic Seemingly Unrelated Regression (DSUR) estimation results showed a positive relationship between tourism and CO_2 emissions. Akadiri et al. (2019) studied the relationship between tourism and CO_2 emission in 15 selected tourist destination countries for the period 1995–2014. The results obtained from the fixed–effect model provided evidence supporting the tourism–induced EKC hypothesis. Nepal et al. (2019) investigated the effects of tourism on CO2 emissions in Nepal for the period 1975–2014. The results obtained from the ARDL model revealed that tourism had a positive effect on CO_2 emissions. Using the Fully Modified Least Squares (FMOLS) method, Balsalobre–Lorente et al. (2020) analyzed the relationship between economic growth, tourism, globalization, energy consumption, and CO_2 emissions in OECD (Organization for Economic Cooperation and Development) for the period 1994–2014. Analysis results proved the tourism–induced EKC hypothesis. Khan et al. (2020) estimated the relationships among tourism, economic growth, energy consumption, and CO_2 emissions in Pakistan and developing countries for the period 1975–2017 by using the ARDL estimator. Given the ARDL estimation, the authors determined a positive relationship between tourism and CO_2 emissions in both short- and long-term. Koçak et al. (2020) investigated the effects of tourism on CO_2 emissions in the most visited countries, using continuously updated FMOLS (CUP–FM) and error correction FMOLS (CUP–BC) methods. Considering the empirical findings, it was determined that the number of tourists positively affected CO_2 emission, while tourism revenues negatively affected CO_2 emissions. Using data of Nigeria belonging to the period of 1995–2016, Bamidele et al. (2021) explored the relationship among tourism, energy, and CO_2 emissions by making use of the ARDL estimator. Given the short- and long-term results, it was found that tourism reduced CO_2 emissions. Within the scope of the tourism-induced EKC hypothesis, Kılavuz et al. (2021)

examined the role of tourism in CO_2 emissions in Turkey for the period 1960–2015. The results obtained from ARDL estimation confirmed the validity of the tourism–induced EKC hypothesis. Using the GMM estimator, Özsoy (2021) examined the effects of tourism on environmental pollution in 22 OECD countries for the period 1995–2014. The results obtained from the analysis revealed an inverse U-shaped relationship between CO_2 and tourism, and the validity of the tourism–induced EKC hypothesis could not be confirmed. Ochoa–Moreno et al. (2022) investigated the relationship between tourism and CO_2 emissions in 20 Latin American countries. In their study employing the annual data of the period 1995–2018, the relationship between variables was estimated using DOLS and Panel DOLS (PDOLS) methods. Estimation results revealed that, in contrast to the tourism–induced EKC hypothesis, there was a U–shaped relationship between tourism and CO_2 emissions. Rej et al. (2022) estimated the relationship between tourism and CO_2 emissions in India for the period 1990–2019 by using ARDL and DOLS estimators. Considering the analysis results, it was found that tourism reduced CO_2 emissions. Using a two–step GMM estimator, Salahodjaev et al. (2022) examined the relationships among tourism, renewable energy, and CO_2 emissions by using the data for the period 1990–2015. Estimation results showed that tourism positively affected CO_2 emissions.

Even though the number is limited, there also are studies investigating the effect of tourism on ECF in literature. Those studies can be summarized as follows:

Using GMM and system GMM estimators, Öztürk et al. (2016) analyzed the validity of the tourism–induced EKC hypothesis in 144 countries, which were groups based on the income levels, for the period 1988–2008. Estimation results showed that, when compared to the lower middle and low–income countries, there was more evidence confirming the tourism–induced EKC hypothesis in upper–middle and high–income countries. In their study, in which they used four indicators (international tourist destinations, tourism revenues, tourism expenditures, and tourism index) representing the tourism, Katırcıoğlu et al. (2018) analyzed the role of tourism in ECF in the top ten international tourist destinations for the period 1995–2014. Using the Panel Random Effects estimation, it was determined that international tourist destinations, tourism revenues, and tourism index negatively affected the ECF and, consequently, the validity of the tourism–induced EKC hypothesis was confirmed. Mikayilov et al. (2019) studied the effects of tourism on ECF in Azerbaijan for the period 1996–2014. As a result of the estimations performed using FMOLS, DOLS, and Temporally Varying Coefficients (TVC) Cointegration Model, it was determined that the tourism–induced EKC hypothesis was valid. Using the data of the years between 1995 and 2018, Anser et al. (2020) analyzed the relationship between tourism and ECF for 130 countries by making use of the Difference GMM estimator. Analysis results confirmed the validity of the tourism–induced EKC hypothesis. For the period 1995–2017, Usman et al. (2020) assessed the relationship between ECF, financial development, energy consumption, and tourism in twenty highest emitting economies by making use of the Augmented Mean Group (AMG) estimator. Estimation results showed that tourism had a negative effect on the ECF. In their study, in which they used ECF and biocapacity variables representing the environment, Alola et al. (2021) tested the relationships among tourism, globalization, economic growth, and the environment in the top ten destination economies by making use of PMG–ARDL method. Test results showed that tourism had a positive effect on ECF and a negative effect on biocapacity. Using the Quantile Regression Model, Lee and Chen (2021) examined the relationships among ECF, economic growth, tourism, and country risk in 123 countries for the period 1992–2016. The results the authors obtained corroborated the validity of the tourism–induced EKC hypothesis. Nathaniel et al. (2021) investigated the factors influencing the ECF in the top ten international tourist destinations for the period 1995–2016. In the study, in which tourism was represented by the number of inbound tourists and the tourism revenues, coefficients were estimated using CUP–FM and CUP–BC estimators. Given the estimation results, it was determined that tourism positively affected ECF. For the period 1995–2016, Adedoyin et al. (2022) examined the relationships among tourism, biocapacity, globalization, and ECF in the top ten international tourist destinations by using PMG–ARDL. Analysis results suggested that tourism had a negative effect on the ECF.

In this study, differing from most studies in the literature, the tourism-induced EKC hypothesis, which is expressed as an inverse U-shaped relationship between tourism and environment, is investigated by employing tourism data directly instead of economic growth. In this sense, it is expected to contribute to the literature.

Tourism-Induced EKC Hypothesis

Environmental Kuznets Curve (EKC) hypothesis mainly depends on the assumption of Kuznets (1955) that there is an inverse U–shaped relationship between income inequality and income. Following Kuznets (1955), Grossman and Krueger (1991) introduced first the EKC hypothesis, which claimed that there was an inverse U–shaped relationship between environmental pollution and income. In the EKC hypothesis, it is assumed that environmental pollution increases until the turning point and, after that level, income increase reduces environmental pollution. On the other hand, Magnani (2001) stated that income was not the only determinant factor and criticized the traditional EKC hypothesis by emphasizing that other factors should also be considered. Hence, the potential relationships of environmental pollution with parameters such as energy, financial development, economic complexity, environmental taxes, and tourism were also scrutinized.

The relationship between environment and tourism was introduced by Pigram (1980) and the linear relationships between environment and tourism were discussed. Following Pigram (1980), Tisdell (1987) also examined the non-linear (parabolic) relationships between environment and tourism. Figure 1 illustrates the linear and parabolic relationships between environmental pollution and tourism activity (Tisdell, 1987).

In Figure 1, lines (a) and (b) represent the negative and positive linear relationships between environmental pollution and tourism activity, respectively. Lines (c) and (d) illustrate the parabolic relationships. Given line (c) representing the U–shaped relationship between environmental pollution and tourism activity, environmental pollution decreases at low levels of tourism activity. When the turning point (τ_2) is exceeded, environmental pollution starts increasing as tourism activity increases. Given line (d) illustrating the inverse U–shaped relationships between tourism activity and environmental pollution, increases in tourism activity increase the environmental pollution until the turning point (τ_1). Then, after this point, increases in tourism activity decrease the environmental pollution (Sherafatian–Jahromi et al., 2017; Tisdell, 1987).

Figure 1. The relationship between environmental pollution and tourism activity
Source: Tisdell (1987, p. 18)

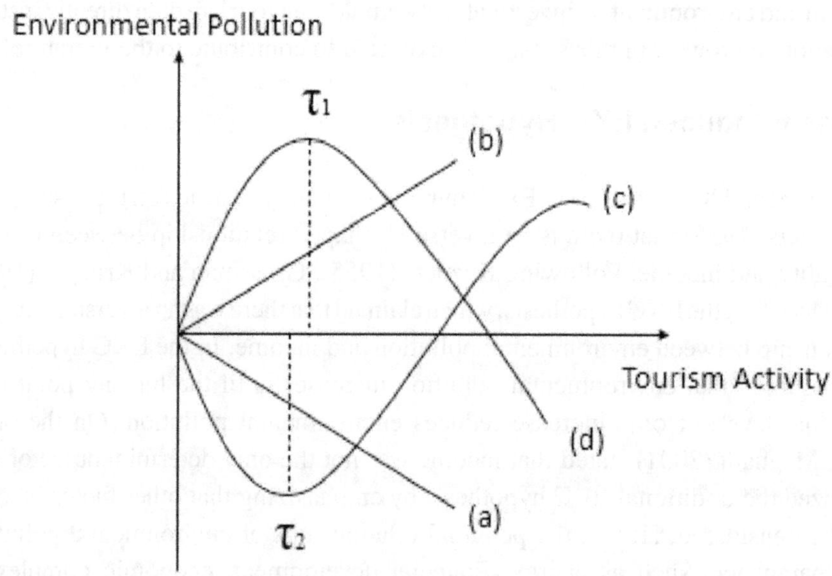

METHODOLOGY

The empirical methodology consists of four steps. In the first step, cross–sectional dependence (CSD) and panel homogeneity are tested. Cross–sectional dependence was tested using Pesaran's (2004) CD_{lm} and Pesaran et al.'s (2008) bias–adjusted" $LM_{adj.}$ tests, while the panel homogeneity was tested using Pesaran and Yamagata's (2008) Delta ($\tilde{\Delta}$) and Delta adjusted $\left(\tilde{\Delta}_{adj.}\right)$ tests. Null and alternative hypotheses of CSD tests are as follows: H_o*: There is no cross-sectional dependence,* H_A*: There is cross-sectional dependence.* $\tilde{\Delta}$ and $\tilde{\Delta}_{adj.}$ tests' null and alternative hypotheses are H_o*: Panel is homogeneous,* H_A*: Panel is heterogeneous.*

In the second step, stationarity characteristics of variables were detected using the bootstrap panel unit root test of Im, Pesaran and Shin (IPS) recommended by Smith et al. (2004). This test is based on the unit root test introduced by Im et al. (2003) and denominated as IPS. In this test, cross-sectional dependence is examined using the bootstrap approach, and $\bar{t}$ test statistics obtained using IPS test's bootstrap version are used. $\bar{t}$ statistic is calculated as shown in Equation (1):

$$\bar{t} = \frac{1}{N}\sum_{i=1}^{N} t_i T \tag{1}$$

In Equation (1), $t_i T$ refers to the individual ADF–t statistics.

In the third step, the long-term cointegration relationship between variables was tested using LM_N^+ test developed by Westerlund and Edgerton (2007) and Durbin–Hausman (DH) test introduced by Westerlund (2008). LM_N^+ test statistic is presented in Equation (2):

$$LM_N^+ = \frac{1}{NT^2}\sum_{i=1}^{N}\sum_{t=1}^{T} \hat{\omega}_i^{-2} S_{it}^2 \tag{2}$$

S_{it} refers to the partial summing process of fully modified z_{it} estimation and $\hat{\omega}_i^2$ refers to long-term estimation variance of u_{it} in terms of Δx_{it}. In presence of CSD, the bootstrap p–value is considered, whereas the asymptotic p-value is used in absence of CSD. DH test's $DH\text{–}_{group}$ and $DH\text{–}_{panel}$ statistics are defined in Equations (3) and (4), respectively:

$$DH -_{grup} = \sum_{i=1}^{n} \hat{S}_i \left(\tilde{\o}_i - \hat{\o}_i\right)^2 \sum_{t=2}^{T} e_{it-1}^2 \tag{3}$$

$$DH -_{panel} = \hat{S}_n \left(\tilde{\o} - \hat{\o}\right)^2 \sum_{i=1}^{n} \sum_{t=2}^{T} e_{it-1}^2 \tag{4}$$

DH–$_{group}$ statistic is valid for heterogeneous panels, whereas DH–$_{panel}$ statistic is valid for homogeneous panels. DH–$_{group}$ and DH–$_{panel}$ statistics' null and alternative hypotheses are as follows: Null hypothesis of DH–$_{group}$ and DH–$_{panel}$ statistics is H_0: $\phi_i=1$ *there is no cointegration relationship for "all units"*. Alternative hypothesis of DH–g$_{roup}$ is H_A: $\phi_i<1$ *there is a cointegration relationship for "so*me units", whereas that of DH–$_{panel}$ statistic is H_A: $\phi_i=\phi$ and $\phi<1$ *there is a cointegration relationship for "all units"* (Westerlund, 2008).

In the fourth step, the long-term coefficients are estimated using the Common Correlated Effects Mean Group (CCE–MG) introduced by Pesaran (2006) and Dynamic Common Correlated Effects (DCCE) developed by Chudik and Pesaran (2015a, b). Having a multivariate error structure, CCE–MG estimator is sensitive to structural breaks and unit roots, it regards CSD and allows the slope coefficient to differ between cross-sections (Pesaran, 2006). Calculated using the simple average of individual CCE estimators, CCE–MG estimator is seen in Equation (5) (Pesaran, 2006):

$$CCE - MG = N^{-1} \sum_{i=1}^{N} \hat{b}_i \tag{5}$$

In Equation (5), $\hat{b}_i$ refers to CCE estimation for each cross–section. Developed as an alternative estimator, the DCCE estimator focuses on CSD and the heterogeneity of the panel. It does not lose its estimation power in presence of structural breaks and offers a good performance for both balanced and unbalanced panels. It can be reliably used even in small sample panels. It significantly eliminates the endogeneity problem and allows for dynamic common correlated effects (Chudik and Pesaran, 2015a, b; Ditzen, 2018). DCCE model is defined in Equation (6):

$$ECF_{it} = \alpha_i ECF_{it-1} + \beta_i X_{it} + \sum_{p=0}^{pT} \gamma_{xip} \bar{X}_{t-p} + \sum_{p=0}^{pT} \gamma_{yip} \bar{Y}_{t-p} + \mu_{it} \tag{6}$$

In Equation (6), ECF_{it} is the dependent variable, ECF_{it-1} is the lag of ECF_{it} as an independent variable, and X_{it} is the independent variable cluster (GDP, TOUR, and $TOUR^2$). pT is the number of lags included in cross-section means, whereas μ_{it} is the error term.

DATASET AND MODEL

In the present study, the validity of the tourism–induced EKC hypothesis in the top nine most visited countries was analyzed for the period 1995–2018 by using ECF. As stated by the United Union's World Tourism Organization, the top ten most visited countries as of 2019 were France (90.2 million), Spain (83.5 million), the USA (79.4 million), China (65.7 million), Italy (64.5 million), Turkey (51.2 million), Mexico (45.0 million), Thailand (39.9 million), Germany (39.6 million), and England (39.4 million), respectively (UNWTO, 2021). On the other hand, France, whose data can be accessed only for the period after 2004, was not involved in the analysis and the analyses were conducted for the remaining nine countries.

Environmental pollution (ECF) was represented by per capita ecological footprint (global hectare), economic growth (GDP) was presented by per capita GDP (prices of the year 2015, USD), and international tourism activity (TOUR) was represented by international tourist arrivals (number of individuals). ECF data was sourced from the Global Footprint Network website (GFN, 2022), whereas GDP and TOUR data were obtained from the World Bank Development Indicator database (WDI, 2022). Data for all variables is transformed into natural logarithms.

The estimation model established following Sherafatian–Jahromi et al. (2017) is presented in Equation (7):

$$lnECF_{it} = \alpha_0 + \beta_1 lnGDP_{it} + \beta_2 lnTOUR_{it} + \beta_3 lnTOUR_{it}^2 + u_{it} \quad (7)$$

In Equation (7), since the increases in GDP will increase the natural resource usage and environmental pressure, β_1 coefficient is expected to be positive. Being statistically significant, positive β_2 coefficient and negative β_3 coefficient suggest that there is an inverse U–shaped relationship between ECF and TOUR and, in other words, the tourism-induced EKC hypothesis is valid.

Considering Equation (7), within the scope of the tourism–induced EKC hypothesis, the relationships between lnECF and lnTOUR are presented as follows:

$\beta_1=\beta_2=0 \Rightarrow$ There is no relationship between lnECF and lnTOUR.
$\beta_1>0, \beta_2=0 \Rightarrow$ There is a linear and positive relationship between lnECF and lnTOUR.

$\beta_1<0, \beta_2=0 \Rightarrow$ There is a linear and negative relationship between lnECF and lnTOUR.
$\beta_1>0, \beta_2<0 \Rightarrow$ There is an inverse U–shaped relationship between lnECF and lnTOUR.
$\beta_1<1, \beta_2>0 \Rightarrow$ There is a U–shaped relationship between lnECF and lnTOUR.

EMPIRICAL RESULTS

The presence of cross-sectional dependence between variables was tested using CD_{lm} and LM_{adj} tests, whereas slope homogeneity was tested using Delta ($\tilde{\Delta}$) and $Delta_{adj}$ ($\tilde{\Delta}_{adj.}$) tests. In Table 1, the cross-sectional dependence and slope homogeneity test results are presented.

Table 1. Cross-sectional dependence and slope homogeneity test results

Variables	CD_{lm} Test	$LM_{adj.}$ Test
lnECF	44.212***	44.017***
lnGDP	61.448***	61.252***
lnTOUR	60.866***	60.670***
$lnTOUR^2$	60.938***	60.743***
lnECF=f(lnGDP, lnTOUR, $lnTOUR^2$)	3.109***	5.532***
Slope homogeneity test	$\tilde{\Delta}$ test	$\tilde{\Delta}_{adj}$ test
lnECF=f(lnGDP, lnTOUR, $lnTOUR^2$)	12.473***	13.945***

Note: *** denotes significance at the 1% level.

Given the CSD test results presented in Table 1, hypothesis H*0* assuming the absence of CSD was rejected for all the variables at the significance level of 1%. Hence, it was found that there was a CSD. The presence of CSD suggests that a shock seen in any of the countries would affect the other countries constituting the panel. On the other hand, based on the assumption that the panel was homogeneous, hypothesis H*0* was rejected at the significance level of 1% and it was found that the slope coefficient varied between the cross–sections and, in other words, the panel was heterogeneous. Then, the stationarity characteristics of variables were tested using the bootstrap IPS panel unit root test and the results are presented in Table 2.

Table 2. Bootstrap IPS unit root test results

Bootstrap IPS	Level	First Difference	Decision
Variables	Test Stat.	Test Stat.	
lnECF	–0.891	–4.960***	I (1)
lnGDP	–1.331	–3.283***	I (1)
lnTOUR	–0.874	–4.081***	I (1)
lnTOUR^2	–0.809	–4.084***	I (1)

Note: *** denotes significance at the 1% level.

In Table 2, given the bootstrap IPS unit root statistics, it was determined that all the variables had unit roots at the level. Taking the first differences of variables, hypothesis H_0 assuming that the panel had unit root was rejected at the significance level of 1% and it was determined that all the variables were stationary. After determining the stationarity characteristics of variables, the long–term relationships were investigated using LM and DH cointegration tests and the results are presented in Table 3.

Table 3. LM and DH cointegration test results

LM Test	Bootstrap p–Value	Asymptotic p–Value
lnECF=*f* (lnGDP, lnTOUR, lnTOUR^2)	1.000	0.333
DH Test	**DH–grup**	**DH–panel**
lnECF=*f* (lnGDP, lnTOUR, lnTOUR^2)	–1.531* (0.063)	–0.061 (0.476)

Note: * denotes significance at the 10% level. () shows p–value.

In Table 3, in the LM test, the bootstrap p-value is considered in presence of CSD, whereas the asymptotic p–value is used in absence of CSD. Since CSD was detected as a result of the CSD test, the decision was made using the bootstrap p–value. Given the bootstrap p–value, hypothesis H_0 which is based on the presence of a cointegration relationship could not be rejected and it was found that there was a cointegration relationship. In the DH test, DH–$_{group}$ statistic is used for heterogeneous panels and DH–$_{panel}$ statistic is employed for homogeneous panels. Since the results of the homogeneity test showed the heterogeneity of the panel, the decision was made considering the DH–$_{group}$ statistic. Given DH–$_{group}$ statistic, hypothesis H_0 which is based on the absence of a cointegration relationship for all the variables was rejected at the significance level of 10%, and it was found that there was a cointegration relationship. In conclusion, it was determined that variables moved together in the long–term and, in other words, they were cointegrated. After determining the presence of cointegration relationships, the coefficients were estimated utilizing CCE–MG and DCCE estimators. Table 4 represents the long-term estimation results of CCE–MG and DCCE.

Table 4. CCE–MG and DCCE coefficient estimation results

lnECF=*f* (lnGDP, lnTOUR, lnTOUR²)	CCE–MG		DCCE	
Variables	**Coefficient**	**t-Stat.**	**Coefficient**	**t-Stat.**
lnGDP	1.188***	8.03	0.490***	6.90
lnTOUR	–13.744***	-2.60	-9.852**	-1.97
lnTOUR²	0.386***	2.63	0.275**	2.01

Note: ** and * denote significance at the 1% and 5% levels, respectively.

Given the CCE–MG and DCCE long–term estimation results presented in Table 4, the lnGDP coefficient was statistically significant and positive. Accordingly, increases in economic activities increase the pressure on the environment and cause environmental pollution. lnTOUR and lnTOUR^2 coefficients were found to be statistically significantly negative and positive, respectively. Given this finding, it was suggested that there was a U–shaped relationship between lnECF and lnTOUR. Thus, it was revealed that increases in the numbers of tourists in leading tourist destination countries did not damage the environment until a turning point but, after the turning point, the increases in the number of tourists started to increase the environmental pollution. The empirical results obtained are in line with the results reported by Mikayilov et al. (2019), Özsoy (2021), and Ochoa–Moreno et al. (2022) but are contrary to the results reported by Paramati et al. (2017), Sherafatian–Jahromi et al. (2017), Anser et al. (2020), Balsalobre–Lorente et al. (2020), and Kılavuz et al. (2021).

CONCLUSION

Together with the effects of globalization, tourism activities have gained speed in the recent period and become one of the driving forces of economic growth. However, tourism activities also have potential negative effects on environmental pollution. Within this context, analyzing the relationship between environmental pollution and tourism activities in the most visited countries within the scope of the tourism–induced EKC hypothesis, the present study aims to contribute to the current discussions. The relationships between the variables were analyzed using the latest panel data test methods. Given the results of LM and DH cointegration tests, environmental pollution, economic growth, tourism activities, and square of tourism activities moved together in the long–term. In other words, these variables are cointegrated. The results obtained from CCE–MG and DCCE estimations showed that the lnGDP coefficient was positive as expected and that environmental pollution increased in parallel with the advancements in the economy. The coefficients of lnTOUR and $lnTOUR^2$ variables were found to be statistically significantly negative and positive, respectively. This finding suggests that there is a U–shaped relationship between environmental pollution and tourism activities and, in other words, the tourism–induced EKC hypothesis could not be confirmed. Accordingly, it can be seen that increases in the number of international inbound tourists would decrease the environmental quality after a specific turning point. In conclusion, it can be stated that the tourism activities in those countries damaged the environment and, in other words, were away from sustainable tourism practices.

The results obtained here revealed that, for those countries, the integration between tourism activities and the environment is a necessity to reduce the ecological footprint and environmental pollution. Accordingly, the factor "environment" should be considered in any decision related with tourism and tourist destinations and sustainable tourism should be promoted by keeping the balance between tourism activities and the environment. The industries closely linked to tourism such as hospitality, transportation, and logistics should be revised in harmony with the environment. Environmental awareness programs should be expanded and involve all the actors in the tourism industry. Besides, priority should be given to measures and practices that increase efficiency in the use of energy and natural resources in all economic activities, especially in tourism activities.

This study has some limitations. First of all, due to the limited tourism arrivals data of the relevant countries, the time spanning started from 1995 and ended in 2018. Second, linear methods were utilized in the analysis, while asymmetric and nonlinear methods were ignored. This study might provide insights for further studies investigating the relationships of tourism activities with carbon footprint, cropland footprint, grazing land footprint, fishing grounds footprint, forest land footprint, and built-up land footprint within the scope of the tourism-induced EKC hypothesis.

REFERENCES

Adedoyin, F. F., Alola, U. V., & Bekun, F. V. (2022). On the nexus between globalization, tourism, economic growth, and biocapacity: Evidence from top tourism destinations. *Environmental Science and Pollution Research International*, *29*(17), 24995–25005. doi:10.100711356-021-17651-8 PMID:34837107

Akadiri, S. S., Lasisi, T. T., Uzuner, G., & Akadiri, A. C. (2019). Examining the impact of globalization in the environmental Kuznets curve hypothesis: The case of tourist destination states. *Environmental Science and Pollution Research International*, *26*(12), 12605–12615. doi:10.100711356-019-04722-0 PMID:30852745

Alola, A. A., Eluwole, K. K., Lasisi, T. T., & Alola, U. V. (2021). Perspectives of globalization and tourism as drivers of ecological footprint in top 10 destination economies. *Environmental Science and Pollution Research International*, *28*(24), 31607–31617. doi:10.100711356-021-12871-4 PMID:33609247

Anser, M. K., Yousaf, Z., Nassani, A. A., Abro, M. M. Q., Zaman, K., & Kabbani, A. (2020). Evaluating ecological footprints through inbound tourism, population density, and global trade. *Polish Journal of Environmental Studies*, *30*(1), 555–560. doi:10.15244/pjoes/122445

Baloch, M. A., Zhang, J., Iqbal, K., & Iqbal, Z. (2019). The effect of financial development on ecological footprint in bri countries: Evidence from panel data estimation. *Environmental Science and Pollution Research International*, *26*(6), 6199–6208. doi:10.100711356-018-3992-9 PMID:30617883

Balsalobre-Lorente, D., Driha, O. M., Shahbaz, M., & Sinha, A. (2020). The effects of tourism and globalization over environmental degradation in developed countries. *Environmental Science and Pollution Research International*, *27*(7), 7130–7144. doi:10.100711356-019-07372-4 PMID:31879881

Bamidele, R., Ozturk, I., Gyamfi, B. A., & Bekun, F. V. (2021). Tourism-induced pollution emission amidst energy mix: Evidence from Nigeria. *Environmental Science and Pollution Research International*, *29*(13), 19752–19761. doi:10.100711356-021-17233-8 PMID:34718978

Ben Jebli, M., Ben Youssef, S., & Apergis, N. (2015). The dynamic interaction between combustible renewables and waste consumption and international tourism: The case of Tunisia. *Environmental Science and Pollution Research International*, *22*(16), 12050–12061. doi:10.100711356-015-4483-x PMID:25874437

Bojanic, D. C., & Warnick, R. B. (2020). The relationship between a country's level of tourism and environmental performance. *Journal of Travel Research, 59*(2), 220–230. doi:10.1177/0047287519827394

Chudik, A., & Pesaran, M. H. (2015a). Large panel data models with cross-sectional dependence: A survey. In B. H. Baltagi (Ed.), *The oxford handbook of panel data* (pp. 3–45). Oxford University Press. doi:10.1093/oxfordhb/9780199940042.013.0001

Chudik, A., & Pesaran, M. H. (2015b). Common correlated effects estimation of heterogeneous dynamic panel data models with weakly exogenous regressors. *Journal of Econometrics, 188*(2), 393–420. doi:10.1016/j.jeconom.2015.03.007

Danish, & Wang, Z. (2018). Dynamic relationship between tourism, economic growth, and environmental quality. *Journal of Sustainable Tourism, 26*(11), 1928-1943.

Ditzen, J. (2018). Estimating dynamic common-correlated effects in stata. *The Stata Journal, 18*(3), 585–617. doi:10.1177/1536867X1801800306

El Menyari, Y. (2021). The effects of international tourism, electricity consumption, and economic growth on CO_2 emissions in North Africa. *Environmental Science and Pollution Research International, 28*(32), 44028–44038. doi:10.100711356-021-13818-5 PMID:33844140

Global Footprint Network. (2022). *National footprint accounts*. Retrieved May 10, 2022, from https://data.footprintnetwork.org/#/

Grossman, G. M., & Krueger, A. B. (1991). Environmental impacts of a north American free trade agreement. *NBER Working Papers Series*, 1–39.

Im, K. S., Pesaran, M. H., & Shin, Y. (2003). Testing for unit roots in heterogeneous panels. *Journal of Econometrics, 115*(1), 53–74. doi:10.1016/S0304-4076(03)00092-7

Işik, C., Kasımatı, E., & Ongan, S. (2017). Analyzing the causalities between economic growth, financial development, international trade, tourism expenditure and/on the CO_2 emissions in Greece. *Energy Sources. Part B, Economics, Planning, and Policy, 12*(7), 665–673. doi:10.1080/15567249.2016.1263251

Katircioglu, S., Gokmenoglu, K. K., & Eren, B. M. (2018). Testing the role of tourism development in ecological footprint quality: Evidence from top 10 tourist destinations. *Environmental Science and Pollution Research International, 25*(33), 33611–33619. doi:10.100711356-018-3324-0 PMID:30276681

Katircioğlu, S. T. (2014). Testing the tourism-induced EKC hypothesis: The case of Singapore. *Economic Modelling, 41*, 383–391. doi:10.1016/j.econmod.2014.05.028

Katircioglu, S. T., Feridun, M., & Kilinc, C. (2014). Estimating tourism-induced energy consumption and CO_2 emissions: The case of Cyprus. *Renewable & Sustainable Energy Reviews*, *29*, 634–640. doi:10.1016/j.rser.2013.09.004

Khan, A., Bibi, S., Ardito, L., Lyu, J., Hayat, H., & Arif, A. M. (2020). Revisiting the dynamics of tourism, economic growth, and environmental pollutants in the emerging economies–sustainable tourism policy implications. *Sustainability (Basel)*, *12*(6), 2533. doi:10.3390u12062533

Kılavuz, E., Oralhan, B., Sarıgül, S. S., & Uluğ, E. E. (2021). The validity of the tourism-induced EKC hypothesis: The case of Turkey. *International Journal of Business and Economic Studies*, *3*(2), 124–138. doi:10.54821/uiecd.1033651

Koçak, E., Ulucak, R., & Ulucak, Z. Ş. (2020). The impact of tourism developments on CO_2 emissions: An advanced panel data estimation. *Tourism Management Perspectives*, *33*(100611), 1–10.

Kuznets, S. (1955). Economic growth and income inequality. *The American Economic Review*, *45*(1), 1–28.

Lee, C. C., & Chen, M. P. (2021). Ecological footprint, tourism development, and country risk: International evidence. *Journal of Cleaner Production*, *279*(123671), 1–24. doi:10.1016/j.jclepro.2020.123671

Lee, J. W., & Brahmasrene, T. (2013). Investigating the influence of tourism on economic growth and carbon emissions: Evidence from panel analysis of the European union. *Tourism Management*, *38*, 69–76. doi:10.1016/j.tourman.2013.02.016

Magnani, E. (2001). The Environmental Kuznets curve: Development path or policy result? *Environmental Modelling & Software*, *16*(2), 157–165. doi:10.1016/S1364-8152(00)00079-7

Mikayilov, J. I., Mukhtarov, S., Mammadov, J., & Azizov, M. (2019). Re-evaluating the environmental impacts of tourism: Does EKC exist? *Environmental Science and Pollution Research International*, *26*(19), 19389–19402. doi:10.100711356-019-05269-w PMID:31073834

Nathaniel, S. P., Barua, S., & Ahmed, Z. (2021). What drives ecological footprint in top ten tourist destinations? Evidence from advanced panel techniques. *Environmental Science and Pollution Research International*, *28*(28), 38322–38331. doi:10.100711356-021-13389-5 PMID:33728610

Nepal, R., Al Irsyad, M. I., & Nepal, S. K. (2019). Tourist arrivals, energy consumption and pollutant emissions in a developing economy–implications for sustainable tourism. *Tourism Management*, *72*, 145–154. doi:10.1016/j.tourman.2018.08.025

Ochoa-Moreno, W. S., Quito, B., Enríquez, D. E., & Álvarez-García, J. (2022). Evaluation of the environmental Kuznets curve hypothesis in a tourism development context: Evidence for 15 Latin American countries. *Business Strategy and the Environment*, *31*(5), 1–13. doi:10.1002/bse.3012

Özsoy, F. (2021). Investigation of the relationship between tourism sector and environmental pollution in the framework of the environmental Kuznets curve hypothesis. *Journal of Abant Social Sciences*, *21*(1), 1–19.

Ozturk, İ., Al-Mulali, U., & Saboori, B. (2016). Investigating the environmental Kuznets curve hypothesis: The role of tourism and ecological footprint. *Environmental Science and Pollution Research International*, *23*(2), 1916–1928. doi:10.100711356-015-5447-x PMID:26408117

Paramati, S. R., Alam, M. S., & Chen, C. F. (2017). The Effects of tourism on economic growth and CO_2 emissions: A comparison between developed and developing economies. *Journal of Travel Research*, *56*(6), 712–724. doi:10.1177/0047287516667848

Pesaran, M. H. (2004). General diagnostic tests for cross section dependence in panels (IZA Discussion Paper No. 1240). *Institute for the Study of Labor (IZA)*, 1-39.

Pesaran, M. H. (2006). Estimation and inference in large heterogeneous panels with a multifactor error structure. *Econometrica*, *74*(4), 967–1012. doi:10.1111/j.1468-0262.2006.00692.x

Pesaran, M. H., Ullah, A., & Yamagata, T. (2008). A bias-adjusted LM test of error cross-section independence. *The Econometrics Journal*, *11*(1), 105–127. doi:10.1111/j.1368-423X.2007.00227.x

Pesaran, M. H., & Yamagata, T. (2008). Testing slope homogeneity in large panels. *Journal of Econometrics*, *142*(1), 50–93. doi:10.1016/j.jeconom.2007.05.010

Pigram, J. J. (1980). Environmental implications of tourism development. *Annals of Tourism Research*, *7*(4), 554–583. doi:10.1016/0160-7383(80)90049-3

Rees, W. E. (1992). Ecological footprints and appropriated carrying capacity: What urban economics leaves out. *Environment and Urbanization*, *4*(2), 121–130. doi:10.1177/095624789200400212

Rej, S., Bandyopadhyay, A., Murshed, M., Mahmood, H., & Razzaq, A. (2022). Pathways to decarbonization in India: The role of environmentally friendly tourism development. *Environmental Science and Pollution Research International*, 29(33), 50281-50302. doi.org/10.1007/s11356-022-19239-2.

Salahodjaev, R., Sharipov, K., Rakhmanov, N., & Khabirov, D. (2022). Tourism, renewable energy and CO_2 emissions: Evidence from Europe and Central Asia. *Environment, Development and Sustainability*, *24*(11), 1–12. doi:10.100710668-021-01993-x

Shakouri, B., Khoshnevis Yazdi, S., & Ghorchebigi, E. (2017). Does tourism development promote CO_2 emissions? *Anatolia*, *28*(3), 444–452. doi:10.1080/13032917.2017.1335648

Sherafatian-Jahromi, R., Othman, M. S., Law, S. H., & Ismail, N. W. (2017). Tourism and CO_2 emissions nexus in Southeast Asia: New evidence from panel estimation. *Environment, Development and Sustainability*, *19*(4), 1407–1423. doi:10.100710668-016-9811-x

Smith, L. V., Leybourne, S., Kim, T. H., & Newbold, P. (2004). More powerful panel data unit root tests with an application to mean reversion in real exchange rates. *Journal of Applied Econometrics*, *19*(2), 147–170. doi:10.1002/jae.723

Solarin, S. A. (2014). Tourist arrivals and macroeconomic determinants of CO_2 emissions in Malaysia. *Anatolia*, *25*(2), 228–241. doi:10.1080/13032917.2013.868364

Solarin, S. A. (2019). Convergence in CO_2 emissions, carbon footprint and ecological footprint: Evidence from OECD countries. *Environmental Science and Pollution Research International*, *26*(6), 6167–6181. doi:10.100711356-018-3993-8 PMID:30617875

Tisdell, C. (1987). Tourism, the environment and profit. *Economic Analysis and Policy*, *17*(1), 13–30. doi:10.1016/S0313-5926(87)50009-1

Tiwari, A. K., Ozturk, I., & Aruna, M. (2013). Tourism, energy consumption and climate change in OECD countries. *International Journal of Energy Economics and Policy*, *3*(3), 247–261.

Uddin, G. A., Salahuddin, M., Alam, K., & Gow, J. (2017). Ecological footprint and real income: Panel data evidence from the 27 highest emitting countries. *Ecological Indicators*, *77*, 166–175. doi:10.1016/j.ecolind.2017.01.003

Ulucak, R., & Lin, D. (2017). Persistence of policy shocks to ecological footprint of the USA. *Ecological Indicators*, *80*, 337–343. doi:10.1016/j.ecolind.2017.05.020

UNWTO. (2020). *International tourism highlights*. Retrieved June 11, 2022, from https://www.e-unwto.org/doi/epdf/10.18111/9789284422456

UNWTO. (2021). *World tourism barometer-statistical annex*. Retrieved June 7, 2022, from https://www.wto.org/english/tratop_e/envir_e/unwto_barom21.pdf

UNWTO. (2022). *UN tourism news*. Retrieved June 4, 2022, from https://www.unwto.org/un-tourism-news-fitur-edition-2022

Usman, M., Kousar, R., & Makhdum, M. S. A. (2020). The role of financial development, tourism, and energy utilization in environmental deficit: Evidence from 20 highest emitting economies. *Environmental Science and Pollution Research International*, *27*(34), 42980–42995. doi:10.100711356-020-10197-1 PMID:32725560

Wackernagel, M., & Rees, W. (1996). Our ecological footprint: Reducing human impact on the earth. New Society Publishers, 1-29.

WDI. (2022). *World development indicators online database*. Retrieved May 21, 2022, from https://databank.worldbank.org/source/world-development-indicators

Westerlund, J. (2008). Panel cointegration tests of the Fisher effect. *Journal of Applied Econometrics*, *23*(2), 193–233. doi:10.1002/jae.967

Westerlund, J., & Edgerton, D. L. (2007). A panel bootstrap cointegration test. *Economics Letters*, *97*(3), 185–190. doi:10.1016/j.econlet.2007.03.003

WTTC. (2022). *World travel & tourism council*. Retrieved April 18, 2022, from https://wttc.org/Research/Economic-Impact

Yilanci, V., & Pata, U. K. (2020). Investigating the EKC hypothesis for China: The role of economic complexity on ecological footprint. *Environmental Science and Pollution Research International*, *27*(26), 32683–32694. doi:10.100711356-020-09434-4 PMID:32514916

KEY TERMS AND DEFINITIONS

Biocapacity: It is an ecosystem's capacity to produce valuable materials and absorb the wastes created by humans. In summary, it is defined as the natural resource supply or reserve.

Cross-Sectional Dependence: It refers to a situation, in which a positive or negative shock in any of the countries constituting the panel affects the other countries.

Ecological Footprint: It depicts the biologically productive area required to sustain the natural values and is a comprehensive indicator measuring the demand pressure of humankind on nature.

Environmental Pollution: It refers to the situation, in which the foreign materials damaging the environment due to human activities intensely mix into air, water, and soil.

International Tourism: It refers to traveling beyond the national borders for a minimum of a day and a maximum of one year for leisure, business, or other purposes.

Service Industry: It refers to the market, in which non-physical products and services such as professional services such as tourism, banking, transportation, communication, wholesale and retail sales, and medicine and governmental services such as defense and justice are traded.

Sustainable Development: It refers to achieving the development goals by preventing the uncontrolled use of natural resources and considering the needs of the next generation. In summary, it means achieving economic development by establishing a balance between humans and nature.

Chapter 7
Ecological Degradation Within the Context of Consumption:
A 30-Year Bibliometric Analysis (1992-2022)

Ece Özer Çizer
https://orcid.org/0000-0002-8597-2073
Yıldız Technical University, Turkey

ABSTRACT

This study examines the evolution of ecological degradation within the context of consumption in the past three decades. The study used the R Studio program to conduct a bibliometric meta-analysis. The data set consists of 2183 articles selected from the Web of Science database by searching the keywords (ALL=("Ecological Degradation") AND ALL=("consum")) OR (ALL=("environmental degradation") AND ALL=("consum*")). Graphics related to annual scientific production, corresponding author's country, most cited countries, most relevant sources, word cloud, word tree, trend topics, and thematic map, which are revealed using the visualization technique, will be included in the study. As a result, this research aims to create a comprehensive framework for articles on ecological degradation within the context of consumption published in the last thirty years.*

DOI: 10.4018/978-1-6684-6727-5.ch007

INTRODUCTION

Environmental degradation is increasing day by day in parallel with social and economic developments (Sarkodie, 2018). Population growth, urbanization, increasing economic activities, and differentiating consumption habits increase the pressure on the environment and natural resources (Khan et al., 2021). While demand and consumption for products and services are increasing around the world, effective use of resources and protection of the environment are among the most difficult issues to meet. Whether the growth is in the economic area or in the population area, when uncontrolled resource consumption occurs in order to grow, it is inevitable to experience problems in terms of resources. On the other hand, resources are the basis of growth, and any problem that may occur with resources directly affects growth.

The natural resources of countries contribute to both the improvement of living standards and the growth of their economies (Erdoğan et al., 2018). However, there is a possibility of environmental degradation when natural resources are misused. To build a sustainable economy, it is essential to manage and use natural resources sustainably. Efficient use of land, water, air, and other essentials stimulates economic growth by increasing production in mining, agriculture, manufacturing, and other industries as well as other sectors. The use of natural resources such as food, biomass, and ecosystems also raises people's standard of living. However, the efficient use of natural resources is associated with environmental problems as well as economic advantages. Natural resources can be depleted rapidly with high consumption levels, and the environment can be damaged by excessive use of the same resources. Consequently, natural resources must be managed sustainably. Using resources effectively, reducing waste, and using renewable resources are essential components of sustainable natural resource management. By using this strategy, environmental problems will be reduced and natural resources will be transferred to future generations. In addition, a balance will be achieved between the sustainable management of natural resources, economic expansion, and environmental sustainability. For this reason, efficient use of natural resources is essential for a functioning economy, and it is also of great importance in terms of reducing environmental problems and transferring natural resources to future generations. Sustainable natural resource management should be applied to meet the needs of future generations by establishing a balance between economic growth and environmental sustainability. Otherwise, it is inevitable to cause environmental degradation while economic growth takes place.

Today, the increase in environmental degradation has come to a point that threatens the life cycle (Xie et al., 2020). Changing climates, destruction of forest areas, increasing desertification and drought, increasing environmental pollution, and decreasing species continue to affect lives on a global scale (Guerrero & Guevara,

2021; Stavi et al., 2021). One of the most important reasons behind environmental degradation is consumption behavior (White et al., 2019; Zhang et al., 2021). Due to both unsustainable consumption patterns and global crises such as global warming and changing climate conditions, the damage to the environment is increasing, as is the resource problem (Joseph et al., 2020). Increasing population and urbanization, along with the technological innovations that emerged with the Industrial Revolution, caused the differentiation of living standards and consumption habits (Zheng et al., 2021). The differentiation of consumption habits and the desire to consume more cause chemical use and waste. This situation also causes ecological deterioration that threatens all living things, especially the depletion of natural resources and climate change (Wassie 2020). Therefore, sustainable and efficient management of resources is an inevitable necessity today. For this, it is necessary to identify unsustainable consumption habits and find solutions for them.

The main causes of unsustainable consumption habits and ecological degradation are climate change, water scarcity, loss of biodiversity, waste management, resource depletion, and unsustainable agricultural practices. For example, as a result of applications such as burning fossil fuels, industrial production, and cutting trees, the amount of greenhouse gases in the atmosphere increases. Climate changes such as sea level rise, global warming, and extreme weather conditions are a result of this. Overuse, pollution of water resources, and climate change are responsible for water scarcity. Health problems, declines in agricultural production, and financial losses occur when there is not enough access to clean water suitable for everyone's needs. In addition to all this, biodiversity is declining. Many species are disappearing as a result of the degradation of their natural habitats due to climate change and human activities. Ecosystem damage and problems with food safety can result from this. On the other hand, more garbage is produced when population and consumption increase rapidly. Health problems and environmental pollution can result from improper handling of garbage. It is essential to adopt actions such as garbage reduction and reuse. However, natural resources are rapidly depleting. Therefore, the use of natural resources such as water, forests, mines, and fossil fuels is very important. Reducing consumption, sustainable production of food resource sources, and equitable distribution of food resources are also of great importance in order to end unsustainable consumption habits.

There are many remarkable studies in the literature on the causes and consequences of environmental degradation. However, there is no bibliometric study that analyzes the common denominator of all these studies. This study, which deals with ecological degradation within the scope of consumption by collecting all these publications on environmental degradation in a common data set, aims to reveal known and unknown consumption patterns that cause ecological degradation. Within the scope of this purpose, first of all, ecological deterioration, types of ecological deterioration, the causes of ecological deterioration, and the concepts of sustainable consumption are included. Then, a systematic literature analysis was applied to 2183 studies, which were obtained from the Web of Science database, to reveal the consumption patterns underlying ecological degradation and which dealt with the issues of ecological degradation and consumption together. In this way, visible and hidden patterns covering the last 30 years have been revealed.

BACKGROUND

Ecological Degradation

The ecological system is being damaged more and more every day due to human activities, especially consumption. Unconscious consumption behavior poses a serious threat to the lives of all living things. The increase in the use of fossil fuels since the industrial revolution is one of the main damages caused by humanity to the environment (Ukaogo et al., 2020; Cheng et al., 2021). While the gradual increase in greenhouse gas density emitted into the atmosphere by the use of fossil fuels causes the planet to warm up more, the increase in global temperature brings along climate change, which is very dangerous for living species (Manowska & Nowrot, 2019).

Emissions play a major role in the increase in global temperatures (Harper et al., 2018). Authorities estimate that global temperature rise will reach 1.5–2 degrees in 2030–2052 if investment in fossil fuels continues (Fahad et al., 2022). Authorities predict that if the temperature increase in the world reaches 2 degrees, the melting glaciers will raise the sea level, some of the continents will be flooded, and agriculture and livestock activities will be seriously affected (DeFries et al., 2019). In the sectoral sense, the largest share of carbon emissions stems from the energy sector and industrial activities (Vieira et al., 2021). While the energy sector is at the forefront of global carbon emissions with approximately 26%, industrial activities have a 20% share, forestry 17%, transportation 14%, and agricultural activities 13% (Tennison et al., 2021; Lamb et al., 2021). In addition to these, the rapid increase in other damages to the environment such as soil, air, and noise pollution, destruction of nature, and unplanned urbanization pose serious risks for the extinction of living species (Kumar et al., 2020).

As a result, environmental pollution destabilizes the ecosystem and paves the way for global warming. Uncontrolled deterioration of the environment is called ecological degradation. Ecological degradation is defined as pollution and consumption of resources such as air, water, and soil; deterioration of natural balance and stable values; destruction of ecosystems; habitat destruction; and extinction of wildlife (Wassie, 2020). It is defined as any change or disorder that is harmful or undesirable to the environment (Mohsin et al., 2019). In the literature, ecological degradation has been handled by dividing it into three main headings: soil, water, and atmospheric degradation (Pata et al., 2022).

Types of Ecological Degradation

Soil Degradation

The quality of the soil is essential for all plants and microorganisms, as well as for the reproduction of food sources for all living things and the survival of life on the planet (Gedamu 2020). However, our soils are seriously polluted due to human intervention. For example, excessive use of fertilizers and pesticides through agricultural practices results in pesticides. Pesticides often contain toxic substances that can contaminate the soil. In addition, excess waste that cannot be destroyed and/or converted as a result of excessive consumption and waste or garbage leaks containing harmful components mix with the soil and pollute it (Saxena et al., 2021).

Water Degradation

Good-quality water is essential for the survival of all living things in nature. Due to human intervention, our water resources are polluted in various ways. Many natural water resources, such as seas, oceans, rivers, lakes, and streams, are polluted because household and industrial waste management is not done properly (Deletic & Wang, 2019).

Atmospheric Deterioration

Atmospheric degradation refers to the degradation of the air we breathe, particulate pollution, and the depletion of the ozone layer. Gas emissions from uncontrolled and overgrowing industries, fuels used to produce energy, and many products that people use daily (including greenhouse gases that contribute to the global warming problem) cause atmospheric degradation. For example, since a significant part of our energy production is done with thermal power plants to produce cheap energy, it causes serious air pollution (Bisht et al., 2020).

Causes of Ecological Degradation

Nature does not deteriorate on its own; environmental degradation occurs entirely through human interventions.

Population Explosion

The human population in the world, which has increased only in the last 50 years, has approached 50% of the total human population that has lived since the beginning of the world (Kopittke et al., 2019). This extremely rapid population growth and density cause huge problems for the world. Excessive industrial production and consumption, excessive domestic and industrial waste, excessive animal and plant food production and consumption, etc. (Sonone et al., 2020). Many reasons lead to the disruption of the ecosystem balance of our planet and serious water, air, and soil pollution (Zwolak et al., 2019).

Deforestation

The destruction of forests poses a great threat to all life, creating great negative effects on the world. The continuous destruction of forests means cutting off the oxygen source, which is the source of life, and decreasing carbon dioxide absorption (Duguma et al., 2019). When oxygen levels decrease, all other sources of life also decrease. When carbon dioxide increases, the effects of global warming are multiplied and accelerated. Another disadvantage that causes deforestation is the excessive consumption and waste of paper products produced from trees. Usually, this paper waste is not recycled and mixes with the soil, causing it to deteriorate (Asamoah et al., 2020). Deforested soils become less nutritious, making soil fertility and reuse difficult (Arias-Ortiz et al., 2021).

Waste Management Issues

Waste leaks: Leakages of harmful substances from waste dumped in landfills cause significant pollution of soil and water resources. In many developing countries, regular waste spills are quite common (Basit et al., 2022).

Waste production: Excessive consumption behaviors also generate enormous amounts of waste. Although waste recycling studies are carried out in developed countries, it is not possible to keep up with excessive consumption. In addition, due to the costs it creates, waste management is quite low in importance in developing and/or underdeveloped countries (Gorrie, 2020).

Illegal waste dumping: Today, one of the causes of environmental degradation is waste that is illegally dumped into the seas, lakes, and rivers. These wastes not only pollute the water resources and make them unusable but also bring the end of the living ecosystem in the water. Apart from water, waste thrown into forests poses a danger to both the environment and living things (Odume, 2022).

Plastic pollution: The amount of plastic garbage produced every year is quite high (Law et al., 2020). For example, it is estimated that around 500 billion disposable plastic cups are used worldwide each year (Foteinis, 2020). Since plastic often needs to be incinerated to get rid of it, this combustion releases harmful gases into the atmosphere. Moreover, some of the plastic litter causes enormous water degradation in our seas and oceans and the death of many marine animals.

Lack of Education

Many people are unaware of how uncontrolled and excessive consumption actions negatively affect the health of the ecosystem (Das & Basu, 2020). Many environmental problems, such as excessive plastic waste, waste of natural resources, excessive consumption behaviors, destruction of forests, and irresponsible use of clean water, leave our planet in the middle of a chain of problems that have become difficult to return to due to the unconscious behavior of people from the past to the present. When the systemic and infrastructural inadequacies of the state administrations are also involved, the negative results increase exponentially. Necessary awareness-raising studies should be carried out to ensure more awareness and ecological sustainability about problems such as more conscious use of natural resources, correct waste management, and elimination of waste, and training should be given to individuals, institutions, and organizations on these issues (Paço et al. 2021).

Agricultural Pollution

Another cause of environmental degradation is agricultural pollution. In agricultural production, the use of large amounts of fertilizers and pesticides and the harmful components in these substances cause soil degradation (Mondal & Tripathy, 2020). Contamination of soils in this way inevitably results in reduced productivity in the fields, which can lead to famine in the long run (Chang et al., 2022). In addition, it is a known fact that these harmful substances pass from the soil to food and from food to all other living things, including humans, negatively affecting their health.

Mining

The environmental impacts of mining can occur on a local, regional, and global scale through direct and indirect mining practices. Impacts can lead to erosion, sinkholes, loss of biodiversity, or contamination of soil, groundwater, and surface waters due to chemicals emitted from mining operations (Barnwell et al., 2020).

Running out of Resources

As the total human population on the planet is increasing day by day and people are not giving up their excessive consumption behaviors for higher living standards, excessive amounts of natural resources are being extracted to meet the demand. This situation leads to the decrease and depletion of many natural resources (Ahmed et al., 2020).

Consumption Behaviors

Today's consumption behavior is a very big problem for the world. The effect of excessive consumption on nature has reached a point where all living resources are no longer sustainable (Longwe, 2022). Consumption of more than is necessary causes overproduction and overuse of resources. In fact, the uncontrolled consumption of limited natural resources adversely affects the natural sustainability of the planet (Bokil & Sinha, 2021). First of all, it causes the destruction of forests, the genocide of living things, environmental (water, soil, and air) pollution, famine, and serious health problems for people and all other living things (Dunlap, 2022).

SUSTAINABLE CONSUMPTION BEHAVIOR

As a result of ecological degradation, two-thirds of the world's underground resources have already been depleted (Politkovskaya et al., 2020). It is predicted that half of the animal and plant species will become extinct in the coming years (Saxena et al., 2021). As long as the world continues to live above its ecological capacity, all these problems will continue to increase. In order to solve all these environmental problems, it will not be enough to change policies alone. There is a need for consumers who have adopted a new lifestyle based on sustainable consumption and are conscious of the environment (Lubowiecki-Vikuk et al., 2021).

The concept of sustainability brings together ecological balance and economic growth (Ahmed et al., 2020). With sustainability, both the effective use of natural resources and the environment are given importance. Sustainable consumption is a type of consumption that can meet the needs of both future generations and today's people. Sustainable consumption takes into account the needs of future generations. Therefore, it offers a consumption style that minimizes the use of natural resources, waste emissions of toxic materials, and environmental pollutants. In this way, it refers to the use of goods and services that aim to leave a better environment for future generations while meeting the basic needs of consumers. It is vital to adopt sustainable consumption behavior in order to prevent ecological degradation (Lin et al., 2018).

METHODOLOGY

This study is aimed at examining the evolution of ecological degradation within the context of consumption in the past three decades. For this purpose, 2183 articles between 1992 and 2022 containing the keywords (ALL=("Ecological Degradation") AND ALL=("consum*")) OR (ALL=("environmental degradation") AND ALL=("consum*")) were selected from the Web of Science database, and bibliometric analysis was applied with the R Studio analysis program. The stages of creating the data set are given in Figure 1. While determining the keywords, both "ecological degradation" and "environmental degradation" degradation keywords were used to increase the number of studies on the scope of ecological degradation. In addition to these keywords, the "consum*" keyword is used to reduce environmental degradation in the consumer context. By limiting the use of relevant keywords while creating the dataset, the dataset was freed from irrelevant studies (Ulucak et al., 2021). In this way, irrelevant studies that were not covered by consumer behavior were not included in the data set. In addition, no time limit was set while creating the data set, and the publication date of the first article in the Web of Science database, which focused on consumer behavior and environmental degradation, was accepted as the beginning. As a result, the analysis is aimed at creating a comprehensive framework for ecological degradation within the context of consumption. The ecological degradation within the context of consumption was discussed under headings within the framework of the main themes that emerged as a result of the bibliometric analysis.

In Figure 2, the 100 most repetitive words in the data set were examined. Among these words, it is seen that words such as "urbanization", "sustainability", "environment", "globalization", and "sustainable development" are common words in articles.

In the word tree shown in Figure 3, the distribution of the frequencies of the words is given in order. In addition to the leading words in the word cloud "economic growth", "environmental degradation", "CO2 emissions", "energy consumption", "renewable energy", "ecological footprint", and "financial development" are frequently discussed in the literature.

Figure 1. PRISMA 2020 flow diagram for new systematic reviews, which included searches of databases, registers, and other sources
** Consider, if feasible to do so, the possibility of reporting the number of records identified from each database or register searched (rather than the total number across all databases/registers).*
*** If automation tools were used, indicate how many records were excluded by a human and how many were excluded by automation tools. Visit: http://www.prisma-statement.org/ (accessed on 4 May 2023)*

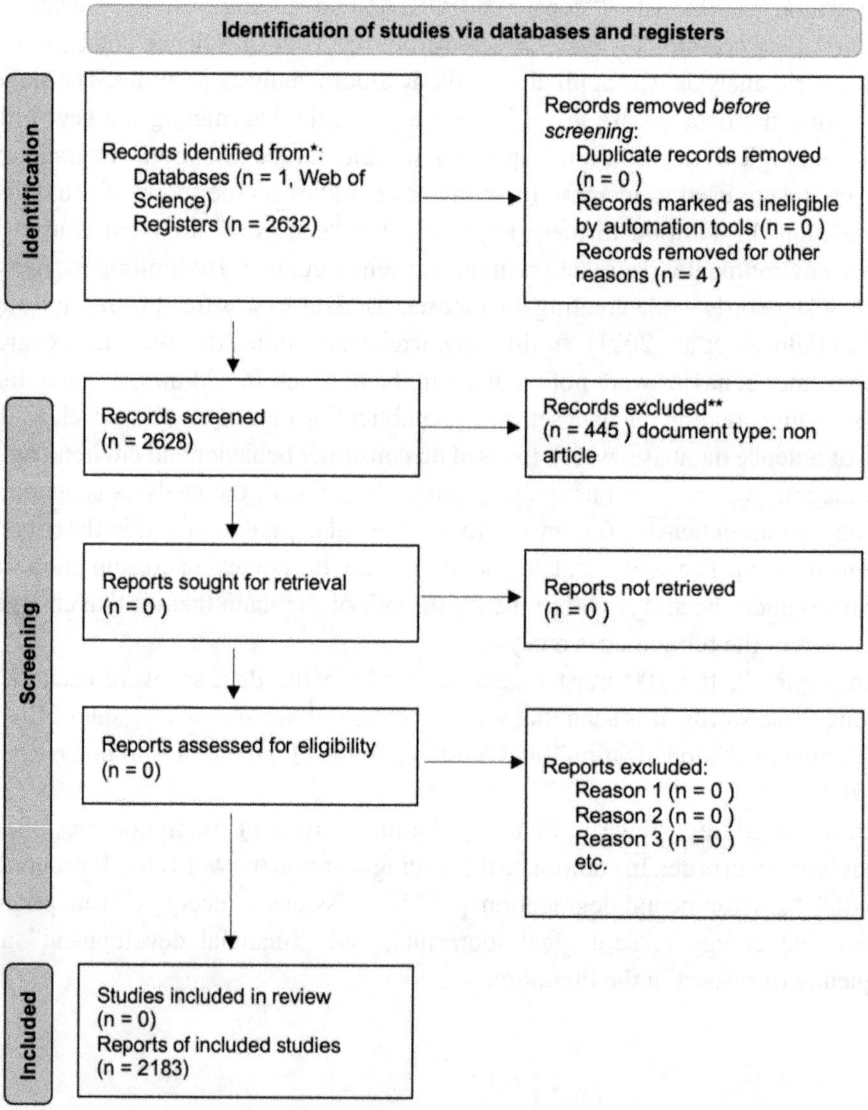

Figure 2. Word cloud

Figure 3. Word tree

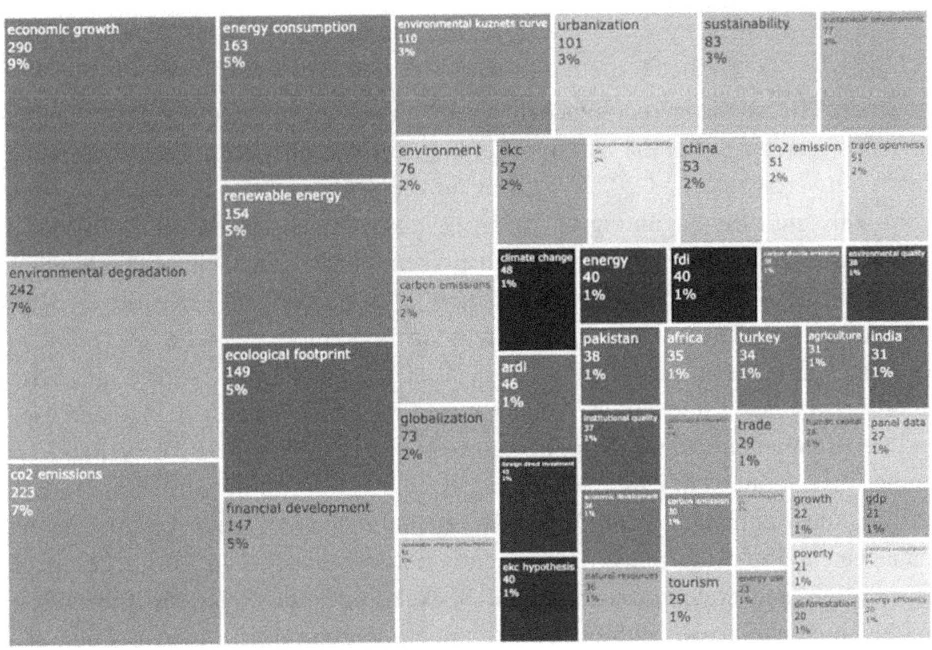

Figure 4. Trend topics

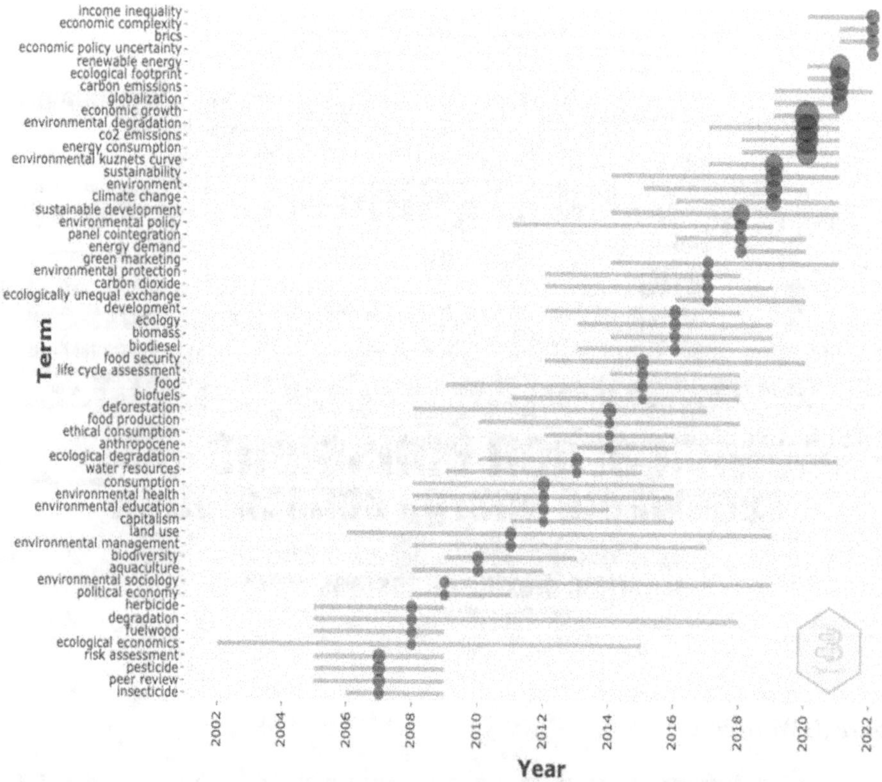

Since 2002, studies on ecological degradation within the context of consumption have been carried out in the literature. In Figure 4, the distribution of the trending topics between 2002 and 2022 is given. In the studies in the literature, starting from 2002 until the beginning of 2015, the main theme studied was "ecological economics". Between 2005 and 2009, it is seen that the concepts of "herbicide", "degradation", "fuelwood", "risk assessment", "pesticide" and "insecticide" have come to the fore. There has been an increase in the number of studies on "political economy" and "environmental sociology" since the beginning of 2009. In 2010, "biodiversity" and "aquaculture" topics gained popularity, and in 2011, "land use" and "environmental sociology" topics gained popularity. After 2012, "consumption" studies in the context of ecological degradation began. In addition to consumption studies, "environmental health", "environmental education" and "environmental management" topics that related to the environment were studied. In 2013, "ecological degradation" and "water resources" topics began being frequently used in the studies. In 2014, publications on "anthropocene", "ethical consumption", "food production", "deforestation", "social media" and "exercise" topics were produced. Between 2015 and 2016, it was seen that "food" and "biomass" related studies accelerated. As of 2017, studies have started to be carried out within the scope of "sustainability", "environmental protection" and "green marketing". In the period from 2019 to the present, studies on "carbon emissions", "renewable energy", "economic growth" and "globalization" have increased. Finally, it has been determined that today, subjects that are discussed from an economic perspective, such as "income inequality", "economic complexity", "brics" and "economic policy uncertainty" are studied.

Figure 5. Thematic map

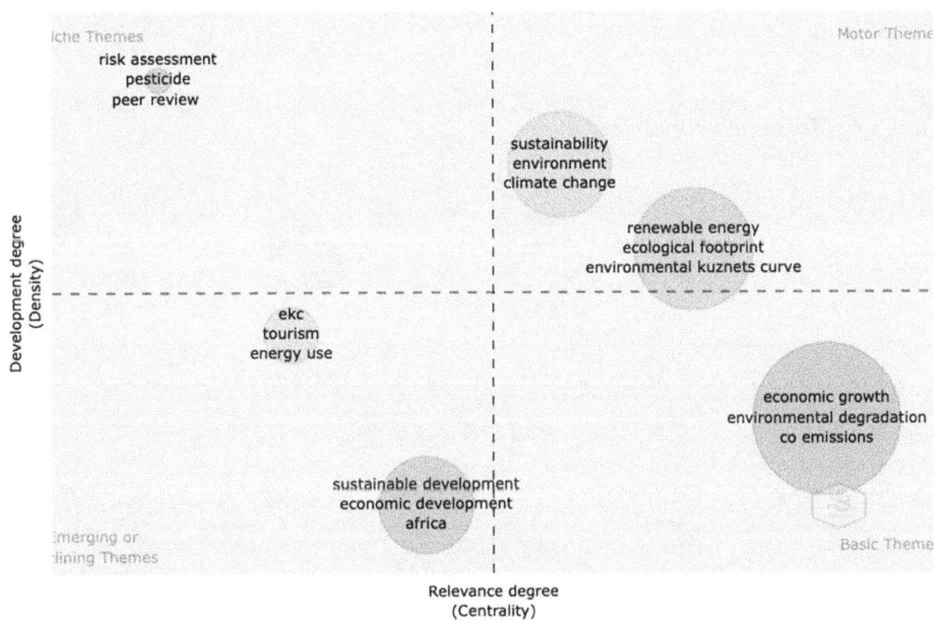

In the study, a thematic map divided into four topological regions based on density and centrality was created, as shown in Figure 5. The upper right region offers "motor" themes indicated with high intensity and centrality. These topics are gathered around two clusters. Topics in each cluster refer to topics that are frequently studied together in the literature. The first cluster includes "sustainability", "environment" and "climate change"; the second cluster includes "renewable energy", "ecological footprint" and "environmental Kuznets curve". These topics in the motor themes need to be further developed given their importance for future research.

In the upper left region, "risk assessment", "pesticide" and "peer revies" topics have high density but low centrality. These topics in this region illustrate specific and underrepresented issues that are areas of rapid development.

Themes in the lower left region are shown with low centrality and density. Two clusters are shown in this region. The first cluster of this region includes "ekc", "tourism" and "energy use" topics, and the second cluster of this region includes "sustainable development", "economic development" and "Africa" topics. These topics in this area of the thematic map now include topics with a downward trend.

Finally, the lower right region contains the core themes shown with high centrality but low intensity. This field includes only one cluster. The cluster of this region includes "economic growth", "environmental degradation" and "CO emissions". These subjects in this region refer to the general importance of the studies to be carried out.

Figure 6. Most cited countries

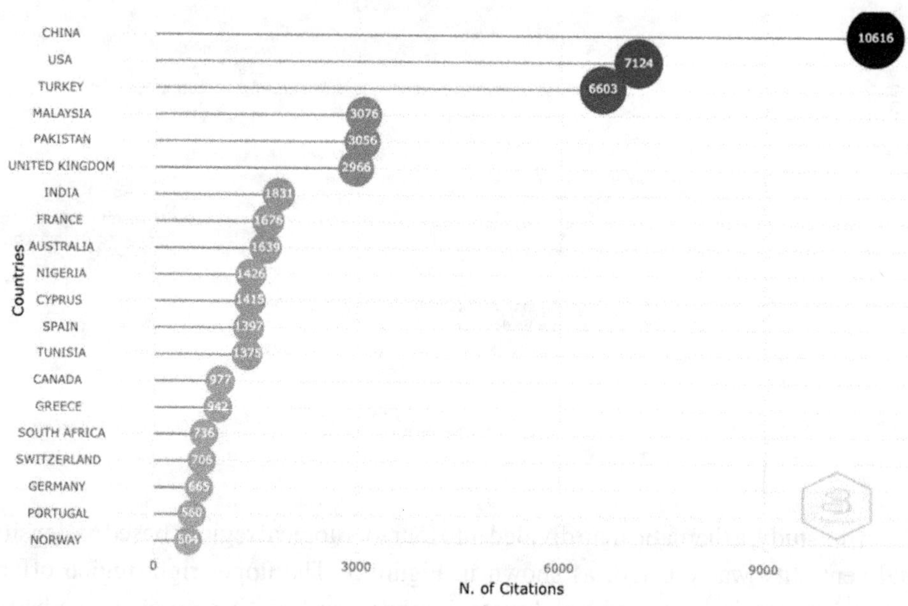

According to Figure 6, it is seen that the most cited publications in studies on ecological degradation within the context of consumption are presented by authors from China, the USA, Turkey, Malaysia, Pakistan, and the United Kingdom, respectively.

Figure 7. Annual scientific production

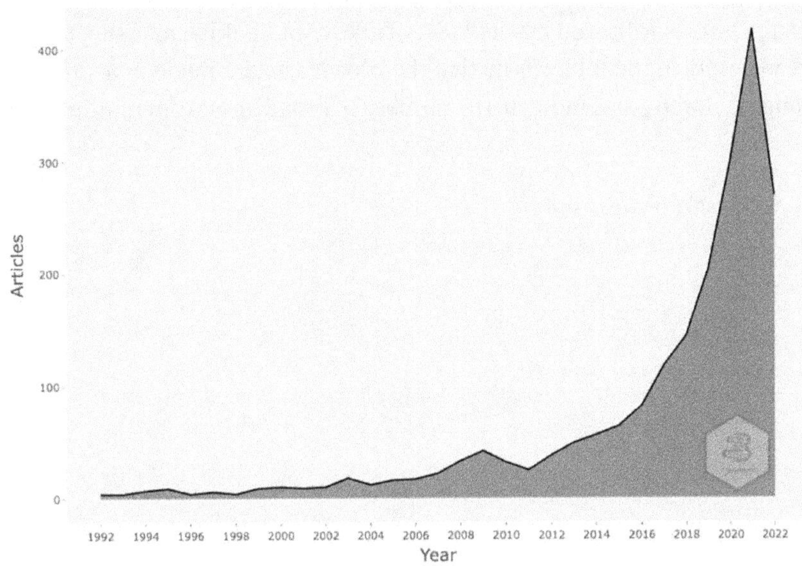

In Figure 7, it is seen that the first studies on ecological degradation within the context of consumption were put forward in 1992. In addition, the number of studies increased rapidly, starting in 2018 and reaching its highest level in 2020.

Figure 8. Corresponding author's country

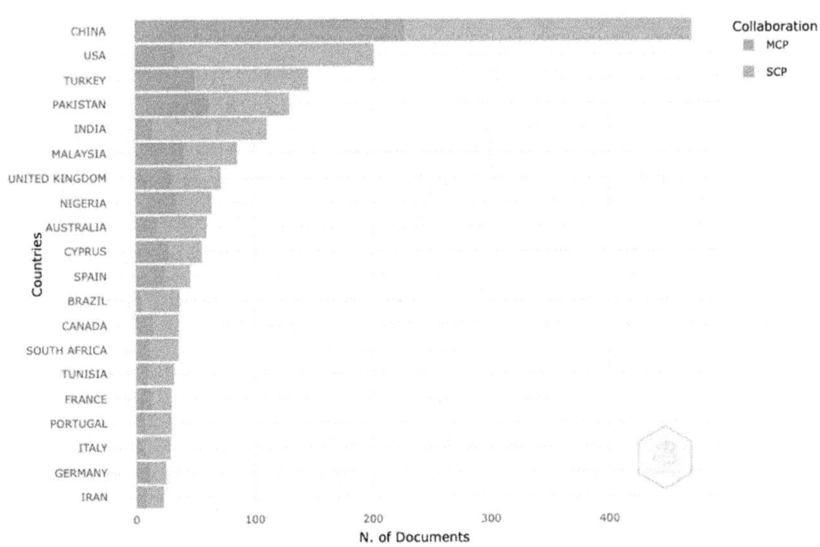

The countries of the responsible authors for the publications were examined and shown in Figure 8. According to Figure 8, the country with the most publications is China. China is followed by the USA, Turkey, and Pakistan, respectively. While the green colors in the table show that the broadcasts are made in a single country, the orange-colored parts indicate the number of broadcasts from multiple countries.

Figure 9. Most relevant sources

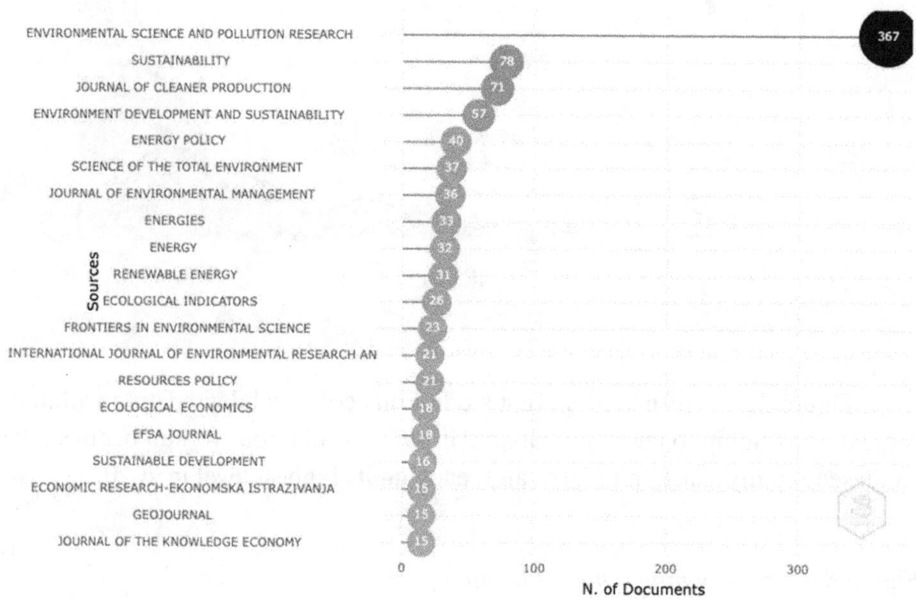

The most relevant journals were searched for studies on ecological degradation within the context of consumption. According to the most relevant sources in Figure 9, "Environmental Science and Pollution Research", "Sustainability", and "Journal of Cleaner Production" are the most popular scientific journals, respectively.

Figure 10 shows the interactivity of the keywords used in the studies in the literature with each other. Keywords are divided into five separate clusters. The bubbles representing economic growth, environmental deterioration, and CO2 emissions were enlarged by the clustering algorithm by positioning them in the co-occurrence network's center, as seen in the Figure 10. Each subset is colored a different color. The red cluster shows the most frequently used word network, and the words in this network are concentrated under the keyword "economic growth", while in the blue cluster it is observed that the keywords are distributed around "CO2 emissions". The words in the purple cluster are concentrated around the keyword "urbanization" and the words in the green cluster are concentrated around the keyword "renewable energy". The last cluster is shown in orange color. Keywords in the orange clusters show a dispersed distribution without centralization.

Figure 10. Co-occurrence network

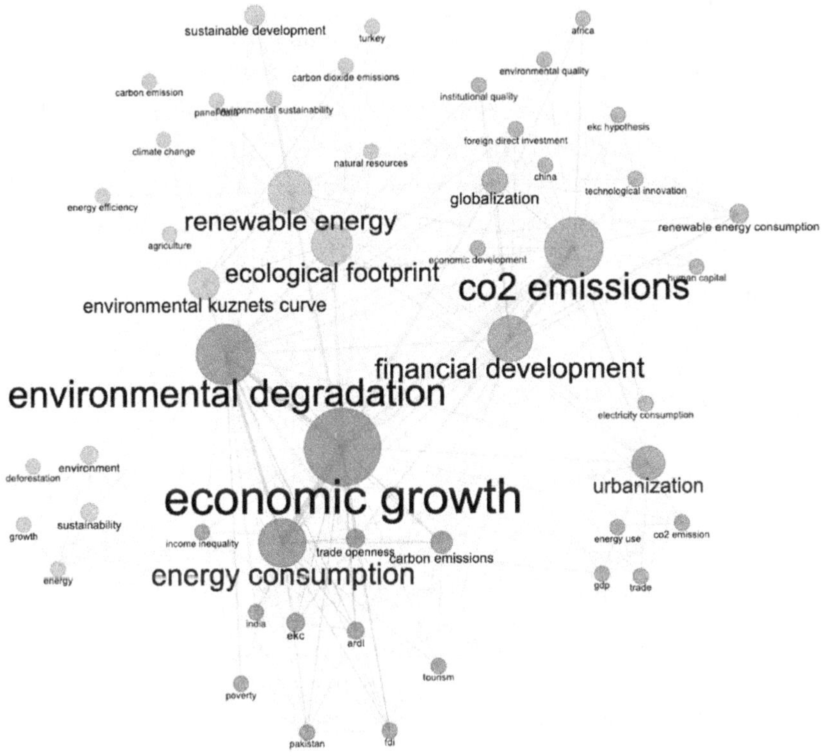

CONCLUSION

In this study, studies on ecological degradation within the scope of consumption in the literature were examined retrospectively. Bibliometric analysis was applied with the R program to 2183 articles published in the Web of Science database in the last 30 years. The analysis results were visualized and interpreted within the scope of the study. The causes of ecological deterioration within the scope of consumption have been revealed. Topics studied over the years, prominent trends and concepts, leading countries, and journals with their contributions to the literature have been revealed.

According to the findings of the study, it is observed that there is a gradual increase in the number of scientific articles produced annually. This shows that the issue of ecological degradation has become increasingly important over the years. Especially in the period after 2018, the number of studies increased rapidly. Another finding of the study is that countries China ranks first among the countries that receive the most citations, by far surpassing its competitors in terms of the number of authors producing the most scientific publications. In the study, it has also been determined that the most prominent word, according to the analysis findings, in the last 30 years has been "economic growth. Studies have shown that as a result of economic growth, consumer behavior changes and triggers ecological deterioration. In recent years, it has been revealed that "income inequality", "economic complexity," and "brics" are among the most popular topics studied in the literature. Finally, based on the thematic map, it has been concluded that, unlike previous years, studies on "sustainable development" have become increasingly important and that more studies should be done on this subject.

REFERENCES

Ahmed, Z., Asghar, M. M., Malik, M. N., & Nawaz, K. (2020). Moving towards a sustainable environment: The dynamic linkage between natural resources, human capital, urbanization, economic growth, and ecological footprint in China. *Resources Policy*, *67*, 101677. doi:10.1016/j.resourpol.2020.101677

Arias-Ortiz, A., Masqué, P., Glass, L., Benson, L., Kennedy, H., Duarte, C. M., Garcia-Orellana, J., Benitez-Nelson, C. R., Humphries, M. S., Ratefinjanahary, I., Ravelonjatovo, J., & Lovelock, C. E. (2021). Losses of soil organic carbon with deforestation in mangroves of Madagascar. *Ecosystems (New York, N.Y.)*, *24*(1), 1–19. doi:10.100710021-020-00500-z

Asamoah, O., Kuittinen, S., Abrefa Danquah, J., Quartey, E. T., Bamwesigye, D., Mario Boateng, C., & Pappinen, A. (2020). Assessing wood waste by timber industry as a contributing factor to deforestation in Ghana. *Forests*, *11*(9), 939. doi:10.3390/f11090939

Avtar, R., Tripathi, S., Aggarwal, A. K., & Kumar, P. (2019). Population–urbanization–energy nexus: A review. *Resources*, *8*(3), 136. doi:10.3390/resources8030136

Barnwell, G. C., Stroud, L., & Watson, M. (2020). "Nothing green can grow without being on the land": Mine-affected communities' psychological experiences of ecological degradation and resistance in Rustenburg, South Africa. *Community Psychology in Global Perspective*, *6*(2), 87-109.

Basit, I., Faizi, F., Mahmood, K., Bilgili, M. S., Yildirim, Y., & Mushtaq, F. (2022). Geospatial alternatives for quantification of bio-thermal influence zone in the vicinity of a solid waste dump. *Waste Management & Research.*

Bisht, A., Kamboj, N., Kamboj, V., & Bisht, A. (2020). A review on the role of emerging anthropogenic activities in environmental degradation and emphasis on their mitigation. *Archives of Agriculture and Environmental Science*, *5*(3), 419–425. doi:10.26832/24566632.2020.0503025

Bokil, Y., & Sinha, D. K. (2021). Green Marketing and Mindful Consumption for Sustainable Development. *Central Asian Journal of Innovations on Tourism Management and Finance*, *2*(9), 4–7.

Chang, J., Fang, W., Liang, J., Zhang, P., Zhang, G., Zhang, H., Zhang, Y., & Wang, Q. (2022). A critical review on interaction of microplastics with organic contaminants in soil and their ecological risks on soil organisms. *Chemosphere*, *306*, 135573. doi:10.1016/j.chemosphere.2022.135573 PMID:35797912

Cheng, Y., Awan, U., Ahmad, S., & Tan, Z. (2021). How do technological innovation and fiscal decentralization affect the environment? A story of the fourth industrial revolution and sustainable growth. *Technological Forecasting and Social Change*, *162*, 120398. doi:10.1016/j.techfore.2020.120398

Das, A., & Basu, T. (2020). Assessment of peri-urban wetland ecological degradation through importance-performance analysis (IPA): A study on Chatra Wetland, India. *Ecological Indicators*, *114*, 106274. doi:10.1016/j.ecolind.2020.106274

DeFries, R. S., Edenhofer, O., Halliday, A. N., Heal, G. M., Lenton, T., Puma, M., ... Ward, B. (2019). *The missing economic risks in assessments of climate change impacts*. Academic Press.

Deletic, A., & Wang, H. (2019). Water pollution control for sustainable development. *Engineering (Beijing)*, *5*(5), 839–840. doi:10.1016/j.eng.2019.07.013

Duguma, L. A., Atela, J., Minang, P. A., Ayana, A. N., Gizachew, B., Nzyoka, J. M., & Bernard, F. (2019). Deforestation and forest degradation as an environmental behavior: Unpacking realities shaping community actions. *Land (Basel)*, *8*(2), 26. doi:10.3390/land8020026

Dunlap, A. (2022). The Self-Reinforcing Cycle of Ecological Degradation and Repression: Revealing the Ecological Cost of Policing and Militarization. In Enforcing Ecocide (pp. 153-176). Palgrave Macmillan.

Erdoğan, S., Gedikli, A., & Genç, S. Y. (2018). An overview of Turkey's national energy policies. *Politico-Economic Evaluation of Current Issues*, 143.

ahad, S., Adnan, M., & Saud, S. (Eds.). (2022). *Improvement of Plant Production in the Era of Climate Change*. CRC Press. doi:10.1201/9781003286417

Foteinis, S. (2020). How small daily choices play a huge role in climate change: The disposable paper cup environmental bane. *Journal of Cleaner Production*, *255*, 120294. doi:10.1016/j.jclepro.2020.120294

Gedamu, M. T. (2020). Soil Degradation and its Management Options in Ethiopia: A Review. *International Journal of Research and Innovations in Earth Science*, *7*(5), 59–76.

Gorrie, M. (2020). Regulating our Consumer Culture: What Role Can the Law Play in Addressing Excessive Consumption? In *Sustainability and Law* (pp. 119-134). Springer.

Guerrero, V. T., & Guevara, F. C. (2021). Influence of Deforestation and Environmental Degradation on Agriculture in Mexico. *Nongxue Xuebao*, *5*(1).

Harper, A. B., Powell, T., Cox, P. M., House, J., Huntingford, C., Lenton, T. M., Sitch, S., Burke, E., Chadburn, S. E., Collins, W. J., Comyn-Platt, E., Daioglou, V., Doelman, J. C., Hayman, G., Robertson, E., van Vuuren, D., Wiltshire, A., Webber, C. P., Bastos, A., ... Shu, S. (2018). Land-use emissions play a critical role in land-based mitigation for Paris climate targets. *Nature Communications*, *9*(1), 1–13. doi:10.103841467-018-05340-z PMID:30087330

Joseph, N., Ryu, D., Malano, H. M., George, B., & Sudheer, K. P. (2020). A review of the assessment of sustainable water use at continental-to-global scale. *Sustainable Water Resources Management*, *6*(2), 1–20. doi:10.100740899-020-00379-7

Khan, I., Hou, F., Le, H. P., & Ali, S. A. (2021). Do natural resources, urbanization, and value-adding manufacturing affect environmental quality? Evidence from the top ten manufacturing countries. *Resources Policy*, *72*, 102109. doi:10.1016/j.resourpol.2021.102109

Kopittke, P. M., Menzies, N. W., Wang, P., McKenna, B. A., & Lombi, E. (2019). Soil and the intensification of agriculture for global food security. *Environment International*, *132*, 105078. doi:10.1016/j.envint.2019.105078 PMID:31400601

Kumar, V., Singh, J., & Kumar, P. (Eds.). (2020). Environmental Degradation: Causes and Remediation Strategies (Vol. 1). Agro Environ Media, Publication Cell of AESA, Agriculture and Environmental Science Academy.

Lamb, W. F., Wiedmann, T., Pongratz, J., Andrew, R., Crippa, M., Olivier, J. G., Wiedenhofer, D., Mattioli, G., Khourdajie, A. A., House, J., Pachauri, S., Figueroa, M., Saheb, Y., Slade, R., Hubacek, K., Sun, L., Ribeiro, S. K., Khennas, S., de la Rue du Can, S., ... Minx, J. (2021). A review of trends and drivers of greenhouse gas emissions by sector from 1990 to 2018. *Environmental Research Letters*, *16*(7), 073005. doi:10.1088/1748-9326/abee4e

Law, K. L., Starr, N., Siegler, T. R., Jambeck, J. R., Mallos, N. J., & Leonard, G. H. (2020). The United States' contribution of plastic waste to land and ocean. *Science Advances*, *6*(44), eabd0288. doi:10.1126ciadv.abd0288 PMID:33127684

Lin, S. T., & Niu, H. J. (2018). Green consumption: Environmental knowledge, environmental consciousness, social norms, and purchasing behavior. *Business Strategy and the Environment*, *27*(8), 1679–1688. doi:10.1002/bse.2233

Longwe, B. (2022). A Comparative Study of Ecological Consumer Behavior Amongst Consumers In India and Pakistan. *Educational Research*, *4*(1), 296–307.

Lubowiecki-Vikuk, A., Dąbrowska, A., & Machnik, A. (2021). Responsible consumer and lifestyle: Sustainability insights. *Sustainable Production and Consumption*, *25*, 91–101. doi:10.1016/j.spc.2020.08.007 PMID:32864405

Manowska, A., & Nowrot, A. (2019). The importance of heat emission caused by global energy production in terms of climate impact. *Energies*, *12*(16), 3069. doi:10.3390/en12163069

Mohsin, M., Abbas, Q., Zhang, J., Ikram, M., & Iqbal, N. (2019). Integrated effect of energy consumption, economic development, and population growth on CO2 based environmental degradation: A case of transport sector. *Environmental Science and Pollution Research International*, *26*(32), 32824–32835. doi:10.100711356-019-06372-8 PMID:31502046

Mondal, T., & Tripathy, B. (2020). Riverbank Erosion and Environmental degradation. *Int J Res*, *7*(7).

Odume, O. N., Onyima, B. N., Nnadozie, C. F., Omovoh, G. O., Mmachaka, T., Omovoh, B. O., Uku, J. E., Akamagwuna, F. C., & Arimoro, F. O. (2022). Governance and Institutional Drivers of Ecological Degradation in Urban River Ecosystems: Insights from Case Studies in African Cities. *Sustainability (Basel)*, *14*(21), 14147. doi:10.3390u142114147

Paço, A., Leal Filho, W., Ávila, L. V., & Dennis, K. (2021). Fostering sustainable consumer behavior regarding clothing: Assessing trends on purchases, recycling and disposal. *Textile Research Journal*, *91*(3-4), 373–384. doi:10.1177/0040517520944524

Pata, U. K., Yilanci, V., Hussain, B., & Naqvi, S. A. A. (2022). Analyzing the role of income inequality and political stability in environmental degradation: Evidence from South Asia. *Gondwana Research, 107*, 13–29. doi:10.1016/j.gr.2022.02.009

Politkovskaya, I., Khvichiya, D., & Artamonova, L. (2020). Economic Support of the Policy of Transition to Sustainable Development of the Territories of the Russian North. In *E3S Web of Conferences* (Vol. 174, p. 04047). EDP Sciences. doi:10.1051/e3sconf/202017404047

Sarkodie, S. A. (2018). The invisible hand and EKC hypothesis: What are the drivers of environmental degradation and pollution in Africa? *Environmental Science and Pollution Research International*, *25*(22), 21993–22022. doi:10.100711356-018-2347-x PMID:29797200

Saxena, R., Rath, R., Gupta, S., & Sood, N. (2021). A review on ecological degradation, its causes and sustainable development in Delhi, India. *Journal of Applied and Natural Science*, *13*(4), 1294–1304. doi:10.31018/jans.v13i4.2978

Sonone, S. S., Jadhav, S., Sankhla, M. S., & Kumar, R. (2020). Water contamination by heavy metals and their toxic effect on aquaculture and human health through food Chain. *Lett. Appl. NanoBioScience*, *10*(2), 2148–2166.

Stavi, I., Paschalidou, A., Kyriazopoulos, A. P., Halbac-Cotoara-Zamfir, R., Siad, S. M., Suska-Malawska, M., Savic, D., Roque de Pinho, J., Thalheimer, L., Williams, D. S., Hashimshony-Yaffe, N., van der Geest, K., Cordovil, C. M. S., & Ficko, A. (2021). Multidimensional food security nexus in drylands under the slow onset effects of climate change. *Land (Basel)*, *10*(12), 1350. doi:10.3390/land10121350

Tennison, I., Roschnik, S., Ashby, B., Boyd, R., Hamilton, I., Oreszczyn, T., Owen, A., Romanello, M., Ruyssevelt, P., Sherman, J. D., Smith, A. Z. P., Steele, K., Watts, N., & Eckelman, M. J. (2021). Health care's response to climate change: A carbon footprint assessment of the NHS in England. *The Lancet. Planetary Health*, *5*(2), e84–e92. doi:10.1016/S2542-5196(20)30271-0 PMID:33581070

Ukaogo, P. O., Ewuzie, U., & Onwuka, C. V. (2020). Environmental pollution: causes, effects, and the remedies. In *Microorganisms for sustainable environment and health* (pp. 419–429). Elsevier. doi:10.1016/B978-0-12-819001-2.00021-8

Ulucak, R., Sari, R., Erdogan, S., & Alexandre Castanho, R. (2021). Bibliometric literature analysis of a multi-dimensional sustainable development issue: Energy poverty. *Sustainability (Basel)*, *13*(17), 9780. doi:10.3390u13179780

Vieira, L. C., Longo, M., & Mura, M. (2021). Are the European manufacturing and energy sectors on track for achieving net-zero emissions in 2050? An empirical analysis. *Energy Policy*, *156*, 112464. doi:10.1016/j.enpol.2021.112464

Wassie, S. B. (2020). Natural resource degradation tendencies in Ethiopia: A review. *Environmental Systems Research*, *9*(1), 1–29. doi:10.118640068-020-00194-1

White, K., Habib, R., & Hardisty, D. J. (2019). How to SHIFT consumer behaviors to be more sustainable: A literature review and guiding framework. *Journal of Marketing*, *83*(3), 22–49. doi:10.1177/0022242919825649

Xie, H., Zhang, Y., Wu, Z., & Lv, T. (2020). A bibliometric analysis on land degradation: Current status, development, and future directions. *Land (Basel)*, *9*(1), 28. doi:10.3390/land9010028

Zhang, L., Godil, D. I., Bibi, M., Khan, M. K., Sarwat, S., & Anser, M. K. (2021). Caring for the environment: How human capital, natural resources, and economic growth interact with environmental degradation in Pakistan? A dynamic ARDL approach. *The Science of the Total Environment*, *774*, 145553. doi:10.1016/j.scitotenv.2021.145553 PMID:33611006

Zheng, S., Wang, R., Mak, T. M., Hsu, S. C., & Tsang, D. C. (2021). How energy service companies moderate the impact of industrialization and urbanization on carbon emissions in China? *The Science of the Total Environment*, *751*, 141610. doi:10.1016/j.scitotenv.2020.141610 PMID:32877794

Zwolak, A., Sarzyńska, M., Szpyrka, E., & Stawarczyk, K. (2019). Sources of soil pollution by heavy metals and their accumulation in vegetables: A review. *Water, Air, and Soil Pollution*, *230*(7), 1–9. doi:10.100711270-019-4221-y

KEY TERMS AND DEFINITIONS

Ecological Footprint: The quantity of land, water, and other resources required to support a population's consumption and waste creation is known as an ecological footprint, which measures the impact of human activities on the environment. The usual unit of measurement is global hectares per capita.

Economic Development: Economic development is the steady expansion and improvement of a nation's economy, which is frequently gauged by changes in GDP, employment, productivity, and standard of life.

Environmental Degradation: Environmental degradation is the term used to describe how human actions, such as pollution, deforestation, and climate change, have caused natural resources, such as air, water, soil, and ecosystems, to deteriorate.

Renewable Energy: Energy sources like solar, wind, hydropower, geothermal, and biomass that replenish themselves naturally and can be sustained over time are referred to as renewable energy sources. Renewable energy sources can be used without depleting the Earth's resources, in contrast to fossil fuels, which are limited and non-renewable.

Sustainable Development: Sustainable development is a development approach that strikes a balance between social justice, environmental conservation, and economic progress. It aims to meet the requirements of the present generation without sacrificing the capacity of future generations to meet their own needs.

Urbanization: The process of population increase and the physical extension of cities and urban regions is referred to as urbanization. It usually entails people moving from rural to urban regions in search of better job prospects and access to facilities and services.

Chapter 8
Convergence in Environmental-Related Technologies Across OECD Countries

Uğur Ursavaş
Zonguldak Bülent Ecevit University, Turkey

ABSTRACT

This study investigates the convergence process of environmental-related technologies patents over the period 1991-2019 across 36 OECD countries. To do so, the club convergence methodology of Phillips and Sul is used, which allows the authors to analyze overall convergence and identify possible convergence clubs. The log-t test results show that OECD countries do not converge to the same equilibrium point in terms of environmental-related technologies patents per capita. However, the club clustering algorithm shows that while 33 OECD countries converge to a common point, Hungary, South Korea, and Lithuania are diverging countries.

INTRODUCTION

In recent years, the consumption of fossil fuels has increased due to several factors, such as rapid economic growth, increasing human development levels, industrialization, etc., which has led to several environmental consequences for countries, including water pollution, global warming, climate change, and energy crisis, etc. One of the ways to cope with these environmental challenges is to increase

DOI: 10.4018/978-1-6684-6727-5.ch008

environmental-related technologies (Yin et al., 2015; Kwon et al., 2017; Mensah et al., 2019). Environmental-related technologies are used in the environmentally friendly production or processing of goods, which can help to decrease environmental degradation in the long run (Mensha et al., 2019; Gu et al., 2019; Danish and Ulucak, 2020; Hussain and Doğan, 2021). The development of new environmental-related technologies significantly affects energy efficiency, boosting the industrial sector's performance in decreasing carbon emissions and reducing environmental degradation (Danish and Ulucak, 2020; Erdoğan et al., 2020; Ahmed, 2020; Sadiq et al., 2022). Progress in environment-related technology raises the proportion of clean energy in the energy mix and decreases energy intensity (Danish and Ulucak, 2020). However, Gu et al. (2019) argue that at different levels of energy technology, the impact of environmental-related technology on environmental degradation may vary. When the level of energy technology is low, energy technological growth leads to a rise in carbon emissions. However, when the level of energy technology surpasses the turning point, the impact of technological progress on carbon emissions reduction becomes significantly greater, which indicates that the direction of technological progress is not straightforward. In other words, technological improvement does not necessarily decrease or raise carbon emissions. Environmental-related technologies can reduce pollution through various channels. For instance, advanced technologies can boost the efficiency of current energy sources, or they may support the production and utilization of new cleaner energy sources, such as wind, hydro, etc. (Ahmad, 2020)

In recent years, numerous studies have investigated the link between environmental-related technology and environmental quality. Gu et al. (2019) identify a U-shaped relationship between carbon emissions and energy technological progress in provinces of China over the period 2005-2016. Danish and Ulucak (2020) show that environmental-related technology positively impacts green growth in BRICS countries over the period 1992-2016 using the Westerlund cointegration methodology. Erdoğan et al. (2020) analyze the relationship between innovation and carbon emissions for G20 countries over the period 1991-2017. The results of CCEMG (Common Correlated Effects Mean Group) and AMG (Augmented Mean Group) show that an increase in innovation in the industrial sector decreases carbon emissions in G20 countries. Using CCEMG and AMG methods, Hussain and Doğan (2021) show the negative relationship between environmental-related technology and the ecological footprint for BRICS countries over the period 1992-2016. Hussain and Khan (2021) find that environment-related technologies negatively affect consumption-based carbon emissions in the top five emitters over the period 1992-2017. Hussain et al. (2022) empirically show that an increase in environmental-related technology decreases carbon emissions in seven emerging countries over the period 1990-2016. Hussain et al. (2022) investigate the link between environmental-related technologies and environmental degradation in BRICS countries for the period 1992-2020. The results of CCEMG and AMG tests show that an increase in environment-related technologies decreases ecological footprint levels in BRICS countries. Sadiq et al. (2022) analyze the relationship between environmental technology and environmental degradation for the top ten countries with the largest ecological footprint levels over the period 1990-2017. The results show that environmental-related technologies have a negative impact on environmental degradation. Sahoo et al. (2022) show that environmental-related technologies have a significant impact on reducing environmental degradation in 14 developing Asian countries.

Although numerous studies have tested the relationship between environmental-related technology and environmental quality, to the author's best knowledge, there is no study testing the convergence in environmental-related technologies. Within this context, this paper investigates the convergence in environmental-related technologies patents per capita across 36 OECD countries over the period 1991-2019. To do so, the nonlinear time-varying factor model proposed by Phillips and Sul (2007) is used. The club convergence methodology of Phillips and Sul (2007) allows us to analyze overall convergence and identify possible sub-convergence clubs. This study focuses on OECD countries for several reasons. First, OECD countries have an important role in global environmental quality. In the past few years, OECD countries have made significant progress in lowering environmental degradation. For instance, in almost all OECD countries, greenhouse gas emissions (GHG) have decreased in recent years due to the economic slowdown that followed the 2008 financial crisis and the implementation of more strict climate regulations. However, improvement is insufficient, and GHG emissions are expected to accelerate again due to recent increases in energy consumption and CO_2 emissions. Besides, as mentioned above, OECD countries have a significant role in global environmental degradation. For instance, by 2020, OECD countries emit nearly 35% of global CO_2 emissions from energy use (OECD, 2020). Second, OECD countries are mostly high-income (developed) and upper-middle-income countries with identical economic characteristics. While most OECD countries are developed countries, it also includes emerging countries. The majority of OECD countries can be classified as post-industrial economies, indicating that they have completed their transition from an industrial economy to a service economy (Yilanci et al., 2019).

This paper contributes to the existing literature on the relationship between environmental-related technologies and environmental degradation on two fronts. First, to the author's best knowledge, this study is the first to test the convergence in environmental-related technologies patents across OECD countries. Second, this study uses a convergence methodology developed by Phillips and Sul (2007), which allows us to test the overall convergence and identify possible convergence clubs. The rest of this paper is organized as follows. Section 2 presents the data and methodological approach of the paper. Section 3 presents the empirical results. Finally, Section 4 offers some concluding remarks.

DATA AND METHODOLOGY

To test the convergence in environmental-related technology, the total number of environmental-related technologies patents per capita is used for 36 OECD countries over the period 1991-2019. Costa Rica and Iceland are excluded due to data availability. The data is obtained from the OECD database. Table 1 shows the descriptive statistics.

Table 1. Descriptive statistics

Variable	Obs	Mean	Std. Dev.
Environmental-related technologies patents per capita	1044	0.0000242	0.0000384

In this paper, we use the econometric methodology developed by Phillips and Sul (2007) to test the hypothesis that OECD countries converge to a common long-run steady-state equilibrium. The starting point of the Phillips and Sul (2007) methodology, which is based on a nonlinear time-varying factor model, is the decomposition of panel data for a total number of environmental-related technologies patents per capita, (X_{it}), into two components:

$$X_{it} = g_{it} + a_{it} \tag{1}$$

where g_{it} is a systematic component and a_{it} is a transitory component. To separate common components from idiosyncratic components, we transform Equation 1 as follows:

$$X_{it} = \left(\frac{g_{it} + a_{it}}{\mu_t}\right)\mu_t = \delta_{it}\mu_t, \tag{2}$$

where μ_t and δ_{it} represent the common component and idiosyncratic component, respectively. δ_{it} measures the distance between the common trend component μ_t and

X_{it}. Since it is impossible to directly estimate the loading coefficients, δ_{it}, without imposing additional structure, the common factor may be removed by constructing the following relative transition paths:

$$h_{it} = \frac{X_{it}}{\frac{1}{N}\sum_{i=1}^{N} X_{it}} = \frac{\delta_{it}}{\frac{1}{N}\sum_{i=1}^{N} \delta_{it}} \tag{3}$$

where h_{it} represents the relative transition parameter which measures the loading coefficient δ_{it} to the panel average at time t. Equation (3) shows the two properties of h_{it}. First, the cross-sectional mean of h_{it} is equal to one. Second, if the factor loading coefficients δ_{it} converge to δ_i, the relative transition parameter h_{it} converges to one. In this case, Equation (4) shows that the cross-sectional variance of the relative transition parameter, H_t, converges to zero asymptotically. The property $H_t \rightarrow 0$ is used to test the null hypothesis of income convergence and to group provinces into convergence clubs.

$$H_t = N^{-1}\sum_{i=1}^{N}(h_{it} - 1)^2 \rightarrow 0 \quad \text{as } t \rightarrow \infty \tag{4}$$

To test the convergence hypothesis, Phillips and Sul (2007) propose the following log-t regression model:

$$\log\left(\frac{H_1}{H_T}\right) - 2\log\left[\log(t)\right] = \alpha + b\log(t) + u_t\,, \tag{5}$$

for $t=[rt], [rt]+1,\ldots,T$ with $r>0$

where $[rt]$ indicates the initial observation in the regression, which implies that the first fraction of the data is discarded. Phillips and Sul (2007) suggest setting r = 0.3 when the sample is small (T ≤ 50), based on Monte Carlo simulations. Phillips and Sul (2007) propose a conventional inferential procedure for Equation (5). Specifically, they recommend a one-sided t-test with heteroskedasticity and autocorrelation-consistent standard errors. The null hypothesis of convergence is rejected if $t_b < -1.65$.

If convergence is not identified for the whole sample, a local clustering procedure is suggested by Phillips and Sul (2007, 2009) to identify local convergence clubs. The procedure has five steps:

1) Cross-sectional ordering: Sort the economies in decreasing order based on their last observation.
2) Core group formation: Select the first k highest economies in the panel to form the subgroup G_k for some $2 \leq k < N$ and apply the log-t regression to obtain the convergence test statistic $t_k=t(G_k)$ for this subgroup. Then, choose the core group size k^* by maximizing t_k over k according to the criterion $k^* argmax_k\{t_k\}$ subject to min $\{t_k\}$ ⟩ – 1.65. t_k indicates the one-side t-statistic that is needed to evaluate the statistical significance of the convergence test. If $t_k > -1.65$ is not valid for k = 2, then the highest economy is dropped from the core group and the algorithm can be repeated again for the rest of the sample.
3) Sieving economies for club membership: The remaining economies are added to the core group G_k one by one and the log-t test is executed again (Eq. (5)). When the economy is added, a new group is formed if the t-statistic is greater than -1.65.
4) Recursion and stopping: Form the new group consisting of all economies that could not be selected in Step 3, and apply log-t test for this subgroup. If $t_k > -1.65$, it indicates that two convergence subgroups exist. Otherwise, if the null hypothesis of convergence is rejected, Steps 1 to 3 are repeated. If no core group is found, then the remaining economies are labeled as divergent and the algorithm stops.
5) Club merging: Run the log-t test for all pairs of initial clubs. The merging procedure is iterative. That is, the log-t test is applied for the initial clubs 1 and 2, and if they fulfill the convergence test jointly, they should be merged into a new one. Repeat this merging procedure for the remaining clubs until the convergence test is rejected.

RESULTS

To analyze the convergence in environmental-related technologies patents per capita across OECD countries, we apply the log-t convergence of Philips and Sul (2007). The results in Table 2 show that the null hypothesis of overall convergence is rejected (-2.6882 < -1.65), which indicates that OECD countries do not converge to common equilibrium point in terms of environmental-related technologies patents per capita.

Table 2. Log-t test results

Variable	Coefficient	SE	T-stat
Environmental-related technologies patents per capita	-0.0557	0.0207	-2.6882

Since the null hypothesis of overall convergence is rejected, we test the existence of local convergence clubs across OECD countries. In Table 3, the results of the clustering algorithm show that there are four convergence clubs and one non-convergent (i.e., divergence) group. Clubs 1, 2, 3 and 4 consist of 8, 9, 12 and 4 countries, respectively. Finally, the non-convergent group includes three countries, Hungary, South Korea, and Lithuania. In contrast with Table 2, Table 3 shows that the null hypothesis of (local) convergence is not rejected as the T-statistic of the log-t test is greater than -1.65 for all clubs.

Table 3. Convergence club classification

Clubs	Provinces	Coefficient	T-Stat
Club 1 [8]	I Chile I Denmark I Finland I France I Germany I Japan I Sweden I I Türkiye I	0.482	13.517
Club 2 [9]	I Austria I Belgium I Greece I Israel I Netherlands I Poland I I Portugal I Switzerland I United States I	0.185	8.718
Club 3 [12]	I Canada I Colombia I Estonia I Ireland I Italy I Latvia I I Luxembourg I Mexico I Norway I Slovenia I Spain I United Kingdom I	0.163	3.501
Club 4 [4]	I Australia I Czech Republic I New Zealand I Slovak Republic I	0.440	3.077
Not convergent Group 5	I Hungary I South Korea I Lithuania I	-1.163	-28.510

The club clustering algorithm tends to overestimate the true number of clubs (Phillips and Sul, 2009; Von Lyncker and Thoennessen, 2017). Phillips and Sul (2009) argue that sequential club merging tests are needed to avoid overestimating the actual number of clubs. Therefore, we apply sequential club merging tests to evaluate whether any of the initially identified clubs can be merged into larger clubs. Table 4 shows the merging test results.

Table 4. Merge test

Clubs	Coefficient	T-Stat
Club 1+2	0.352	12.357
Club 2+3	0.075	2.312
Club 3+4	0.156	3.272
Club 4+Group 5	-1.167	-25.324

According to the results of merging tests, the number of convergence clubs has been reduced from four to one. Table 5 shows the final club classification after the merging test. For Club 1, the T-statistic is greater than -1.65. Thus, the null hypothesis of (within) convergence is not rejected. The first convergence club includes 33 out of 36 countries. The second group includes three countries, Hungary, South Korea, Lithuania, that are diverging. It is observed from Table 5 that with the exception of Hungary, South Korea, and Lithuania, 33 out of the 36 OECD countries are converging to a common equilibria

Table 5. Final convergence clubs

Clubs	Provinces	Coefficient	T-Stat
Club 1 [33]	I Australia I Austria I Belgium I Canada I Chile I Colombia I I Czech Republic I Denmark I Estonia I Finland I France I I Germany I Greece I Ireland I Israel I Italy I Japan I Latvia I I Luxembourg I Mexico I Netherlands I New Zealand I Norway I I Poland I Portugal I Slovak Republic I Slovenia I Spain I I Sweden I Switzerland I Türkiye I United Kingdom I I United States I	0.094	4.057
Not convergent Group 2	I Hungary ISouth Korea I Lithuania I	-1.163	-28.510

Figure 1 shows the relative transition paths of Club 1 and three divergent countries. As expected, since Club 1 includes almost all of the countries in the sample, the relative transition path of Club 1 is almost on the panel average. Figure 1 shows that the relative transition path of South Korea is largely above the panel average, which indicates that the number of environmental-related technologies per capita in South Korea is converging to a higher level. However, the relative transition paths of Hungary and Lithuania are below the panel mean, which indicates that the number of environmental-related technologies in these countries is converging to a lower level.

CONCLUSION

This paper investigates the convergence in environmental-related technologies patents per capita across 36 OECD countries over the period 1991-2019. To do so, the Phillips and Sul (2007;2009) convergence methodology is used, which allows us to analyze overall convergence and identify possible convergence clubs. The club convergence methodology also has other advantages compared to other methods. First, it is not necessary to make any assumptions regarding trend stationarity or stochastic nonstationarity. Second, the log-t regression test solves the problem of biased and inconsistent estimation induced by endogeneity and omitted variables in the augmented Solow regression model (Du, 2017).

Figure 1. Relative transition paths

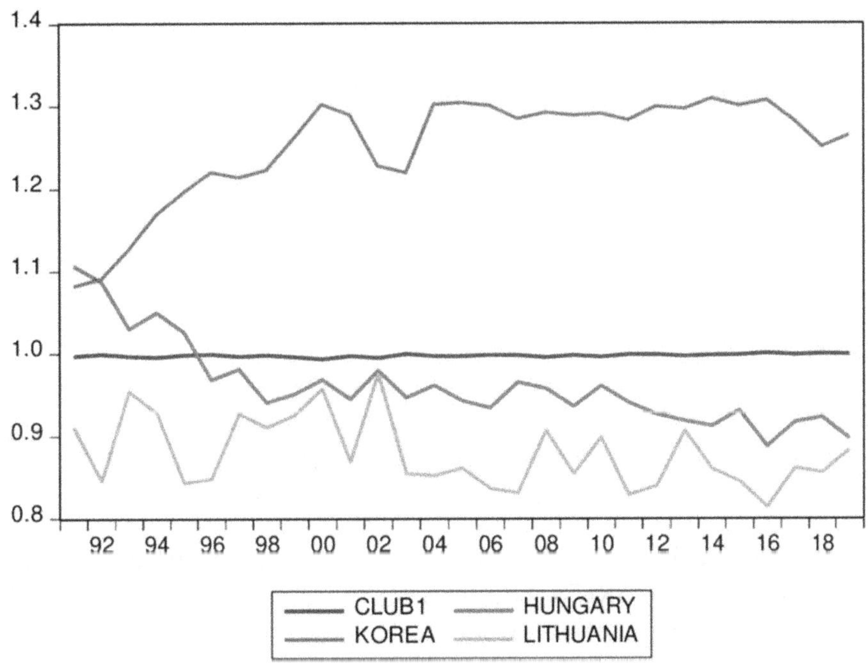

The results of the log-t test show that there are four convergence clubs in terms of environmental-related technologies patents per capita. As proposed by Phillips and Sul (2009), the club clustering algorithm tends to overestimate the true number of clubs. Therefore, the merging test is applied. After the merging test, the results indicate that there is one convergence club which includes 33 out of the 36 OECD countries. The countries located in Club 1 are converging to a single equilibrium point, whereas Hungary, South Korea, and Lithuania are diverging.

Environmental-related technologies are significantly related to environmental degradation. Since environmental-related technologies may help decrease environmental degradation, OECD countries may enhance investment in producing environmental-related technologies. However, OECD does not consist of homogeneous countries in terms of development level, technology, environmental quality, etc. As mentioned above, technological improvement does not necessarily decrease or raise carbon emissions. The effect of technological change on the environment may differ at different levels of technology. Therefore, countries should consider their own characteristics when formulating policies. Furthermore, governments should implement environmental policies and regulations that promote environmental-related technologies.

Disclosure Statement

The author declares no conflict of interest/competing interests.

Funding

There is no funding information.

REFERENCES

Ahmed, K. (2020). Environmental policy stringency, related technological change and emissions inventory in 20 OECD countries. *Journal of Environmental Management, 274*, 111209. doi:10.1016/j.jenvman.2020.111209 PMID:32798847

Danish & Ulucak, R. (2020). How do environmental technologies affect green growth? Evidence from BRICS economies. *Science of the Total Environment, 712*, 136504. doi:10.1016/j.scitotenv.2020.136504

Du, K. (2017). Econometric convergence test and club clustering using Stata. *The Stata Journal, 17*(4), 882–900. doi:10.1177/1536867X1801700407

Erdoğan, S., Yıldırım, S., Yıldırım, D. Ç., & Gedikli, A. (2020). The effects of innovation on sectoral carbon emissions: Evidence from G20 countries. *Journal of Environmental Management, 267*, 110637. doi:10.1016/j.jenvman.2020.110637 PMID:32349957

Gu, W., Zhao, X., Yan, X., Wang, C., & Li, Q. (2019). Energy technological progress, energy consumption, and CO_2 emissions: Empirical evidence from China. *Journal of Cleaner Production, 236*, 117666. doi:10.1016/j.jclepro.2019.117666

Hussain, M., Arshad, Z., & Bashir, A. (2022). Do economic policy uncertainty and environment-related technologies help in limiting ecological footprint? *Environmental Science and Pollution Research International, 29*(31), 1–8. doi:10.100711356-022-19000-9 PMID:35171421

Hussain, M., & Dogan, E. (2021). The role of institutional quality and environment-related technologies in environmental degradation for BRICS. *Journal of Cleaner Production, 304*, 127059. doi:10.1016/j.jclepro.2021.127059

Hussain, M., & Khan, J. A. (2021). The nexus of environment-related technologies and consumption-based carbon emissions in top five emitters: Empirical analysis through dynamic common correlated effects estimator. *Environmental Science and Pollution Research International, 30*(10), 1–10. doi:10.100711356-021-15333-z PMID:34254244

Hussain, M., Mir, G. M., Usman, M., Ye, C., & Mansoor, S. (2022). Analysing the role of environment-related technologies and carbon emissions in emerging economies: A step towards sustainable development. *Environmental Technology, 43*(3), 367–375. doi:10.1080/09593330.2020.1788171 PMID:32579488

Kwon, D. S., Cho, J. H., & Sohn, S. Y. (2017). Comparison of technology efficiency for CO2 emissions reduction among European countries based on DEA with decomposed factors. *Journal of Cleaner Production, 151*, 109–120. doi:10.1016/j.jclepro.2017.03.065

Mensah, C. N., Long, X., Dauda, L., Boamah, K. B., Salman, M., Appiah-Twum, F., & Tachie, A. K. (2019). Technological innovation and green growth in the Organization for Economic Cooperation and Development economies. *Journal of Cleaner Production, 240*, 118204. doi:10.1016/j.jclepro.2019.118204

Organization for Economic Cooperation and Development. (2020). *Environment at a glance indicators*. Retrieved July, 2022, from https://www.oecd.org/environment/environment-at-a-glance/Climate-Change-Archive-June-2020.pdf

Phillips, P. C., & Sul, D. (2007). Transition modeling and econometric convergence tests. *Econometrica, 75*(6), 1771–1855. doi:10.1111/j.1468-0262.2007.00811.x

Phillips, P. C., & Sul, D. (2009). Economic transition and growth. *Journal of Applied Econometrics, 24*(7), 1153–1185. doi:10.1002/jae.1080

Sadiq, M., Wen, F., & Dagestani, A. A. (2022). Environmental footprint impacts of nuclear energy consumption: The role of environmental technology and globalization in ten largest ecological footprint countries. *Nuclear Engineering and Technology, 54*(10), 3672–3681. Advance online publication. doi:10.1016/j.net.2022.05.016

Sahoo, B., Behera, D. K., & Rahut, D. (2022). Decarbonization: Examining the role of environmental innovation versus renewable energy use. *Environmental Science and Pollution Research International, 29*(32), 1–16. doi:10.100711356-022-18686-1 PMID:35199265

Von Lyncker, K., & Thoennessen, R. (2017). Regional club convergence in the EU: Evidence from a panel data analysis. *Empirical Economics*, *52*(2), 525–553. doi:10.100700181-016-1096-2

Yilanci, V., Gorus, M. S., & Aydin, M. (2019). Are shocks to ecological footprint in OECD countries permanent or temporary? *Journal of Cleaner Production*, *212*, 270–301. doi:10.1016/j.jclepro.2018.11.299

Yin, J., Zheng, M., & Chen, J. (2015). The effects of environmental regulation and technical progress on CO2 Kuznets curve: An evidence from China. *Energy Policy*, *77*, 97–108. doi:10.1016/j.enpol.2014.11.008

Chapter 9
Disaggregating Renewable and Nonrenewable Energy Consumption in the Energy Growth Nexis:
Evidence From the Panel Frequence Domain Approach in OECD Countries

Veli Yilanci
Canakkale Onsekiz University, Turkey

Murat Aslan
Independent Researcher, Turkey

Önder Özgür
Independent Researcher, Turkey

ABSTRACT

The environmentalists argue that climate change is a real threat, and therefore, they believe both developed and developing countries should shift their energy sources from fossil-based to renewable alternatives, including solar, wind, geothermal, hydropower, bio-oils, and ocean power to sustain economic growth. In this context, this study investigates the causal relationship between two sets of pairs—(i) renewable energy consumption and economic growth and (ii) nonrenewable energy consumption—and economic growth for 16 OECD countries for the 1971-2015 period. To fulfill this objective, the study employs a recently introduced frequency domain approach in

DOI: 10.4018/978-1-6684-6727-5.ch009

both single-country form and multi-country (panel) form. The study indicates some dissimilarities on renewable and nonrenewable energy consumption and economic growth nexus in OECD countries. Although the results of single country frequency domain test mostly support neutrality hypothesis, multi-country frequency domain test finds strong evidence supporting bidirectional causality among these sets of pairs.

1. INTRODUCTION

After 2nd WW, the economic development through rapid industrialization has caused immense environmental distress. Solid evidence presented by natural scientists showing that the increase in energy use, primarily fossil fuels, has caused extensive irreversible environmental damage and furthermore, the climate change will be expected to cause an enormous long-term threat to national economies and possibly will endanger global ecosystem (Hansen and Skinner, 2005). In order to mitigate and even to reverse the environmental threat, some concrete measures have been taken by international community, including the Kyoto Protocol and Paris Agreement. Improving energy efficiency is one of the operational tools outlined in these agreements for reducing CO2 emission and greenhouse gases alongside promoting energy conservation policies, stimulating low-carbon energy production (renewables) and promoting capture and storage (CCS) technology.[1] Renewable energy-based alternatives are the most probable solution in this matter since they are green, sustainable, and affordable (Bulut and Muratoglu 2018).

OPEC shocks, volatility in oil prices, global climate change and increasing public awareness about environmental problems have significantly influenced the way economists think about the issues related to energy and environment which in turn triggered more research being conducted in economic literature. Understanding and analyzing possible links and/or causality among economic growth, energy use and environmental hazard have been very vibrant research area in the economic field for several reasons. Renewable energy consumption is still limited (i.e. constituting about %15 of global energy market), in spite of having an upward trend, due to its relatively high costs and technological barriers in many countries, while non-renewable energy consumption has been increasing considerably in comparison due mainly to its relatively low costs and ease of operation (Nguyen and Kakinaka 2019). Therefore, in order policymakers to evaluate the usefulness of energy efficiency and conservation policies as well as the viability of alternative forms of energy production, in-depth analysis about the direction of causality among these variables is critical. Volatility, as well as upward trend in energy prices, also have great implications.

In-depth understanding about the links/causality is vital for the analysis of indirect impacts of volatility in energy prices (or fluctuations in energy costs) in the growth process of the nations and their magnitude (Tsani and Menegaki 2018).

The empirical studies examining the links between energy-growth relations cover a broad array of research questions, but the direction of causality has been a focal point since Kraft and Kraft (1978). After this seminal work, a large number of studies using different methodological tools and/or different data coverage have aimed at testing causal relation between economic growth and energy consumption; albeit conflicting results. Based on the results of these studies, four distinct hypotheses have been established; namely, (i) growth hypothesis, (ii) the conservation hypothesis, (iii) the feedback hypothesis, and (iv) neutrality hypothesis (Shiu and Lam, 2004; Squalli, 2007; Ozturk, 2010). Each hypothesis has a distinct connotation regarding energy conservation policies.

According to the growth hypothesis, energy consumption, regarded as a complementary factor to labor and capital in the production process, contributes to economic growth through both direct and indirect channels. Therefore, the growth hypothesis anticipates a unidirectional causality running from energy consumption to economic growth which implies that any energy conservation measures may have adverse effects over economic performance. The conservation hypothesis, on the other hand, postulates that expanded economic activity tends to require more energy which in turn implies that there exists a unidirectional causality running from economic growth to energy consumption and therefore policies with the aim of effectively curbing energy consumption may not lead to deterioration in economic growth performance. Intuitively, the hypothesis is more in tune with countries that are not energy dependent. The feedback hypothesis rests on the mechanism working in a bidirectional manner; that is, economic growth and energy consumption tend to feed each other simultaneously, and therefore causality runs in both directions. For this case, therefore, energy conservation policies may have adverse effects on economic growth. The last hypothesis is the neutrality hypothesis, and it postulates that there is a little or no interdependency between energy consumption and growth. To be more exact, the neutrality hypothesis predicts that energy conservative policies have little or no adverse effects on real economic activity.

According to environmentalists, the climate change is a real threat, and therefore they believe both developed and developing countries should shift their energy sources from fossil-based to renewable alternatives, including solar, wind, geothermal, hydropower, bio-oils and ocean power. International initiatives (i.e., Kyoto Protocol and Paris Agreement) have intensified the debate on the implementation of energy conservative policies and switching to alternative forms of energy production, moving away from carbon-emitting fossil fuels (Tsani and Menegaki 2018). The debate has induced scholars to understand the implications of such actions for economic growth better. Recently, therefore, we have observed that in the empirical field, the focus shifted to investigating the causality between renewable energy consumption and economic growth. This tendency naturally stimulates enthusiasm that looks at energy issues from broader perspective by incorporating both the conventional and renewable energy sources in the causality analysis.

The current study is motivated by two distinct points. Although many studies, using different data sets (covering different countries or regions) and different methodological tools or approaches have investigated the causal links over growth and energy nexus, there are some areas with a potential room for contribution. The existing studies in the empirical field have generally employed the conventional causality tests in a time domain, and the number of studies utilizing the frequency domain approach is very rare. Secondly, although the literature on "conventional" energy and growth nexus are rather large, the renewable energy-based studies are still scarce (Bhattacharya et al. 2016).

This study aims at analyzing the direction of causality between economic growth and energy consumption/renewable energy consumption for 16 OECD countries over the period between 1971-2015. The pairwise causality is tested for two sets: (i) conventional energy consumption and economic growth and (ii) renewable energy consumption and economic growth. In the literature, studies concentrating on the direction of causality between economic growth and renewable energy consumption is very rare and therefore the first contribution of this study would result from incorporating renewable energy in examining the direction of causality among these pairs for OECD countries. In exploring the direction of causality for these pairs, the study utilizes the Granger causality tests in the frequency domain a recently introduced methodological tool. According to our best knowledge, there is no study utilizing frequency domain analysis with three distinct frequency alternatives (short, medium and long term) in addressing the direction of causality between renewable energy and economic growth for OECD countries. Since the pairwise causality is explored by using two different sub-methodological configurations including (i) time domain and frequency domain and (ii) single-country and multi-country (panel), the study allows comparing the results of alternative methodological setups. In other words, using alternative causality models (time and frequency domain analysis) with single and multi-country settings, this study will contribute the existing stock of knowledge by showing the importance of methodological construct in exploring the causality issue.

In order to address energy security as well as climate change problem, most OECD member countries, governments have invested large amounts of public money to support renewable energy development (OECD 2012). Although some studies exist having examined the direction of causality for energy-related variables for OECD countries, we believe that there is room for additional research to be performed using methodological tools equipped with superior and flexible features. The frequency domain causality test is first proposed by Breitung and Candelon (2006) on a single country basis and then developed by Croux and Reusens (2013) for panel data. The conventional time domain causality methods, in general, produce a single test statistic for the entire data span. However, the direction and strength of Granger causality may vary over different frequencies -short-run, medium-run and long-run (Lemmens et al. 2008). That is, unlike the time domain causality test, frequency domain causality test allows us to decompose causality by different time horizons, and hence, could detect causal links between variables at certain cycle lengths even when the time domain causality test may fail to pick up any causality (Gupta et al. 2015). In other words, instead of setting a single measure for the entire relationship, Granger causality test in frequency domain calculates the causality relationship for each individual frequency component separately. Therefore, the frequency domain analysis allows us to differentiate any changes in the direction and/or the strength of the causality in short/medium- and long-term between energy consumption and economic growth for selected OECD countries.

The organization of this article is as follows. In the following section, the selected literature will be reviewed. Section 3 explains the econometric methodology. Section 4 first explains the data and then discusses the empirical findings. Section 5 reviews the conclusion.

2. LITERATURE REVIEW

Since Kraft and Kraft's seminal study in 1978, there is a vast literature aimed at exploring the direction of causality between growth and energy consumption, albeit conflicting results. The use of different econometric methods, the examination of different periods, the phase of economic development, the energy consumption patterns and the heterogeneity of country-specific factors seem to contribute the conflicting results in these studies (Ozturk 2010). Based on our research question, we classify the existing literature into three distinct subcategories. The first category consists of the studies concentrating on the nexus between conventional energy consumption-economic growth; the second category focuses on analyzing the causal link between renewable energy sources and economic growth. The third group includes economic growth, conventional energy consumption, and renewable energy consumption and explores the direction of causality among these pairs.

Table 1. Nomenclature list

Acronyms			
ADF	Augmented Dickey-Fuller	**IPS**	Im-Pesaran-Shin
CADF	Cross-Sectionally ADF	**LM**	Lagrange Multiplier
CD	Cross-Section Dependence	**REN**	Renewable Energy Consumption
CIPS	Cross-Sectionally Augmented IPS	**VAR**	Vector Auto Regression
CO_2	Carbon Dioxide	**VECM**	Vector Error Correction Model
EC	Energy consumption	**TY**	Toda-Yamamoto causality test
EK	Emirmahmutoglu and Kose panel causality test	**Y**	GDP per capita

Much of the existing literature on the causal relationship between energy consumption and economic growth particularly pay attention to conventional fossil energy sources (See Table-2). Of these earlier studies, Yu and Hwang (1980) examine the way of causality in US economy by using Sim's causality test covering the period between 1947 and 1979 and their results confirm no causal relationship between energy consumption and GNP. By extending the period Stern (1993) employs Granger causality test over 1947 to 1990 in the US economy, and the results exhibit unidirectional causality running from energy consumption to GDP.

Altınay and Karagol (2004) examine the case in Turkey over the period between 1950 and 2000. They use Hsiao version of Granger causality test and confirm that there is no evidence of causality between energy consumption and economic growth based on a detrended data. By employing standard Granger causality test, Paul and Bhattacharya (2004) provide a unidirectional causal relation running from energy consumption to economic growth in India during 1950 and 1996. Through a panel cointegration technique, Lee et al. (2008) aim to analyze the key role of the capital stock and applies multivariate panel causality test in 22 OECD countries. This study finds that there is unidirectional causality from energy consumption to capital stock. Narayan and Prasad (2008) analyze the electricity consumption-real GDP causality nexus through a bootstrapped causality test for 30 OECD countries from 1971 to 2000. They conclude that electricity conservation policies negatively affect real GDP in eight of these countries while their empirical findings support neutrality hypothesis for the remaining countries. Ozturk et al. (2010) extend the analysis by covering 51 low-income, lower-middle, and upper-middle-income countries and utilizes panel causality test. Their study concludes that although there is long-run Granger causality running from GDP to energy consumption for low-income countries, there is bidirectional causality between energy consumption and GDP for middle-income countries.

Alam et al. (2011) use the multivariate framework proposed by Toda and Yamamoto (1995) and examine the causal relationship between energy consumptions, CO_2 emissions and economic growth in India covering the period 1971 and 2006. Empirical findings of their study imply that there is a bidirectional Granger causality between energy consumption and CO2 emissions in the long-run. However, their study does not provide supportive evidence for the causal relationship between energy consumption and real income. By employing the more recent technique, Dergiades et al. (2013) consider non-linearity and perform both linear and non-linear causality analysis. Their empirical findings support both linear and nonlinear causal link between energy consumption and economic growth in Greece.

Pala (2016) uses panel Granger causality and panel vector error correction model (VECM) from 1995 to 2013 in OECD countries. This study suggests that there is bi-directional causality between energy consumption and economic growth in the short-term whereas empirical findings do not provide supportive evidence for long-run causality. More recently, Ranjbar et al. (2017) examine symmetric and asymmetric causal relationships between energy consumption and economic growth in time and frequency domains in South Africa between 1965 and 2012. They suggest that the asymmetric effects of energy consumption on economic growth do not differ. They find that positive or negative changes in energy consumption neither Granger cause real GDP, nor real GDP growth. Furthermore, their findings through time and frequency domains imply that the causality is not stable, implying that relationship changes over time.

The second strand of the literature on the causality issue is carried out on causality between renewable energy and economic growth. In their sectoral analysis of renewable energy consumption economic growth nexus, Bowden and Payne (2010) provide supportive evidence of unidirectional causality in residential nonrenewable energy consumption and real output level and no causal relationship in commercial and industrial sectors in US economy.

Apergis and Payne (2011) use panel VECM from 1980 to 2006 in six Central American countries. The results reported by the study suggest that there is bi-directional causality between energy consumption and economic growth both in the short-run and long-run. Considering the nonlinearities in the series, Nazlıoglu et al. (2014) utilize both linear and non-linear Granger causality test to identify the causal relationship between electricity consumption and economic growth in Turkey over 1967 to 2007. Their findings for the linear causality test reveals bidirectional causality both in the short-run and long-run while nonlinear Granger causality analysis supports the neutrality hypothesis. Alper and Oguz (2016) utilize asymmetric Granger causality test for eight new European Union member countries over 1990 to 2009. Empirical findings of this study present that there is no causal relationship between renewable energy consumption and economic growth in Cyprus, Estonia, Hungary,

Poland, and Slovenia whereas an increase in economic growth causes an increase in energy consumption in the Czech Republic. The only country that implies a causal relationship running from energy consumption to economic growth in Bulgaria.

In the third strand of the literature, there exist studies performing empirical analysis to investigate the causal relationship between both renewable and conventional energy consumption and economic growth. Of these studies, Apergis and Payne (2012) utilize short-run VECM, and long-run causality test in 80 countries and empirical findings of the panel error correction model exhibits bidirectional causality between renewable and non-renewable energy consumption and economic growth both in the short-run and long-run. Also, they find a bidirectional short-run causality between renewable and nonrenewable energy consumption which represents the substitutability of the energy sources. Similarly, Tugcu et al. (2012) employ an asymmetric causality test in exploring the relative importance of renewable and non-renewable energy sources on economic growth in G7 countries. Empirical findings provide evidence of bidirectional causality for all countries in case of the classical production function. More recently, Destek and Aslan (2017) also analyze the relative performance of renewable and non-renewable energy consumption on economic growth in 17 emerging countries over 1980-2012. They use bootstrap panel causality that allows both cross-sectional dependency and country-specific heterogeneity across countries. Their findings provide supportive evidence for each hypothesis on energy consumption and growth nexus and present mixed results in each country.

The overall empirical results on energy-growth nexus suggest that there is no single pattern of causality among these variables because (a) the causality among these variables is very complex, (b) heterogeneity; that is different countries may possess different conditions, depending for example, on their stage of development, their industrial base, legal structure on environmental issues, and (c) different methodological tools or time periods that a particular study employs.

3. DATA AND ECONOMETRIC METHODOLOGY

3.1. Data

In this study, the yearly data consisting of 16 OECD countries (Australia, Austria, Canada, Chile, Denmark, Finland, France, Germany, Greece, Iceland, Israel, Mexico, Portugal, Sweden, Turkey, and the USA) covers the period 1971-2015 are used. The data utilized in this study include three variables:

Table 2. Review of the selected studies

Authors	Period	Country	Causality Relationship
Conventional Energy Consumption and Economic Growth			
Yu and Hwang (1980)	1947-1979	US	GDP- - -EC
Stern (1993)	1947-1990	US	EC→GDP
Altınay and Karagol (2004)	1950-2000	Turkey	GDP- - -EC
Paul and Bhattacharya (2004)	1950-1996	India	EC→GDP
Lee et al. (2008)	1960-2001	22 OECD Countries	GDP- - -EC
Narayan et al. (2008)	1971-2000	30 OECD Countries	Mixed results
Ozturk et al. (2010)	1971-2005	51 Countries	Mixed results
Alam et al. (2011)	1971-2006	India	GDP- - -EC
Dergiades (2013)	1960-2008	Greece	EC→GDP
Pala (2016)	1995-2013	OECD Countries	EC→GDP (SR) GDP- - -EC (LR)
Ranjbar et al. (2017)	1965-2012	South Africa	Mixed in time and frequency domains
Renewable Energy Consumption and Economic Growth			
Bowden and Payne (2010)	1949-2006	US	Mixed with respect to sectors
Apergis and Payne (2011)	1980-2006	6 Central American Countries	REN→GDP (SR and LR)
Nazlıoglu et al. (2014)	1967-2007	Turkey	REN→GDP (Linear Test) GDP→REN (Linear Test) GDP- - -REN (Nonlinear)
Alper and Oguz (2016)	1990-2009	New EU Countries	Mixed results
Conventional and Renewable Energy Consumption and Economic Growth			
Apergis and Payne (2012)	1990-2007	80 Countries	EC(REN)→GDP GDP→EC(REN) EC→REN REN→EC
Tugcu et al. (2012)	1980-2009	G7 Countries	REN→GDP GDP→REN
Destek and Aslan (2017)	1980-2012	17 Emerging Countries	Mixed results

Note: The "→" represents the direction of causality. "- - -" represents no causality.

(i) Conventional energy consumption per capita (EC) measured in kilotons of oil equivalent
(ii) the renewable energy consumption per capita (REN) measured in kilotons of oil equivalent
(iii) Real GDP per capita (Y) in constant 2010 US dollars.

EC and REN are obtained from International Energy Agency (IEA) Global Energy Data, and GDP comes from the World Bank-World Development Indicators (WDI). All the variables are transformed into natural logarithm form before exposing them to econometric tests and analysis, and the above abbreviations are modified this way: lnY, lnEC, and lnREN.

3.2. Econometric Methodology

The empirical studies in energy literature generally mainly use causality tests in time-domain to examine the causality between energy consumption and economic growth. In this study, we implement both the causality test in time and frequency domain. Moreover, for each case, we carry out both the single-country test and the multi-country (panel) test. The assumptions and outline of frequency domain causality test procedure for individual and multi-country settings are explained below.

The single country causality test in the frequency domain is based on the conventional VAR model. Let's consider the following vector autoregressive model (VAR(p)):

$$X_t = \sum_{i=1}^{p} \beta_{11,i} X_{t-i} + \sum_{i=1}^{p} \beta_{12,i} Y_{t-i} + \varepsilon_{1t} \quad (1)$$

$$Y_t = \sum_{i=1}^{p} \beta_{21,i} X_{t-i} + \sum_{i=1}^{p} \beta_{22,i} Y_{t-i} + \varepsilon_{2t} \quad (2)$$

To test the null of no causality from Y to X at frequency w by using Model 1, we test the following linear restrictions by a standard F test that has a distribution as $F(2, T-2p)$ where p shows the lag length of VAR and T indicates the number of observations:

$$\sum_{i=1}^{p} \beta_{12,i} \cos(iw) = 0, \quad (3)$$

$$\sum_{i=1}^{p} \beta_{12,i} \sin\left(iw\right) = 0. \tag{4}$$

Croux and Reusens (2013) show that incremental R^2, that is the difference between the R^2 of unrestricted and restricted models can be employed as a measure of causality and the null can be tested at the level α when

$$Incremental\ R^2 > F_{(2,T-2p,1-\alpha)} \frac{2}{T-2p}\left(1-R^2\right) \tag{5}$$

Where $F_{(2,T-2p,1-\alpha)}$ shows the α upper critical value of F.

The panel data version of frequency domain causality analysis is formulated by Croux and Reusens (2013). In order to incorporate frequency domain analysis into panel setting, they estimate the following seemingly unrelated regression equations:

$$X_{j,t} = \sum_{i=1}^{p} \alpha_{i,j} X_{j,t-i} + \sum_{i=1}^{p} \beta_{i,j} Y_{j,t-i} + \varepsilon_{j,t} \quad \text{for } j=1,\ldots,G \tag{6}$$

where $X_{j,t}$ and $Y_{t,t}$ are the variables of the country j at time t and G shows the number of countries. The null hypothesis of no causality from X_t to Y_t can be tested, examining the following restrictions:

$$\sum_{i=1}^{p} \beta_{i,j} \cos\left(iw\right) = 0 \quad \text{for } j=1,\ldots,G \tag{7}$$

$$\sum_{i=1}^{p} \beta_{i,j} \sin\left(iw\right) = 0 \quad \text{for } j=1,\ldots,G \tag{8}$$

These restrictions can also be tested using *Incremental* $R^2(R^2\text{-}R^{2*})$; where R^2 and R^{2*} are the McElroy R squared values of the Unrestricted and restricted SUR model). The null is rejected at the level α if,

$$Incremental\ R^2 > F_{(2G,G(T-2p),1-\alpha)} \frac{2G}{G\left(T-2p\right)}\left(1-R^2\right) \tag{9}$$

Here, $F_{(2G,G(T-2p),1-\alpha)}$ shows the α critical value of F distribution.

4. EMPIRICAL RESULTS AND DISCUSSION

Before implementing causality tests, we determine the integration levels of the variables using unit root tests. However, in the case of existence of cross-sectional dependence, the usage of the first-generation panel unit root tests can produce biased results. So, we first test the existence of cross-sectional dependence and the results of the tests are presented in Table 3.

Table 3. Cross-sectional dependence test results

	Breusch-Pagan LM	Pesaran-Scaled LM	Bias-Corrected Scaled LM	Pesaran CD
lnY	4936.683	309.882	309.701	70.211
	(0.00)	(0.00)	(0.00)	(0.00)
lnEC	2493.382	152.168	151.986	15.669
	(0.00)	(0.00)	(0.00)	(0.00)
lnREN	2225.569	134.881	134.699	3.970
	(0.00)	(0.00)	(0.00)	(0.00)

Note: The numbers in parentheses show the p-values.

The results in Table 3 show that there exists cross-sectional dependency in all variables in the analysis; and therefore, the panel unit root test(s) with second-generation characteristics is more appropriate for the data. Due to the presence of cross-sectional dependency in the panel, we carry out cross-sectionally augmented Dickey-Fuller (CADF), and the cross-sectionally augmented Im, Pesaran and Shin (or CIPS) test. The results of both CADF and CIPS are reported in Table-4.

The results of the CIPS panel unit root test show that only lnEC is stationary at level whereas both lnREN and lnY are difference-stationary. On the other hand, the CADF unit root tests which help us to determine the integration levels of the variables individually, and it shows that for the majority of the countries both Model-A (trend only specification) and Model-B (level and trend specification) generate presence of unit root for all the variables. Note also that for both Model-A and Model-B the first difference of both lnEC and lnREN tend to revert back to their respective long-term trend; that is, the first difference in these variables are stationary. However, the results for lnY is not very clear. For Finland, Germany, Greece, and Portugal, CADF test with alternative specifications (Model-A and Model-B and also in level and first difference) was not able to find a clear mean reversion. For Chile, France, and Sweden, CADF test for Y detects mixed results.

Table 4. CADF unit root test results

CADF UNIT ROOT TEST												
	lnEC				lnREN				lnY			
	Model-A		Model-B		Model-A		Model-B		Model-A		Model-B	
Series	**lnEC**	**ΔlnEC**	**lnEC**	**ΔlnEC**	**lnREN**	**ΔlnREN**	**lnREN**	**ΔlnREN**	**lnY**	**ΔlnY**	**lnY**	**ΔlnY**
Australia	-2.67	-6.58**	-2.77	-6.46**	-4.96**	-8.37**	-5.26**	-8.29**	-2.02	-6.35**	-3.08	-6.26**
Austria	-3.73*	-7.37**	-4.36*	-7.32**	-2.26	-6.79**	-2.36	-6.65**	-0.92	-5.02**	-1.61	-5.74**
Canada	-1.74	-6.85**	-1.73	-7.27**	-2.56	-5.58**	-3.58	-6.03**	-2.29	-4.08**	-2.54	-4.01*
Chile	-2.29	-4.94**	-1.39	-5.34**	-1.62	-6.03**	-2.26	-6.08**	-2.88	-3.56*	-3.61	-3.76
Denmark	0.51	-5.13**	-1.45	-5.12**	-2.53	-5.22**	-3.68	-5.03**	-1.77	-4.14*	-2.05	-5.06**
Finland	-1.85	-7.69**	-2.92	-7.54**	-1.96	-6.29**	-2.54	-6.31**	-1.66	-3.02	-2.49	-3.06
France	-2.44	-6.48**	-2.37	-6.39**	-1.06	-6.71**	-3.82*	-6.65**	-1.39	-3.81*	-2.00	-4.31*
Germany	0.12	-5.24**	-1.74	-5.23**	-4.49**	-6.16**	-4.27*	-6.08**	-1.95	-3.26	-2.45	-3.23
Greece	-2.60	-5.82**	-2.35	-5.86**	-1.48	-3.84*	-2.39	-4.09*	-1.83	-3.06	-3.60	-3.07
Iceland	-2.37	-5.94**	-1.83	-6.27**	-1.17	-6.59**	-3.24	-6.51**	-1.29	-5.58**	-1.16	-5.77**
Israel	-0.57	-5.12**	-1.77	-5.11**	-3.05	-5.59**	-3.00	-5.54**	-1.87	-4.56**	-2.12	-4.48*
Mexico	-0.38	-6.52**	-2.85	-6.69**	-2.31	-6.18**	-2.09	-5.65**	-2.20	-4.90**	-2.28	-5.49**
Portuguese	-2.69	-3.81*	-1.77	-4.09*	-2.43	-5.68**	-2.78	-6.84**	-1.44	-3.05	-1.60	-3.23
Sweden	-1.93	-7.45**	-2.02	-7.39**	-1.50	-6.56**	-3.42	-6.46**	-1.54	-3.62*	-1.76	-3.56
Turkey	-0.52	-6.77**	-3.11	-6.86**	-0.42	-8.29**	-4.16*	-8.15**	-0.65	-5.72**	-2.41	-5.81**
USA	-1.72	-3.99*	-1.94	-4.03*	-2.57	-5.17**	-2.50	-5.77**	-1.56	-4.01**	-1.02	-4.27*
CIPS Panel Unit Root Test												
Panel	-1.68	-5.98*	-2.27	-6.06*	-2.27**	-6.19*	-3.22*	-6.26*	-1.97	-4.24*	-2.24	-4.45*

Notes:

(1): Model-A is CADF test incorporating level variable only while Model-B is CADF test incorporating both level and trend variable.

(2) Critical values for CIPS test with intercept case are -2.36, -2.2, -2.11 and with the intercept and trend case are -2.85, -2.71 and -2.63 for the 1%, 5%, 10% levels respectively.

(3) Critical values for the CADF test with intercept case are; -3.99, 3.28, 2.94, and with the intercept-only model (Model-A) and for intercept and trend case (Model-B) are -4.52, -3.79 and -3.4463 for the 1%, 5%, 10% levels respectively. * and ** denotes the rejection of the null of a unit root at the 1% and 5% levels respectively.

Having established the unit root characteristics of the data, in the second step we examine the direction of the causality between two sets of our pairs: (i) lnY-lnEC and (ii) lnY-lnREN, in both time domain and frequency domain.

We start with individual country-based time domain approach; in doing so, we apply Toda-Yamamoto (TY) causality test. An advantage of the Toda-Yamamoto test is that it is not necessary to difference the series to make them stationary. We first, determine the VAR model lag (m) then add the extra lags considering the maximum integration levels of the variables(dmax) and estimate the VAR(m+dmax) with the levels and then implement F-constraint test for the first m lags. The results of the TY causality test are reported in Table 5.

Table 5. Individual (TY) and panel (EK) causality tests

	H_0: lnREN ↛ lnY	H_0: lnY ↛ lnREN	H_0: lnEC ↛ lnY	H_0: lnY ↛ lnEC
Australia	0.168	1.054	0.839	1.623
	(0.682)	(0.305)	(0.360)	(0.203)
Austria	0.565	0.590	2.845*	0.051
	(0.452)	(0.443)	(0.092)	(0.822)
Canada	0.033	0.016	0.480	1.159
	(0.855)	(0.900)	(0.488)	(0.282)
Chile	1.436	0.116	1.854	0.016
	(0.231)	(0.733)	(0.173)	(0.900)
Denmark	1.902	0.938	1.704	0.29
	(0.593)	(0.816)	(0.192)	(0.590)
Finland	0.087	0.012	0.853	4.432
	(0.768)	(0.914)	(0.837)	(0.218)
France	0.023	0.175	1.795	0.004
	(0.879)	(0.676)	(0.180)	(0.949)
Germany	4.135*	2.561	0.008	1.162
	(0.042)	(0.109)	(0.927)	(0.281)
Greece	0.350	0.165	2.607	19.485*
	(0.554)	(0.684)	(0.626)	(0.001)
Iceland	0.491	1.060	8.872*	3.616
	(0.483)	(0.303)	(0.031)	(0.306)
Israel	0.357	0.906	1.448	0.239
	(0.550)	(0.341)	(0.229)	(0.625)

continues on following page

Table 5. Continued

	H_0: lnREN ↛ lnY	H_0: lnY ↛ lnREN	H_0: lnEC ↛ lnY	H_0: lnY ↛ lnEC
Mexico	0.283	0.928	6.033*	0.541
	(0.595)	(0.335)	(0.014)	(0.462)
Portuguese	1.454	4.907*	4.428	8.115*
	(0.228)	(0.027)	(0.219)	(0.044)
Sweden	0.120	0.046	0.673	1.055
	(0.729)	(0.830)	(0.412)	(0.304)
Turkey	0.427	0.013	0.001	0.709
	(0.514)	(0.908)	(0.974)	(0.400)
USA	1.323	1.041	0 .000	0.186
	(0.250)	(0.308)	(0.999)	(0.666)
Panel EK Causality Test				
Test Statts → **Critical Values ↓**	26.329	28.28	43.231	43.946
1%	61.284	60.473	64.777	62.86
5%	50.971	50.635	52.464	51.829
10%	46.208	45.931	47.234	46.694

Note: Number in parentheses Show the p-value. * denotes the significance at the 10% level.

On the basis of the individual results between lnEC and lnY, 11 out of 16 series are in line with the neutrality hypothesis; no statistically significant causality is detected for these countries. Moreover, the feedback hypothesis is not valid for none of these 16 countries. Unidirectional causality running from lnEC to lnY is detected for Austria, Iceland, and Mexico suggesting that conservation policy may cause some adverse consequences for these countries. The reverse causality running from lnY to lnEC seems to be in tune with Greece and Portugal.

The time domain causality test for the pair of lnREN and lnY detects unidirectional causality only for 2 out of 16 countries. The unidirectional causality running from lnREN to lnY is detected only for Germany while the reverse causality (that is unidirectional causality running from lnY to lnREN) is relevant only for Portugal. Moreover, there is no evidence in support of the feedback hypothesis between renewable lnREN and lnY. In other words, on the basis of the results of time domain analysis, the causal links between lnREN and lnY are quite weak, majority of our sample is in support of naturality hypothesis.

Having established fairly weak unidirectional or bidirectional causal links over the pairs by using time domain individual test, this study then proceeds to apply a panel causality test of Emirmahmutoglu and Kose (2011) (EK hereafter). Note that EK is a panel test compatible with time domain configuration. By applying a time domain panel causality test, the current study aims at providing an additional benchmark for comparison.

We use EK test by combining the p-values of TY causality test via the Fisher method:

$$EK = -2\sum_{j=1}^{G} \ln\left(p_j\right) \tag{10}$$

where p_j shows the p-value of TY causality test for the jth country. We test the null hypothesis of non-causality. We obtain the critical values of the EK test via bootstrap simulations.

The results of the EK test are presented at the bottom of Table-5. Similar to the results of the single country causality test with time domain property, the results of the EK panel causality test are also in favoring naturality hypothesis. In other words, the results of both single country causality test and panel causality test in time domain analysis seem to support neutrality hypothesis mostly.

Finally, the frequency domain causality test is utilized. In this step, the direction of causality among the pairs is analyzed at three distinct frequencies: 0.5, 1.5 and 2.5. While the frequency level 0.5 represents long term (permanent causality), the frequency level of 2.5 is regarded as short term and is interpreted as temporary causality, and between these two extremes, the frequency level of 1.5 represents a medium (or intermediate) run causality.

The results of both single country and multi-country Granger causality in the frequency domain tests for the pair of lnREN and lnY are presented at Table-6. Single country causality in the frequency domain test seems to pick up slightly more unidirectional cases relative to those of the time-frequency test. The test produces results providing evidence of one-way causality from lnREN to lnY in short and middle-term for Austria, in middle-term for Iceland, and long-term for Germany, Portugal, and Sweden. In other words, renewable energy stimulating policies may contribute to economic growth performance for these countries. On the other spectrum, we find only temporary (or short term) causal relationship from lnY to lnREN for France. Similar to the results we obtained from time-frequency analysis, no bidirectional causality is found in the single-country analysis. Overall, the results for single country Granger causality in the frequency domain for lnY and lnREN are mainly in supporting neutrality hypothesis.

When compared the results of individual country time domain causality test with the results of frequency domain test for lnY and lnREN, we notice slight differences. The number of unidirectional causality cases detected in frequency domain analysis is slightly more than those of time domain analysis. In terms of the unidirectional causality running from lnY to lnREN, both time and frequency domain tests pick up only one country, but each test identifies different country: Portugal for the former and France for the latter. However, in terms of the direction of causality running from lnREN to lnY, the number of cases identified by the frequency domain test is more than that of time-frequency test. Put it differently, although the time domain analysis finds only Germany, the frequency domain approach identifies two or three cases (depending on the selection of frequency). Specifically, the results favoring the growth hypothesis detected by the frequency domain approach are: (i) for the long run three cases (Germany, Portugal, and Sweden); (ii) for the medium run two cases (Australia and Iceland) and (iii) for short term one case (Australia).

Table 6. Frequency domain test results between REN and Y

	H_0: lnREN ↛ lnY				**H_0: lnY ↛ lnREN**			
	LR w=0.5	**MR w=1.5**	**SR w=2.5**	**c.v. %10**	**LR w=0.5**	**MR w=1.5**	**SR w=2.5**	**c.v. %10**
Australia	0.1144	0.1281*	0.1579*	0.1170	0.0337	0.0461	0.0168	0.1301
Austria	0.0005	0.0005	0.0014	0.0056	0.0961	0.0336	0.0270	0.1179
Canada	0.0578	0.0261	0.0573	0.1194	0.0097	0.0051	0.0037	0.1269
Chile	0.0082	0.0017	0.0009	0.0959	0.0177	0.0140	0.0014	0.1259
Denmark	0.0019	0.0109	0.0096	0.0754	0.0347	0.0221	0.0074	0.1276
Finland	0.1039	0.0448	0.0103	0.1224	0.0173	0.0058	0.0065	0.1015
France	0.0165	0.0106	0.0440	0.1180	0.0524	0.0888	0.1393*	0.1104
Germany	0.1218*	0.0489	0.0562	0.1202	0.0687	0.0817	0.0406	0.1067
Greece	0.0031	0.0051	0.0053	0.1355	0.0005	0.0004	0.0005	0.0035
Iceland	0.0702	0.1345*	0.1154	0.1211	0.0697	0.0771	0.0185	0.1019
Israel	0.0301	0.0137	0.0173	0.1358	0.0324	0.0750	0.0277	0.1246
Mexico	0.0350	0.0326	0.0091	0.1279	0.0535	0.0399	0.0195	0.1258
Portuguese	0.1334*	0.0532	0.0865	0.1072	0.0479	0.0609	0.0499	0.0887
Sweden	0.1279*	0.0943	0.0407	0.1127	0.0213	0.0263	0.0137	0.1213
Turkey	0.0719	0.0634	0.0138	0.1060	0.0690	0.0648	0.0248	0.1289
USA	0.0065	0.0004	0.0066	0.0951	0.0352	0.0352	0.0354	0.1153
Panel	0.1263*	0.1371*	0.1031*	0.0477	0.1182*	0.0361	0.0682*	0.0447

Notes: * denotes the rejection of the null hypothesis at least at 10% significance level.
SR: Short Run, MR: Medium Run, LR: Long Run.

Nevertheless, the major differences between time and frequency domain are due to panel level analysis. While the results of the test for the overall panel for the time domain analysis support neutrality hypothesis between lnREN and lnY, the results of frequency domain test suggest that there exists statistically significant bi-directional causality between the two. In other words, by the results we obtain from the multi-country level analysis in the frequency domain approach for lnREN and lnY, we infer that both renewable energy consumption and economic growth drive each other.

The results of Granger causality in the frequency domain between lnEC and lnY (in both single country level and multi-country level) are displayed in Table 7. Overall, the results of the single country test for alternative frequencies seem to be in support of neutrality hypothesis. In each frequency level, roughly two-thirds of the results are in line with the neutrality hypothesis. The results of the individual test indicate unidirectional causality from lnEC to lnY (or growth hypothesis) in the middle and short term for Finland, in the middle -, and long-term Greece, in short, and long term for Portugal. There also exists unidirectional causality from lnY to lnEC in middle-, and short term in Denmark, in short, middle-, and long-term for Iceland and in the short term for Mexico. Finally, the feedback hypothesis is supported for middle-term frequency for Mexico. Overall, as in the previous tests for single country level, the single country analysis in the frequency domain approach for lnEC and lnY, in general, produces results mainly in support of the neutrality hypothesis.

The results of panel causality test in frequency domain show that there exists bidirectional causality between lnEC and lnY. The results of Granger causality test in frequency domain for the whole panel have some important implications, particularly for policymakers. Better economic performance needs more energy, and at the same time, more energy consumption may stimulate economic growth. For our 16 OECD countries, the results suggest that energy consumption and economic growth seem to complement each other and therefore energy conservation measures may adversely affect economic growth. However, the existence of the bidirectional causality between lnREN and lnY identified by the frequency domain panel analysis (see Table-6) provide an important alternative and quite suitable policy recommendation. The existence of bidirectional causality between lnREN and lnY suggest that higher renewable energy consumption may induce growth and vice versa. Therefore, these countries may be recommended to implement measures stimulating the production of renewable energy sources.

Table 7. Frequency domain test results (EC and GDP)

	H_0: lnEC ↛ lnY				H_0: lnY ↛ lnEC			
	LR w=0.5	**MR w=1.5**	**SR w=2.5**	**c.v. %10**	**LR w=0.5**	**MR w=1.5**	**SR w=2.5**	**c.v. %10**
Australia	0.0015	0.0039	0.0023	0.0081	0.0063	0.0116	0.0124	0.1343
Austria	0.0003	0.0013	0.0044	0.0113	0.1191	0.0580	0.0429	0.1145
Canada	0.0116	0.0018	0.0060	0.0522	0.0202	0.0192	0.0173	0.1252
Chile	0.0008	0.0009	0.0009	0.0023	0.0642	0.0822	0.0826	0.1167
Denmark	0.0041	0.0033	0.0007	0.0319	0.0125	0.2059*	0.2130*	0.1026
Finland	0.0069	0.0171*	0.0162*	0.0165	0.0399	0.0209	0.0162	0.0988
France	0.0070	0.0289	0.0227	0.0321	0.0144	0.0372	0.0314	0.1254
Germany	0.0144	0.0009	0.0184	0.0187	0.0877	0.0306	0.0189	0.1059
Greece	0.0092*	0.0084*	0.0080*	0.0035	0.0012	0.0002	0.0002	0.0034
Iceland	0.0003	0.0005	0.0012	0.0040	0.1848*	0.1779*	0.1765*	0.0875
Israel	0.0004	0.0001	0.0010	0.0087	0.0168	0.0573	0.0584	0.1269
Mexico	0.0066	0.0196*	0.0160	0.0161	0.0250	0.1422*	0.1610*	0.0943
Portuguese	0.0023*	0.0008	0.0021*	0.0012	0.0296	0.0011	0.0316	0.0722
Sweden	0.0030	0.0027	0.0053	0.0199	0.0306	0.0435	0.0401	0.1185
Turkey	0.0008	0.0004	0.0001	0.0054	0.0020	0.0045	0.0052	0.1372
USA	0.0006	0.0029	0.0089	0.0171	0.0043	0.0020	0.0017	0.1196
Panel	0.01195	0.0165*	0.0106*	0.0028	0.082*	0.1041*	0.1038*	0.0467

Notes: * denotes the rejection of the null hypothesis at least at 10% significance level.
SR: Short Run, MR: Medium Run, LR: Long Run.

5. CONCLUSION

This study analyzes the direction of causality for two sets of pairs: (i) the economic growth-conventional energy consumption and (ii) economic growth and renewable energy consumption for 16 OECD countries during 1971-2015. In examining the causality nexus among these pairs, this study adopts a recently developed frequency domain causality method. In order to address the direction of causality among these pairs, this study utilizes both the single country and the multi-country Granger causality tests in the frequency domain. Furthermore, in order to compare our results, the study also utilizes the conventional (time domain) Granger causality tests in both single country level and panel level.

The results of both country level and panel level causality tests in time domain analysis for our variables (economic growth, energy consumption, and renewable energy consumption) are in general favoring neutrality hypothesis. The number of cases for either unidirectional or bidirectional causality identified by the time domain analysis for single country analysis is very few. We also fail to find any evidence supporting the unidirectional or bidirectional causality between the pairs in the multi-country level time domain causality test. Therefore, for our sample, the time domain analysis generally suggests that these pairs do not have causal links.

We find mixed results for frequency domain analysis. According to the results of single country frequency domain causality test, the number of unidirectional and bidirectional causality cases is slightly more than those of time domain analysis. However, the overall results for single country analysis for frequency domain causality test still favor neutrality hypothesis. Although the results vary concerning the selection of specific frequency parameter level and with respect to our pairs, around %80-90 of the results are in favoring no-causality.

The results of multi-country (or panel) frequency domain causality tests for our pairs are quite different. Although time-domain approach in both single-country and multi-country levels support neutrality hypothesis in two pairs (i.e., lnEC-lnY and lnREN-lnY), the frequency domain approach in panel level produces statistically significant bidirectional causality between both lnEC-lnY and lnREN-lnY. In other words, higher economic growth tends to stimulate energy consumption and simultaneously higher energy consumption cause better economic performance. Furthermore, the results of panel frequency domain test show that higher economic growth may induce renewable energy consumption and therefore these two variables may complement each other. Overall assessment of our results in panel level suggests that countries may need more energy for their economic progress, but they can implement some energy friendly measures aimed at promoting use of renewable energy sources.

For almost all countries, energy will remain an important issue for years to come. Energy efficiency and the use of renewable energy sources will be pivotal points in the delicate balance between economic growth and environment. This study finds mixed results in addressing the causal nexus between economic growth and conventional energy consumption and renewable energy consumption. The factors influencing the production and consumption of both energy and renewable energy are very diverse, and we believe that further studies should concentrate on the details, that is they should endogenize the factors in both supply and demand side of the renewable energy. We also believe that the transmission channels (the mechanisms that the consumption of energy and renewable energy are transmitted to economic growth or the mechanisms that economic growth is transmitted towards consumption of these sources) are also an important area for further research.

REFERENCES

Alam, M. J., Begum, I. A., Buysse, J., Rahman, S., & Van Huylenbroeck, G. (2011). Dynamic modeling of the causal relationship between energy consumption, CO2 emissions and economic growth in India. *Renewable & Sustainable Energy Reviews*, *15*(6), 3243–3251. doi:10.1016/j.rser.2011.04.029

Alper, A., & Oguz, O. (2016). The role of renewable energy consumption in economic growth: Evidence from asymmetric causality. *Renewable & Sustainable Energy Reviews*, *60*, 953–959. doi:10.1016/j.rser.2016.01.123

Altinay, G., & Karagol, E. (2004). Structural break, unit root, and the causality between energy consumption and GDP in Turkey. *Energy Economics*, *26*(6), 985–994. doi:10.1016/j.eneco.2004.07.001

Apergis, N., & Payne, J. E. (2011). The renewable energy consumption–growth nexus in Central America. *Applied Energy*, *88*(1), 343–347. doi:10.1016/j.apenergy.2010.07.013

Apergis, N., & Payne, J. E. (2012). Renewable and non-renewable energy consumption-growth nexus: Evidence from a panel error correction model. *Energy Economics*, *34*(3), 733–738. doi:10.1016/j.eneco.2011.04.007

Bhattacharya, M., Paramati, S. R., Ozturk, I., & Bhattacharya, S. (2016). The effect of renewable energy consumption on economic growth: Evidence from top 38 countries. *Applied Energy*, *162*, 733–741. doi:10.1016/j.apenergy.2015.10.104

Bowden, N., & Payne, J. E. (2010). Sectoral analysis of the causal relationship between renewable and non-renewable energy consumption and real output in the US. *Energy Sources. Part B, Economics, Planning, and Policy*, *5*(4), 400–408. doi:10.1080/15567240802534250

Breitung, J., & Candelon, B. (2006). Testing for short- and long-run causality: A frequency-domain approach. *Journal of Econometrics*, *132*(2), 363–378. doi:10.1016/j.jeconom.2005.02.004

Bulut, U., & Muratoglu, G. (2018). Renewable energy in Turkey: Great potential, low but increasing utilization, and an empirical analysis on renewable energy-growth nexus. *Energy Policy*, *123*, 240–250. doi:10.1016/j.enpol.2018.08.057

Croux, C., & Reusens, P. (2013). Do stock prices contain predictive power for the future economic activity? A Granger causality analysis in the frequency domain. *Journal of Macroeconomics*, *35*, 93–103. doi:10.1016/j.jmacro.2012.10.001

Dergiades, T., Martinopoulos, G., & Tsoulfidis, L. (2013). Energy consumption and economic growth: Parametric and non-parametric causality testing for the case of Greece. *Energy Economics*, *36*, 686–697. doi:10.1016/j.eneco.2012.11.017

Destek, M. A., & Aslan, A. (2017). Renewable and non-renewable energy consumption and economic growth in emerging economies: Evidence from bootstrap panel causality. *Renewable Energy*, *111*, 757–763. doi:10.1016/j.renene.2017.05.008

Emirmahmutoglu, F., & Kose, N. (2011). Testing for Granger causality in heterogeneous mixed panels. *Economic Modelling*, *28*(3), 870–876. doi:10.1016/j.econmod.2010.10.018

Gupta, R., Gil-Alana, L. A., & Yaya, O. S. (2015). Do sunspot numbers cause global temperatures? Evidence from a frequency domain causality test. *Applied Economics*, *47*(8), 798–808. doi:10.1080/00036846.2014.980575

Hansen, C., & Skinner, R. (2005). *Nuclear Power and Renewables: Strange Bedfellows?* Oxford Institute for Energy Studies.

Kraft, J., & Kraft, A. (1978). On the relationship between energy and GNP. *The Journal of Energy and Development*, 401-403.

Lee, C. C., Chang, C. P., & Chen, P. F. (2008). Energy-income causality in OECD countries revisited: The key role of capital stock. *Energy Economics*, *30*(5), 2359–2373. doi:10.1016/j.eneco.2008.01.005

Lemmens, A., Croux, C., & Dekimpe, M. G. (2008). Measuring and testing Granger causality over the spectrum: An application to European production expectation surveys. *International Journal of Forecasting*, *24*(3), 414–431. doi:10.1016/j.ijforecast.2008.03.004

Narayan, P. K., & Prasad, A. (2008). Electricity consumption–real GDP causality nexus: Evidence from a bootstrapped causality test for 30 OECD countries. *Energy Policy*, *36*(2), 910–918. doi:10.1016/j.enpol.2007.10.017

Nazlioglu, S., Kayhan, S., & Adiguzel, U. (2014). Electricity consumption and economic growth in Turkey: Cointegration, linear and nonlinear Granger causality. *Energy Sources. Part B, Economics, Planning, and Policy*, *9*(4), 315–324. doi:10.1080/15567249.2010.495970

Nguyen, K. H., & Kakinaka, M. (2019). Renewable energy consumption, carbon emissions, and development stages: Some evidence from panel cointegration analysis. *Renewable Energy*, *132*, 1049–1057. doi:10.1016/j.renene.2018.08.069

OECD. (2012). *Linking Renewable Energy to Rural Development, OECD Greeen Growth Studies*. OECD Publishing.

Ozturk, I. (2010). A literature survey on energy–growth nexus. *Energy Policy, 38*(1), 340–349. doi:10.1016/j.enpol.2009.09.024

Ozturk, I., Aslan, A., & Kalyoncu, H. (2010). Energy consumption and economic growth relationship: Evidence from panel data for low and middle-income countries. *Energy Policy, 38*(8), 4422–4428. doi:10.1016/j.enpol.2010.03.071

Pala, A. (2016). Which energy-growth hypothesis is valid in OECD countries? Evidence from panel Granger causality. *International Journal of Energy Economics and Policy, 6*(1), 28–34.

Paul, S., & Bhattacharya, R. N. (2004). Causality between energy consumption and economic growth in India: A note on conflicting results. *Energy Economics, 26*(6), 977–983. doi:10.1016/j.eneco.2004.07.002

Ranjbar, O., Chang, T., Nel, E., & Gupta, R. (2017). Energy consumption and economic growth nexus in South Africa: Asymmetric frequency domain approach. *Energy Sources. Part B, Economics, Planning, and Policy, 12*(1), 24–31. doi:10.1080/15567249.2015.1020120

Shiu, A., & Lam, P. (2004). Electricity consumption and economic growth in China. *Energy Policy, 32*(1), 47–54. doi:10.1016/S0301-4215(02)00250-1

Squalli, J. (2007). Electricity consumption and economic growth: Bounds and causality analyses of OPEC members. *Energy Economics, 29*(6), 1192–1205. doi:10.1016/j.eneco.2006.10.001

Stern, D. I. (1993). Energy and economic growth in the USA: A multivariate approach. *Energy Economics, 15*(2), 137–150. doi:10.1016/0140-9883(93)90033-N

Toda, H. Y., & Yamamoto, T. (1995). Statistical inference in vector autoregressions with possibly integrated processes. *Journal of Econometrics, 66*(1-2), 225–250. doi:10.1016/0304-4076(94)01616-8

Tsani, S., & Menegaki, A. N. (2018). The Energy-Growth Nexus (EGN) Checklist for Authors. *The Economics and Econometrics of the Energy-Growth Nexus*, 347-376. . doi:10.1016/B978-0-12-812746-9.00011-0

Tugcu, C. T., Ozturk, I., & Aslan, A. (2012). Renewable and non-renewable energy consumption and economic growth relationship revisited: Evidence from G7 countries. *Energy Economics, 34*(6), 1942–1950. doi:10.1016/j.eneco.2012.08.021

Yu, E. S. H., & Hwang, B. K. (1984). The relationship between energy and GNP: Further results. *Energy Economics*, *6*(3), 186–190. doi:10.1016/0140-9883(84)90015-X

ENDNOTE

1 Note that a list and the definition of abbreviations used in this study are given on Table 1.

Chapter 10
Green Consumption Behavior and Environmental Sustainability Considerations

F. Gül Aygen Ispahi
https://orcid.org/0000-0001-6980-7919
Independent Researcher, Turkey

ABSTRACT

The purpose of this chapter is to review the factors that have an impact on green consumption behavior of individual consumers, as they relate to environmental sustainability considerations. After a brief review of the evolution of environmentally related research and relevant terminology, factors expected to have an effect on pro-environmental behaviors of individuals, inclusive of various demographic, individual, social-cultural, and contextual/situational considerations are assessed. A distinction between high- and low-income countries with respect to their reasons for engaging in environmentally friendly behaviors is also made in the discussion section. The chapter ends with a number of suggestions for future research and some concluding remarks.

INTRODUCTION

The rapidly increasing world population, high rates of urbanization, changing lifestyles and consumer demands in many parts of the globe, affluence and over-consumption in some, poverty and under-consumption in still other markets, advances in science and technology which result in both a better life for us all, on the one hand, but environmental worries such as global warming, air pollution, depletion and pollution

DOI: 10.4018/978-1-6684-6727-5.ch010

of natural resources, irresponsible deforestation, widespread desertification, toxic wastes, plant and animal species loss, on the other... These are only a few of the facts mankind is confronted with and may be worrying about to varying degrees in today's world. Yet, if one thinks about tomorrow's world, the concept of sustainability comes into scene which entails caring about the lives of future generations, as well.

The Global Development Research Center (GDRC, 2021) points out that there are over one hundred definitions of sustainability and sustainable development but that the best known belongs to the United Nation's (UN) World Commission on Environment and Development (WCED). Accordingly, one may talk about sustainable development where it "meets the needs of the present without compromising the ability of future generations to meet their own needs" (WCED, 1987, p.24) and it is based on three fundamental pillars: Economic, social, and environmental (Purvis et al., 2019). While the economic dimension suggests a system of production which satisfies current levels of consumption without endangering future needs, the social dimension entails concepts like quality of life, education, community development, equal opportunity, cultural identity, accessibility, participation, empowerment, and institutional stability (Mensah, 2019). Last but not the least, the environmental dimension is concerned with the natural environment and its productiveness together with its resiliency to support human life (Mensah, 2019). Likewise, it addresses different issues such as global warming, energy consumption, water consumption, carbon emissions, the use of chemicals, waste minimization, and noise pollution (Erol et al., 2009). As such, it encompasses topics like resource management, habitat restoration and preservation, and environmental protection (Mensah, 2019). The foregoing implies that for sustainable development to take place, there is a need to have a coordinated development of nature, society, economy, and science and technology (Wang, 2022).

In the rather affluent parts of the world, overconsumption is one of the major factors contributing to environmental deterioration. This situation not only harms the natural environment but also the well-being of consumers and quality of life, in turn. Besides this, due to their huge population sizes, emerging economies of the world, be it China, India, or Brazil, put additional stress on the ecosystem. Hence, both the developed and the developing nations are urged to find ways so as to decrease and may be reverse the negative effects of their burden on the environment by promoting pro-environmental behaviors (Felix & Braunsberger, 2016). Likewise, the impact of firm activities on the environment has an utmost influence on people's lives and well-being; their non-sustainable deeds have quite adverse effects, globally. When talking about firm activities, it should not be forgotten that the different technologies businesses use to produce goods and services both influence and are influenced by societies, substantially. Through technology, productivity, efficiency, and cost savings may be increased, product wastes and use of chemicals may be reduced, resources may be protected, and progress may be measured, analyzed, and tracked over time so as to help bring the impact of the industrial sector on the environment to a minimum level (Schwartz, 2021).

Given these facts, one might be tempted to think that it is the responsibility of governments and big corporations to deal with environmental issues because of their huge impact on the environment. Nevertheless, based on the studies of Grunert and Chen & Chai, conducted in 1993 and 2010, respectively, Chekima et al. (2016) point out that between 30 to 40% of environmental worsening is a direct result of individuals' non-sustainable consumption. It is also stated by Wijekoon and Sabri (2021) that consumer household purchases are responsible for 40% of environmental damage.

Based on the foregoing, it is important to realize that not only governments and the business world but also individual consumers, worldwide, are all responsible for considering the coming generations and leaving them a safe and sound planet on which to live. Even though sustainable development cannot be achieved by any one of these parties alone, each has a role to play in accomplishing this important deed and it should not be forgotten that each step taken in the betterment of the planet, whether small or big, will surely make a difference.

As such, the purpose of this chapter is to look at the situation from the perspective of consumers and review literature pertaining to factors expected to have an effect on green consumption behavior. A distinction between high and low income countries with respect to their reasons for engaging in environmentally friendly behaviors is also made.

BACKGROUND

Environmental deterioration has been considered to be a problem since the Industrial Revolution; nevertheless, Rachel Carson's book titled the 'Silent Spring' (1962) may be considered to be the pioneer publication in which the environment was first thought as a contemporary issue in the business domain (Kilbourne & Beckmann,1998).

Yet it was not until the early 1970s that the issue began to be of interest to researchers; even so, over the decades, there has been an ebb and flow of social concern about the environment and hence academic interest in the subject: Whereas the 1980s saw a decline, there was a resurgence of interest in environmental matters in the 1990s which made a peak in the 2000s and 2010s due mainly to factors such as the use of advanced technology, stricter governmental regulations, closer scrutiny from various environmental organizations and the media, and continuous rise of growing global concern about environmental quality (Chamorro et al., 2009;

Kilbourne & Beckmann, 1998; Lee, 2008, Leonidou & Leonidou, 2011; McDonagh & Prothero, 2014). It seems that 2020s will receive increasing attention given the world's prevailing situation.

Looking at the earlier years of environmentally related research, one can see that until the second half of the 1990's, the individual consumer was the unit of analysis and research was mainly conducted at the micromarketing level (Leonidou & Leonidou, 2011). The main purpose of these studies was to define an environmentally concerned market segment based on the individual characteristics of consumers which, in turn, was expected to help companies develop their marketing strategies, accordingly (Kilbourne & Beckmann, 1998). Beginning in 1995, however, rather macro considerations such as the dominant social paradigm (DSP) -of industrialized western countries-, value systems, the morality of marketing, consumerism, consumers' involvement in voluntary simplicity, sustainable consumption, and/or anti-consumption behaviors (Chamorro et al., 2009; Kilbourne & Beckmann, 1998; McDonagh & Prothero, 2014) also began to be of interest to researchers.

According to McDonagh and Prothero (2014), "what happens at the individual level will always remain important" (p.1196) yet it is also of utmost relevance to explore consumption and consumption practices, taking into consideration macro issues, as well. Hence as Kilbourne and Beckmann argued back in 1998, "macro related research combined with research at the micromarketing level affords a much greater opportunity for the creation of effective and enduring environmental policy than can either level of research by itself" (p.525).

'Green consumption behavior', entails consumption behaviors that people perceive to have either no impact or a minimum or reduced impact on the environment (Johnstone & Tan, 2015) while 'pro-environmental behavior' can be defined as "behavior that consciously seeks to minimize the negative impact of one's actions on the natural and built world" through minimization of resource and energy consumption, use of non-toxic substances, and reduction of waste production (Kollmuss & Agyeman, 2002, p.240). Inclusive in such behavior are purchase of environmentally friendly products, recycling -of paper and paper products, cardboard, plastic bottles and containers, aluminum/tin cans and containers, used glass bottles and containers, and used batteries-, careful in-house usage of water, electricity and electrical appliances, avoidance of product purchases that harm the environment or cause environmental damage, engagement in mindful transportation -such as the use of public vehicles rather than private cars-, and taking part in profit or not-for-profit environmental organizations -by donating money to charities involved in environmental causes, boycotting products or companies based on their negative environmental impact, joining a group that supports environmental protection, volunteering for pro-environmental activities, taking part in a protest about an environmental issue and the like (Kalamas et al., 2014). As such, green products aim at reducing waste, being less harmful to the environment and/or reducing consumption of natural resources (Majid & Russell, 2015). Likewise, the 'green consumer' has been defined by Webster in 1975 "as a consumer (who) takes into account the public consequences of his/her private consumption or who tries to use his/her purchasing power to promote social change" (Gonçalves et al., 2016, p.1485).

FACTORS EXPECTED TO HAVE AN EFFECT ON GREEN CONSUMPTION BEHAVIOR

Much research has been conducted to identify 'the green consumer' and investigate the various factors that have an impact on pro-environmental behavior of individuals. These factors, shown in Figure 1, may be broadly categorized as follows:

Figure 1. Factors expected to have an effect on green consumption behavior
Source: Author's elaboration

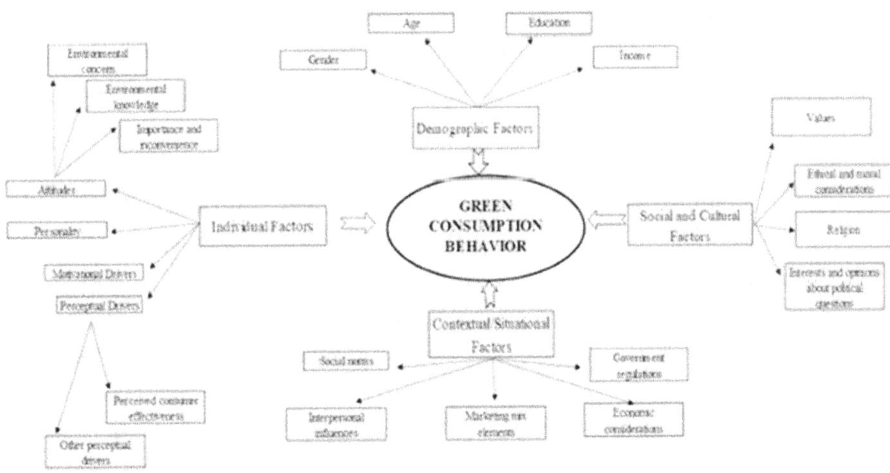

Demographic Factors

One of the most frequently used variables in this regard seems to be the demographic characteristics inclusive of variables such as gender, age, education, and income (Chekima et al., 2016; Chen et al. 2013; Gonzales et al., 2015; Johnstone & Lindh, 2018; Lin & Wu, 2018; Robert & James, 1999; Samdahl & Robertson, 1989). Nevertheless, in spite of their popularity, findings of these studies have been rather 'inconsistent/contradictory' and hence 'inconclusive' most of the time (Diamantopoulos et al. 2003; Leonidou et al. 2010; Roberts,1996; Shrum et al., 1995; Straughan & Roberts, 1999).

With respect to gender, women were found to show more pro-environmental behaviors compared to men in several studies (Chen et al., 2013; Diamantopoulos et al., 2003; Jain & Kaur, 2006; Roberts,1995; Roberts,1996). On the other hand, men were found to be more likely to purchase environmentally friendly/organic products compared to women in some cases (Roitner-Schobesberger et al., 2008). Still some other researchers found no gender influence on pro-environmental actions or purchase of environmentally friendly/organic products (Singh & Verma, 2017).

In case of age, either younger (Chen et al., 2013; Singh & Verma, 2017) or relatively older (Roberts 1996; Roitner-Schobesberger et al., 2008; Tung et al., 2012) consumers were found to be more concerned about environmental issues, performed more pro-environmental behaviors, and/or purchased more environmentally friendly/organic products, in different studies. In some cases it was found that age had no effect on buying green products (Tilikidou & Delistavrou, 2012). There were again some researchers reporting no clear relationship between age and pro-environmental actions (Clark et al., 2003; Diamantopoulos et al., 2003; Jain & Kaur, 2006).

Coming to income, whereas some authors found middle and high income consumers to show more pro-environmental behaviors compared to lower-income consumers (Dimitri & Dettmann, 2012; Roitner-Schobesberger et al., 2008; Singh & Verma, 2017), another study found a negative relationship between income and ecologically conscious consumer behavior (Roberts, 1996); still some other researchers concluded that income was not a predictor for pro-environmental behavior (Chan 1996; Jain & Kaur, 2006; Tilikidou & Delistavrou, 2012).

Surprisingly, similar inconsistent results were obtained in case of education. Even though one might expect the level of education to be a better predictor of environmental concern and pro-environmental behaviors, there were studies reporting either a positive (Chekima et al., 2016; Dimitri & Dettmann, 2012; Roitner-Schobesberger et al., 2008; Singh & Verma, 2017) or a negative (Samdahl and Robertson, 1989) relationship between the two. In other cases, no clear relationship could be established between the level of education and purchase of green products (Straughan & Roberts, 1999) or else education was found not to be a significant predictor of ecologically conscious consumer behavior (Kassarjian, 1971).

Hence, it has been suggested by various authors that, even though they are important, there is not much relevance in the use of demographic characteristics in identifying the green consumer and understanding green consumer behavior (Diamantopoulos et al. 2003; Kinnear et al. 1974; Peattie, 2001; Roberts, 1996). Some other authors have further argued that demographics are less important than knowledge, values, and/or attitude in explaining ecologically friendly behavior

(Laroche et al. 2001). Kalamas et al. (2014) point out that psychographics may be more influential in explaining pro-environmental behaviors compared to standard demographic variables. Moreover, according to Diamantopoulos et al. (2003), it is not possible to talk about an accurate profile of the green consumer unless attention is paid to all aspects of environmental consciousness; namely, knowledge about green issues, attitudes toward environmental quality, and environmentally sensitive behavior. Besides these arguments, a very recent study by Wijekoon and Sabri (2021) points out that while demographics do not essentially influence green purchase behavior, the use of socio-demographic variables may be considered to play an 'auxiliary role' in academic studies. Furthermore, Testa et al.'s (2021) systematic review of 113 articles published between the years of 2000 and 2018 on 'drivers to green consumption' identifies six demographic variables (gender, age, income, education level, family composition, other demographics) and specifies them as being the least relevant category of independent variables with respect to their driving effect and they further state that socio-demographic variables are considered as control variables in most of the reviewed articles. As such, in a very recent study, Dangelico et al. (2021) also used the socio-demographic factors of gender, age, education, and income as 'control' variables.

Despite these suggestions, however, it seems that the dilemma of whether the use of sociodemographic variables for explaining green purchases is effective or not will continue in the years to come: In their recent study, Witek and Kuzniar (2021) found that gender, age, education level, personal financial situation, and the number of children in the family all had an impact on both awareness and purchases of green products, in an emerging market. Likewise, in their recent study, Riskos et al. (2021) argue that demographic factors (i.e. gender, age, education, occupation, and marital status) have a significant effect on green product buying behavior.

Individual Factors

These factors are related to internal influences that affect individual consumers:

Personality Measures

As for personality measures, based on Hines et al.'s meta-analysis of 128 pro-environmental behavior research studies conducted in 1986, Kollmuss and Agyeman (2002) have listed the variables associated with responsible pro-environmental behavior, one of which is 'locus of control', the extent to which an individual perceives to have an ability to make some change through his/her own effort and behavior. Those who have a strong internal locus of control think that they can change things by their deeds (Kollmuss & Agyeman, 2002, p.243) while those with an external locus of control feel that, not themselves, but powerful others can bring about change and hence are much less likely to engage in environmentally friendly behavior.

Another personality variable identified in this context is 'individual sense of responsibility'. Accordingly, those individuals who have a higher sense of personal responsibility are expected to exhibit environmentally responsible behavior (Kollmuss & Agyeman, 2002).

Still another personality variable that has received attention is 'alienation', a feeling of being isolated from community, society, or culture (Do Paço & Raposo, 2009); yet, with respect to identifying the ecologically concerned consumers, this variable has produced conflicting results (Schwepker & Cornwell, 1991).

Other personality measures such as conservatism, dogmatism (Schlegelmilch et al., 1996), personal competence (Kollmuss & Agyeman, 2002; Lu et al. 2015), cosmopolitanism (Schwepker & Cornwell,1991), aggression, anxiety, self-esteem, understanding, tolerance, harm avoidance (Kinnear et al., 1974), dominance (Webster, 1975), and egoism (De Pelsmacker et al., 2005; Kollmuss & Agyeman, 2002; Lu et al., 2015), have also been analyzed as to their effect on green purchases, over the years.

Attitudes

Attitudes, the enduring positive or negative feelings about a person, an object, or an issue (Kollmuss & Agyeman, 2002) are believed to be one of the major factors that guide human behavior, in cognitive psychology (Johnstone & Tan, 2015). Research regarding attitudes has mainly relied on the Theory of Reasoned Action (TRA) developed by Fishbein and Ajzen in 1975 and the Theory of Planned Behavior (TPB) presented by Ajzen in 1985, mainly because of their clarity and simplicity (Johnstone & Tan, 2015). In the earlier versions of these models, 'intention' is identified as a major predictor of behavior, which, in turn, is a function of the attitudes towards the behavior and subjective norms (Testa et al., 2015). Accordingly, based on these theories, several studies have been undertaken by researchers to investigate the internal factors influential in the environmentally friendly behavior of individuals through the use of 'environmental concerns' (EC) and 'environmental knowledge' (EK) (Testa et al., 2015).

Environmental Concern

Do Paço and Raposo (2010) define 'environmental concern' (EC) "as an attitude that is related to environmental consequences" (p.430) and that it is influenced by an individual's direct experiences, by the experiences of others, and by communication produced by the media. Many studies have argued that the more concerned the consumers are about the environment, the stronger is their intention to purchase a green product (Do Paço et al., 2013; Lin & Wu, 2018; Newton et al., 2015); nevertheless, empirical findings do not support this argument much (Newton et al., 2015; Testa et al., 2015). Straughan and Roberts (1999) also note that environmental concern does not play an vital role in environmentally conscious consumer behavior (ECCB), in spite of the fact that this role is significant.

Environmental Knowledge

Environmental knowledge (EK), which refers to how much an individual knows about environmental issues (Chan,1999), has been another factor that has received attention in exploring green attitudes and behavior since it is frequently assumed to lead to green consumption behavior (Johnstone & Tan, 2015). Even though environmental knowledge has been considered to be an influential factor in every step of the buying decision process, findings of studies regarding the influence of environmental knowledge on behavior are contradictory (Do Paço & Raposo, 2010; Johnstone & Tan, 2015; Laroche et al., 2001; Tanner & Kast, 2003) and it is mostly agreed by researchers that just a small portion of pro-environmental behavior can be explained by EK and environmental awareness (Kollmuss & Agyeman, 2002).

Moreover, an average correlation of 0.30 was found between EK and behavior in a meta-analysis of 128 studies by Hines et al. conducted in 1986 and 1987 (Johnstone & Tan, 2015; Laroche et al., 2002). Based on a study by Kempton et al. conducted in 1995, Kollmuss and Agyeman (2002) also indicate that EK is not a prerequisite for pro-environmental behavior as lack of knowledge is found to be equally strong among environmentalists and non-environmentalists, in the aforementioned study. Furthermore, as Kollmuss and Agyeman (2002) put, many studies have revealed that very detailed technical knowledge does not enhance pro-environmental behavior, either. With respect to EK, Tanner and Kast (2003) have also made a distinction between knowledge about facts ("knowledge about definitions and causes / consequences of environmental problems") and knowledge about actions ("information about possible actions") (p. 886) and have argued that action-related knowledge is more likely to affect behavior compared to factual knowledge.

Importance and Inconvenience

According to Laroche et al. (2002), attitudes are the most significant predictors of individuals' willingness to pay more for ecologically favorable products and the two most influential attitudes in this respect are the 'importance' and the 'inconvenience' of being environmentally friendly. Importance relates to the extent to which consumers think environmentally friendly behaviors as important to themselves or the society as a whole while inconvenience refers to the extent to which behaving in an environmentally friendly manner is perceived to be inconvenient for the individual.

As such, even if a deed is found to be important by an individual, it may not be accomplished due to its inconvenience. Furthermore, whether or not individuals believe an environmentally friendly behavior to be important, its inconvenience has a greater influence on their actions.

Based on a number of studies, Tanner and Kast (2003) indicate that when engaging in environmental behavior, individuals make judgments about products or behaviors rather than judgments about environmental problems and that specific consumer beliefs rather than general environmental concern predict environmentally friendly consumer behavior more accurately. Likewise, in their point of view, positive attitudes toward environmental protection, fair trade, and local production are all powerful predictors of green purchases.

Nevertheless, as Kollmuss and Agyeman (2002) note, environmental attitudes have been found to have a rather minute and varying impact on pro-environmental behavior.

Motivational Drivers

As for motivational drivers, Johnstone and Tan (2015) cite various studies that have found factors such as 'ecological affect' (Chan's study conducted in 2001), 'personal circumstances'-such as allergies- (Soler's study conducted in 1996), 'one's level of involvement with nature'-living close to nature, being dependent on the natural environment- (Soler's study conducted in 1996), and 'his/her emotional affinity towards nature' (Kals et al.'s study conducted in 1999; van Tonder et al., 2020) to affect green consumption behavior.

As has been mentioned by Lu et al. (2015), some health-related factors (mainly health concern (Yin et al.'s study conducted in 2010), food safety concern, and health consciousness (Michaelidou and Hassan's study conducted in 2008; Singh and Verma, 2017)) have also been of interest to researchers, as one of the driving forces in green consumption.

Perceptual Drivers

Perceived Consumer Effectiveness

Perceived consumer effectiveness (PCE), which can be explained as the extent to which an individual believes that his/her actions will make a difference in solving a problem (Kinnear et al., 1974), seems to be one of the most promising variables in this respect (Do Paço & Raposo, 2010; Jaiswal & Kant, 2018; Kinnear et al., 1974; Lee, 2008; Roberts, 1996; Stadelmann & Schubert, 2018). Accordingly, even if an individual is concerned with the environment, he/she will not undertake ecologically conscious consumer behavior (ECCB) unless he/she believes that his/her actions will be effective (Straughan & Roberts, 1999). As such, PCE has received considerable attention in several studies and has been found to have the highest explanatory power among the factors that act as antecedents to environmentally friendly behavior (González et al. 2015).

Other Perceptual Drivers

Besides PCE, several studies attempt to figure out the effect of other perceptual drivers -such as 'price perceptions', 'quality perceptions', 'risk perceptions', 'consumers' perceptions of sales promotions', 'perception of trust', 'perceived performance', and 'pro-social status perception' (as cited in Johnstone & Tan, 2015) on green consumption behavior. Likewise, Laroche et al. (2002) mention 'perception of severity of environmental problems' and Lee (2008) points out to 'perceived environmental responsibility' as other perceptual drivers affecting green consumption behavior.

Social and Cultural Factors

Consumers are also influenced by a number of factors in their social and cultural settings:

Values

As Schwartz has defined back in 1994, values are desirable goals which serve as guiding principles in people's lives and they vary in importance (Laroche et al., 2001). An individual's values are most influenced by his 'social net' (family, neighbors, peers), the 'exo-system' (the media and political organizations), and the 'macro-system' (the cultural context he lives in), in descending order (Kollmuss & Agyeman, 2002). As González et al. (2015) point out, research on values has concentrated both on the direct and the indirect effect of values on environmentally friendly behaviors, especially in high income countries. With respect to their direct effect, past research has revealed that values influence behaviors differently and that they are not all equally relevant in explaining different pro-environmental behaviors of consumers (González et al., 2015). In case of their indirect effect, on the other hand, values have an influence on behaviors via beliefs, personal norms, attitudes, and environmental consciousness (González et al., 2015).

Besides, 'biospheric' (concerns about the protection of non-human elements of the environment), social/altruistic (concern for the welfare of others), and egoistic (concern about oneself) values, as were proposed by Stern et al. in 1999, all influence environmentally friendly behaviors, the former two positively, the latter, negatively (González et al., 2015).

Furthermore, cultural value orientations such as 'collectivism', 'individualism', 'long-term orientation' (LTO), and 'man-nature orientation' (MNO) have been studied by various researchers so as to reveal their effects on environmental concerns and green behavior (Chan, 2001; Do Paço et al, 2013; Laroche et al., 2001; Leonidou et al., 2010). With respect to 'collectivism', Leonidou et al. (2010) point out to several studies that have confirmed a positive association between collectivism and environmentally friendly attitudes. Besides, based on the findings of other researchers, Laroche et al. (2001) state that whereas collectivist individuals behave more environmentally friendly, individualistic people tend not to do so. Sreen et al. (2018) also indicate that collectivism is an antecedent of green consumption.

LTO can be explained as the degree to which a person has a practical, future-oriented viewpoint. Long term oriented people were found to develop attitudes related to protecting the natural environment both in their homes and within the society in which they lived (Ghazali et al., 2017; Leonidou et al., 2010; Sreen et al., 2018).

MNO, individuals' friendly deeds toward nature because of a big affection for nature and also of a great passion to protect it (Chekima et al., 2016), is suggested to have an effect on their ecological affect, environmental knowledge, and their commitment to purchase green products, positively (Do Paço et al., 2013). Furthermore, a positive relationship was found between MNO and green purchase intention (Chekima et al, 2016).

Another culture-related proposal belongs to Kollmuss and Agyeman (2002): They state that cultures in small countries with large populations (such as Switzerland) tend to be more considerate of resources compared to cultures in large countries with plentiful of resources (such as the USA).

Personal values, such as security, fun/enjoyment, and self-gratification (Jansson et al., 2010; McCarty & Shrum, 1994), 'materialism' (Banerjee & McKeage, 1994; De Pelsmacker et al., 2005; Gu et al., 2020; Kilbourne & Pickett, 2008), and 'hedonic value' (Jaini et al., 2020; Papista & Krystallis, 2013; van Riper et al., 2020) have also been studied by various researchers as potential predictors of or barriers to environmentally responsible behavior.

Ethical and Moral Considerations

As green purchase behaviors aim at sustaining natural resources and protecting the environment, they are considered to be a type of ethical consumer behavior. Ethical and moral considerations inclusive of moral concern, moral attitudes, moral responsibilities, ethical motives, ethical obligation, and ethical judgment were investigated by various researchers, over the years (Arvola et al., 2008; Dean et al., 2008; Honkanen et al., 2006; Lu et al., 2015; Rettie et al., 2012; Wang & Wang, 2016).

Tanner and Kast (2003) point out to several studies which have revealed that environmental behavior is enhanced by personal norms (feelings of moral obligation) and that "environmentally friendly behavior may be characterized as morally demanding" (p.886).

Among the major ethical factors expected to have an influence on green purchase behavior are 'deontology' and 'law obedience' (Leonidou et al., 2010). Accordingly, 'deontology' refers to "whether the intention and/or consequences of a certain act is right or wrong" (p.1325). As such, people who hold strong deontological views believe they need to protect the environment so as to fulfill their moral duty towards humanity (Leonidou et al., 2010). The other factor, law obedience, refers to "the extent to which an individual respects the laws of his/her country" (p.1326) and a law-obedient individual is believed to be ready to make personal sacrifices in order to protect the environment. Anyhow, taking Carrigan and Attala's study conducted in 2001 as a base, Rettie et al. (2014) state that even though consumers may have good intentions, actual buying behavior is not affected by ethical concerns much. Johnstone and Tan (2015) also refer to Carrigan and Attala's study (2001) and argue that purchase decisions do not necessarily rest on ethical considerations. Moreover, as Rettie et al. (2014) point out, in a study conducted by Szmigin et al. in 2009, it was found that even very ethically conscious consumers were inconsistent in their purchase behavior.

Religion

Religions, being fundamental belief systems, encompass consumers' core values yet they have not been studied in environmental research much, until the last decade. Unfortunately, these recent studies yielded mixed findings; in some cases it was found that religious consumers could be more sustainable than less/non-religious consumers (Leary et al., 2016; Minton et al., 2015; Minton et al., 2018b) and that having religious values positively affected green purchase behavior (Ghazali et al., 2018) whereas in some other cases, the opposite was found to be true (Minton, 2013; Wolkomir et al., 1997). Minton et al. (2018b) suggest that as sustainability is being encouraged more in religious institutions nowadays, it is possible that the more recent research reveals a rather more positive influence of religion on sustainability. In a very recent study on the influence of religiosity on sustainable behavior across three different cultures, Kaplan and Iyer (2021) found that in cultures that were collectivist with a long-term orientation, religiosity had a significant influence on sustainable behavior.

When talking about 'religion', besides religiosity, religious affiliation has to be taken into consideration, as well. The primary world religions fall into the two categories of Abrahamic religions (like Christianity, Judaism, and Islam) and Indian religions (such as Hinduism, Buddhism, and Sikhism, among others) (Vaughan, 2022). Christianity forms the major religious group in the Western world and it is widely believed by Christians that "God gave humans dominion over the Earth and all of its creatures" (Clements et. al., 2014, p. 87) and that nature exists to serve mankind. As such, Western Christianity has been characterized as being at odds with environmentalism and ecological values (Clements et. al., 2014). The second largest world religion is Islam and with respect to the tenets of Islam, all resources upon which life depends have been created by God and it is the right and privilege of all people and species to make use of these resources. As such, mankind should be very careful in ensuring the interests and rights of all other living beings on earth since they are equal partners and should not abuse, misuse, or distort the natural resources; each generation can only benefit from these resources but cannot own them (Bagader et al., 2011). In Asian cultures, apart from Islam and also no-religion populations, one can see the influence of different religions, mainly Hinduism and Buddhism in the Southwest and Confucianism and Taoism in the East. Most Hindus believe that all living things are sacred because they are part of God, as is the natural world, and many believe in showing respect for all life, humans, animals, and vegetables (BBC, 2022). Most Buddhists, on the other hand, believe people need to live simply and respect the cycle and balance in nature so everything can continue for future generations (BBC, 2022). Likewise, Confucianism and Taoism posit the view that man and nature are interconnected and that there is a need for balance and harmony between the two (Polonsky et. al., 2014).

As mentioned above, with respect to different religious thoughts, Minton et al. (2018b) state that Christians participate in fewer sustainable behaviors than people with other belief systems (i.e. Eastern religions and Atheists). Likewise, Kalamas et al. (2014) indicate that Eastern religions are more eco-friendly than Christianity. They also refer to the Sarıgöllü study conducted in 2009 and point out that "the emphasis on fate in Muslim cultures would imply a more passive rather than a proactive stance towards the environment" (p.374). Besides, in contrast to Christians, highly religious Buddhists are more likely to participate in sustainable behaviors (Felix & Braunsberger, 2016).

Interests and Opinions About Political Questions

With respect to interests and opinions about political questions, it is generally believed that environmental issues are "part of a 'liberal' (left-leaning) political agenda" and that those who have a liberal political background tend to have a stronger commitment to the green movement compared to those with more conservative political views (Do Paço et al., 2009, p.19; Watkins et al., 2016).

Several other studies have also been cited by Leonidou et al. (2010) and Liu et al. (2014) revealing citizens with stronger liberal political ideology as being more concerned about the environment.

Contextual/Situational Factors

Various environmental considerations also play an important role in influencing green consumption behavior.

In their systematic analysis of 113 articles published between the years of 2000 and 2018 on the topic of 'drivers to green consumption', Testa et al. (2021) state that contextual factors –"i.e., the particular situation and physical environment that surround the consumer at the moment of product selection" (p.4842)- can influence an individual's purchase choice. Yet, according to them, such factors are contingent in nature and they can affect momentary decisions but not an individual's basic system of values or habits. As such, contextual factors may be studied under different groups; namely, product access and shop stimuli, social norms, economic benefits, and media and advertising. As for product access and shop stimuli, considerations such as distance, parking access, the convenience of its location, shop's environment -inclusive of factors such as space and product distribution, lightning, music, the presence of directions, and other visual stimuli- are all influential on product choice. Coming to social norms, presence of certain other people in a given setting or, else, particular social contexts are contextual factors on which social norms depend and both peers, parents, and general social pressure are considered to be key factors influencing an individual's preference for green products. One other factor that is found to affect purchases to a great extent is media attention on environmental issues or scandals related to specific brands (Testa et al., 2021). In a similar vein, marketing mix elements are considered to be one of the contextual factors affecting consumers' purchase decisions in other studies, as well (ElHaffar et al., 2020).

Technologies used to produce goods and services may also be thought of as a contextual factor that has an effect on green consumption behavior. As such, use of sustainable technology may result in a preference for the resulting goods and services that would hamper the adverse effects of environmental deterioration, provided that they are perceived to be useful and easy to use.

Economic constraints, lack of choice and availability (Gleim et al., 2013; Johnstone & Tan, 2015; Tanner & Kast, 2003), social pressures, and opportunities to choose different actions (Hines et al.'s 1986, 87 study as cited in Kollmuss and Agyeman, 2002), have also been investigated as situational factors by various researchers which act as either potential inhibitors or drivers of pro-environmental purchase behavior.

In line with this, together with several studies, Ertz et al. (2016) state that consumers are influenced by both intra-personal factors and also contextual or situational factors -such as interpersonal influences, government regulations, availability of recycling facilities, quality of public transport, and pricing regimes- capable of hampering or easing pro-environmental behaviors. Hence, in their point of view, it is a more fruitful approach to consider contextual factors together with intra-psychic factors in evaluating pro-environmental behaviors rather than using these two sets of factors exclusively.

DISCUSSION

Based on the foregoing, it will not be wrong to quote Kollmuss and Agyeman's (2002) words regarding the factors expected to have an effect on green consumption behavior: "The question of what shapes pro-environmental behavior is such a complex one that it cannot be visualized through one single framework or diagram" (p.239). As such, it is not quite possible to make generalizations and reach conclusions with respect to how an individual will behave under different circumstances. Nevertheless, in spite of this complexity and ambiguity, even being aware of this fact would be a good starting point in dealing with environmental issues.

At this point, it may be useful to make a distinction between high and low income countries with respect to their reasons for engaging in environmentally friendly behaviors. As has been revealed above, a vast array of factors seem to have an effect on environmentally friendly behavior of consumers. However, majority of the research conducted on this specific topic has concentrated on high income countries -at least till the last decade. Apart from cultural considerations (inclusive of religions and cultural orientations such as individualism-collectivism, long-term orientation, or man-nature orientation), what happens in high income countries need not necessarily take place in the rather less-developed parts of the world- especially when their immediate concerns are taken into account.

Depending also on the different stages of development they are at, reasons for engaging in environmentally friendly behaviors in developing countries may be quite different from those in the developed ones. In some instances, environmentally friendly behavior in developing countries may not, in fact, be due to well-entrenched environmental values but rather to a "traditional custom of savings and frugality" (González et al., 2015, p. 288), as poverty is a specific predictor of lifestyles in these countries; consumers may be trying to satisfy their basic needs so as to survive and may be thinking of their own lives, rather than acting in a more altruistic manner taking into consideration coming generations, as well. Associated with this is a desire of some of the low-income consumers to 'save money' when they engage in 'reducing', 'reusing', and 'recycling' activities. Besides, there are also some more affluent segments in such countries who engage in environmentally friendly behaviors mainly because of their aspirations and willingness to lead comparable lives to those of high-income country citizens (González et al., 2015).

Anyhow, large numbers of people in the world live on a few dollars a day which makes it impossible for them to act pro-environmentally. These people do not really lack environmental concern but environmental issues are not considered to be one of their most pressing problems (Kollmuss & Agyeman, 2002). It will sure take time –much time – for a large part of the world to engage in more 'caring' and pro-environmental lifestyles. In this sense, there is a long way to go…

Nonetheless, in most instances, huge populations of emerging economies, such as China, India, and Brazil, have a heavy burden on the ecosystem just as overconsumption in high-income countries contributes to environmental deterioration. Hence, as has been mentioned above, there is a need to find ways to decrease these negative effects, on the part of both the developing and the developed world, which necessitate rather macro considerations.

Another important consideration is that even though green consumption practices are necessary for environmental sustainability, it should not be forgotten that if used too much, they, as well, may lead to overconsumption of resources; hence, it seems that there will be a need for some sort of a systemic change if mankind is to find enduring solutions to increasing environmental problems. That would again take one to rather macro considerations.

One of the main issues related to consumer behavior seems to be the so-called 'attitude-behavior gap', a notion which implies that consumers do not always do the things they say they would like to do. Hence, there results a disparity between what they say (they intend to do) and what they actually do. This is also the case with green consumption behavior; people say they are greatly concerned about environmental problems, in most instances; yet, when it comes to action, there is a halt! This, in fact, may be the most outstanding problem to be dealt with in environmental sustainability considerations. The gap has to be diminished somehow so as to give way to more pro-environmental behaviors.

FUTURE RESEARCH DIRECTIONS

Besides the 'attitude-behavior gap', there may also be a gap between reported behavior and actual behavior, as most of the studies conducted on this topic measure self-reported behavior which may involve social desirability bias, to an extent. Hence, future research may aim at studying actual behavior, as well, so as to grasp a clearer picture of how consumers behave in reality.

As has been mentioned above, various studies have been conducted on cultural value dimensions of 'individualism /collectivism', 'man-nature orientation', and 'long-term orientation' as they relate to green purchasing behavior. Yet, other cultural value dimensions such as 'power distance', 'masculinity / femininity', 'uncertainty avoidance', and 'indulgence', which have not been studied much in the field of green consumer behavior may also be analyzed by researchers so as to examine their effects on environmental sustainability.

Likewise, cross-cultural works are also of much importance in making comparisons between the different societies of the world with respect to their environmental understanding and sustainability considerations. Furthermore, cross-cultural research may be conducted to recognize the similarities and differences in the decision making processes of consumers living in the different parts of the world (Sharma, 2021) and also to understand why sustainability related initiatives succeed in some countries but fail in others (Minton et al., 2018a).

There is a need to find ways of increasing the mass market preference for green products; hence, apart from studies directed at the green consumer segment, it would be desirable to conduct studies targeted at the mass market (Wymer & Polonsky, 2015).

Besides, use of mixed methods -of both quantitative and qualitative techniques in the same study- (Sharma, 2021) and longitudinal studies that capture changes in consumer behavior over time would also be fruitful, as they have not been used in green marketing literature to study the attitudes and purchase behavior of consumers with respect to sustainable behavior (Ghazali et al., 2018; Sharma, 2021).

Emotions (especially, warm glow, emotional value, neophobia, nostalgia, and emotions, in general) have a potential role in influencing customer green purchase behavior; therefore, future research should also concentrate on the influence of different types of emotions on consumer green purchase behavior, as they have not been researched much up until the present time (Wijekoon & Sabri, 2021). Studies related to the influence of other specific emotions, such as guilt, generativity (Joshi & Rahman, 2015), pleasure, and humor on green purchase behavior may also be recommended (ElHaffar et al., 2020; Joshi & Rahman, 2015).

Liu et al. (2014) refer to a "technofix worldview", stating that young people tend to spend most of their time indoors and hence they cannot connect with nature and the environment (p.88). Likewise, the increasing use of smart phones and social media tools throughout the globe, also leads to an alignment and disconnection from nature -as well as from all face-to-face communication. As such, research on the "technofix" youth may also yield important considerations as to how to deal with these problematic realities of our day which actually impede the 'man-nature' relationship.

Future studies may also concentrate on the impact of technology, especially of sustainable technology, on green consumption behavior of individuals, as such research seems to be quite scarce.

CONCLUSION

The aim of this chapter was to review the factors that have an impact on green consumption behavior of individual consumers, as they relate to environmental sustainability considerations.

It can be stated that, at the micro level, environmentally friendly purchases of consumers and the factors that influence their behaviors will always be an important consideration for companies, even if these factors are quite numerous and complex in nature so as to make generalizations. Besides, as Leonidou et al. (2010) replicate from Roberts and Bacon's study (1997), "consumers' uncontrollable consumption of goods and irresponsible treatment of the environment make it necessary to understand their motives, attitudes, behavior, and actions with regard to ecological issues so as to alleviate at least part of the problems associated with the environment" (p. 1320). Hence, although macro considerations are of utmost importance, it should not be forgotten that research at the micro level is also quite invaluable and as has been stated above, macro related research together with micro level research will prove to be more fruitful in creating an effective and long-lasting environmental policy, than can any of these streams of research, individually (Kilbourne & Beckmann, 1998).

One of the issues not to be overlooked is related to the distinction between the developed versus the developing country citizens with respect to their reasons for engaging in environmentally friendly behaviors. These should be analyzed separately and necessary actions should be taken based on this distinction.

Besides the environment, each and every other living thing on the world, is influenced by environmentally responsible or irresponsible behaviors, to a great extent. Hence, it is important to realize that even small steps matter in the betterment of humanity's future and accordingly, efforts should be directed at encouraging these steps. Moreover, in encouraging these steps, effective use of technology -especially of sustainable technology- may foster green consumption behavior of individual consumers worldwide.

The Mother Earth is a gift for us all... Let's try to find ways to save her before it is too late...

REFERENCES

Arvola, A., Vassallo, M., Dean, M., Lampila, P., Saba, A., Lahteenmaki, R., & Shepherd, R. (2008). Predicting intentions to purchase organic food: The role of affective and moral attitudes in the theory of planned behavior. *Appetite*, *50*(2-3), 443–454. doi:10.1016/j.appet.2007.09.010 PMID:18036702

Bagader, A., El-Sabbagh, A., Al-Glayand, M., & Samarrai, M. (2011). *Environmental protection in Islam: A general introduction.* Retrieved June 17, 2022, from https://www.islamreligion.com/ articles/307/viewall/environmental-protection-in-islam/

Banerjee, B., & McKeage, K. (1994). How green is my value: Exploring the relationship between environmentalism and materialism. *Advances in Consumer Research. Association for Consumer Research (U. S.)*, *21*, 147–152.

BBC. (2022). *Bitesize / Environment.* Retrieved February 25, 2022, from http://www.bbc.co.uk >revision

Chamorro, R., Rubio, S., & Miranda, F. J. (2009). Characteristics of research on green marketing. *Business Strategy and the Environment*, *18*(4), 223–239. doi:10.1002/bse.571

Chan, R. Y. K. (1999). Environmental attitudes and behavior of consumers in China. *Journal of International Consumer Marketing*, *11*(4), 25–52. doi:10.1300/J046v11n04_03

Chan, R. Y. K. (2001). Determinants of Chinese consumers' green purchase behavior. *Psychology and Marketing*, *18*(4), 389–413. doi:10.1002/mar.1013

Chan, T.-S. (1996). Concerns for environmental issues and consumer purchase preferences: A two-country study. *Journal of International Consumer Marketing*, *9*(1), 43–55. doi:10.1300/J046v09n01_04

Chekima, B., Wafa, S. A., Igau, O. A., Chekima, S., & Sondoh, S. L. (2016). Examining green consumerism motivational drivers: Does premium price and demographics matter to green purchasing? *Journal of Cleaner Production*, *112*, 3436–3450. doi:10.1016/j.jclepro.2015.09.102

Chen, X., Peterson, M. N., Hull, V., Lu, C., Hong, D., & Liu, J. (2013). How perceived exposure to environmental harm influences environmental behavior in urban China. *Ambio*, *42*(1), 52–60. doi:10.100713280-012-0335-9 PMID:22821144

Clark, C. F., Kotchen, M. J., & Moore, M. R. (2003). Internal and external influences on pro-environmental behavior: Participation in a green electricity program. *Journal of Environmental Psychology*, *23*(3), 237–246. doi:10.1016/S0272-4944(02)00105-6

Clements, J. M., McCright, A. M., & Xiao, C. (2014). Green Christians? An empirical examination of environmental concern within the U.S. general public. *Organization & Environment*, *27*(1), 85–102. doi:10.1177/1086026613495475

Dangelico, R. M., Nonino, F., & Pompei, A. (2021). Which are the determinants of green purchase behaviour? A study of Italian consumers. *Business Strategy and the Environment*, *30*(5), 2600–2620. doi:10.1002/bse.2766

De Pelsmacker, P., Driesen, L., & Rayp, G. (2005). Do consumers care about ethics? Willingness to pay for fair-trade coffee. *The Journal of Consumer Affairs*, *39*(2), 363–385. doi:10.1111/j.1745-6606.2005.00019.x

Dean, M., Raats, M. M., & Shepherd, R. (2008). Moral concerns and consumer choice of fresh and processed organic foods. *Journal of Applied Social Psychology*, *38*(8), 2088–2107. doi:10.1111/j.1559-1816.2008.00382.x

Diamantopoulos, A., Schlegelmilch, B. B., Sinkovics, R. R., & Bohlen, G. M. (2003). Can socio-demographics still play a role in profiling green consumers? A review of the evidence and an empirical investigation. *Journal of Business Research*, *56*(6), 465–480. doi:10.1016/S0148-2963(01)00241-7

Dimitri, C., & Dettmann, R. (2012). Organic food consumers: What do we really know about them? *British Food Journal*, *114*(8), 1157–1183. doi:10.1108/00070701211252101

Do Paço, A. M. F., Alves, H., Shiel, C., & Filho, W. L. (2013). Development of a green consumer behavior model. *International Journal of Consumer Studies*, *37*(4), 414–421. doi:10.1111/ijcs.12009

Do Paço, A. M. F., & Raposo, M. L. B. (2009). Green segmentation: An application to the Portuguese consumer market. *Marketing Intelligence & Planning*, *27*(3), 364–379. doi:10.1108/02634500910955245

Do Paço, A. M. F., & Raposo, M. L. B. (2010). Green consumer market segmentation: Empirical findings from Portugal. *International Journal of Consumer Studies*, *34*(4), 429–436. doi:10.1111/j.1470-6431.2010.00869.x

Do Paço, A. M. F., Raposo, M. L. B., & Filho, W. L. (2009). Identifying the green consumer: A segmentation study. *Journal of Targeting. Measurement and Analysis for Marketing*, *17*(1), 17–25. doi:10.1057/jt.2008.28

ElHaffar, G., Durif, F., & Dube, L. (2020). Towards closing the attitude-intention-behavior gap in green consumption: A narrative review of the literature and an overview of future research directions. *Journal of Cleaner Production*, *275*(122556), 1–20. doi:10.1016/j.jclepro.2020.122556

Erol, I., Cakar, N., Erel, D., & Sari, R. (2009). Sustainability in the Turkish retailing industry. *Sustainable Development (Bradford)*, *17*(1), 49–67. doi:10.1002d.369

Ertz, M., Karakaş, F., & Sarıgöllü, E. (2016). Exploring pro-environmental behaviors of consumers: An analysis of contextual factors, attitudes, and behaviors. *Journal of Business Research*, *69*(10), 3971–3980. doi:10.1016/j.jbusres.2016.06.010

Felix, R., & Braunsberger, K. (2016). I believe therefore I care. The relationship between religiosity, environmental attitudes, and green product purchase in Mexico. *International Marketing Review*, *33*(1), 137–155. doi:10.1108/IMR-07-2014-0216

GDRC (Global Development Research Center). (2021). *Sustainable development*. Retrieved April 29, 2021, from http://www.gdrc.org/sustdev/definitions

Ghazali, E. M., Mutum, D. S., & Ariswibowo, N. (2018). Impact of religious values and habit on an extended green purchase behaviour model. *International Journal of Consumer Studies*, *42*(6), 639–654. doi:10.1111/ijcs.12472

Gleim, M. R., Smith, J. S., Andrews, D., & Cronin, J. J. Jr. (2013). Against the green: A multi-method examination of the barriers to green consumption. *Journal of Retailing*, *89*(1), 44–61. doi:10.1016/j.jretai.2012.10.001

Gonçalves, H. M., Lourenço, T. F., & Silva, G. M. (2016). Green buying behavior and the theory of consumption values: A fuzzy-set approach. *Journal of Business Research*, *69*(4), 1484–1491. doi:10.1016/j.jbusres.2015.10.129

González, E. M., Felix, R., Carrete, L., Centeno, E., & Castaño, R. (2015). Green shades: A segmentation approach based on ecological consumer behavior in an emerging economy. *Journal of Marketing Theory and Practice*, *23*(3), 287–302. doi:10.1080/10696679.2015.1032395

Gu, D., Gao, S., Wang, R., Jiang, J., & Xu, Y. (2020). The negative associations between materialism and pro-environmental attitudes and behaviors: Individual and regional evidence from China. *Environment and Behavior*, *52*(6), 611–638. doi:10.1177/0013916518811902

Honkanen, P., Verplanken, B., & Olsen, S. O. (2006). Ethical values and motives driving organic food choice. *Journal of Consumer Behaviour*, *5*(5), 420–430. doi:10.1002/cb.190

Jain, S. K., & Kaur, G. (2006). Role of sociodemographics in segmenting and profiling green consumers: An exploratory study of consumers in India. *Journal of International Consumer Marketing*, *18*(3), 107–117. doi:10.1300/J046v18n03_06

Jaini, A., Quoquab, F., Mohammad, J., & Hussin, N. (2020). Antecedents of green purchase behavior of cosmetics products: An empirical investigation among Malaysian consumers. *International Journal of Ethics and Systems*, *36*(2), 185–203. doi:10.1108/IJOES-11-2018-0170

Jaiswal, D., & Kant, R. (2018). Green purchasing behavior: A conceptual framework and empirical investigation of Indian consumers. *Journal of Retailing and Consumer Services*, *41*, 60–69. doi:10.1016/j.jretconser.2017.11.008

Jansson, J., Marell, A., & Nordlund, A. (2010). Green consumer behavior: Determinants of curtailment and eco-innovation adoption. *Journal of Consumer Marketing*, *27*(4), 358–370. doi:10.1108/07363761011052396

Johnstone, L., & Lindh, C. (2018). The sustainability-age dilemma: A theory of (un) planned behavior via influencers. *Journal of Consumer Behaviour*, *17*(1), 127–139. doi:10.1002/cb.1693

Johnstone, M.-L., & Tan, L. P. (2015). Exploring the gap between consumers' green rhetoric and purchasing behavior. *Journal of Business Ethics*, *132*(2), 311–328. doi:10.100710551-014-2316-3

Joshi, Y., & Rahman, Z. (2015). Factors affecting green purchase behavior and future research directions. *International Strategic Management Review*, *3*(1-2), 128–143. doi:10.1016/j.ism.2015.04.001

Kalamas, M., Cleveland, M., & Laroche, M. (2014). Pro-environmental behaviors for thee but not for me: Green giants, green Gods, and external environmental locus of control. *Journal of Business Research*, *67*(2), 12–22. doi:10.1016/j.jbusres.2013.03.007

Kaplan, B., & Iyer, E. S. (2021). Motivating sustainable behaviors: The role of religiosity in a cross-cultural context. *The Journal of Consumer Affairs*, *55*(3), 792–820. doi:10.1111/joca.12376

Kassarjian, H. H. (1971). Incorporating ecology into marketing strategy: The case of air pollution. *Journal of Marketing*, *35*(3), 61–65. doi:10.1177/002224297103500310

Kilbourne, W. E., & Beckmann, S. C. (1998). Review and critical assessment of research on marketing and the environment. *Journal of Marketing Management*, *14*(6), 513–532. doi:10.1362/026725798784867716

Kilbourne, W. E., & Pickett, G. (2008). How materialism affects environmental beliefs, concern, and environmentally responsible behavior. *Journal of Business Research*, *61*(9), 885–893. doi:10.1016/j.jbusres.2007.09.016

Kim, Y., & Choi, S. M. (2005). Antecedents of green purchase behaviour: An examination of collectivism, environmental concern and PCE. *Advances in Consumer Research. Association for Consumer Research (U. S.), 32*, 592–59.

Kinnear, T. C., Taylor, J. R., & Ahmed, S. A. (1974). Ecologically concerned consumers: Who are they? *Journal of Marketing, 38*, 20–24.

Kollmuss, A., & Agyeman, J. (2002). Mind the gap: Why do people act environmentally and what are the barriers to pro-environmental behavior. *Environmental Education Research, 8*(3), 239–260. doi:10.1080/13504620220145401

Laroche, M., Bergeron, J., & Barbaro-Forleo, G. (2001). Targeting consumers who are willing to pay more for environmentally friendly products. *Journal of Consumer Marketing, 18*(6), 503–520. doi:10.1108/EUM0000000006155

Laroche, M., Tomiuk, M.-A., Bergeron, J., & Barbaro-Forleo, G. (2002). Cultural differences in environmental knowledge, attitudes, and behaviors of Canadian consumers. *Canadian Journal of Administrative Sciences, 19*(3), 267–283. doi:10.1111/j.1936-4490.2002.tb00272.x

Leary, R. B., Minton, E. A., & Mittelstaedt, J. D. (2016). Thou shall not? The influence of religion on beliefs of stewardship and dominion, sustainable behaviors, and marketing systems. *Journal of Macromarketing, 36*(4), 457–470. doi:10.1177/0276146715626219

Lee, K. (2008). Opportunities for green marketing: Young consumers. *Marketing Intelligence & Planning, 26*(6), 573–586. doi:10.1108/02634500810902839

Leonidou, C. N., & Leonidou, L. C. (2011). Research into environmental marketing/management: A bibliographic analysis. *European Journal of Marketing, 45*(1/2), 68–103. doi:10.1108/03090561111095603

Leonidou, C. N., Leonidou, L. C., & Kvasova, O. (2010). Antecedents and outcomes of consumer environmentally friendly attitudes and behavior. *Journal of Marketing Management, 26*(13-14), 1319–1344. doi:10.1080/0267257X.2010.523710

Li, Q., Long, R., & Chen, H. (2017). Empirical study of the willingness of consumers to purchase low-carbon products by considering carbon labels: A case study. *Journal of Cleaner Production, 161*, 1237–1250. doi:10.1016/j.jclepro.2017.04.154

Lin, B., & Wu, W. (2018). Why people want to buy electric vehicle: An empirical study in first-tier cities of China. *Energy Policy, 112*, 233–241. doi:10.1016/j.enpol.2017.10.026

Liu, X., Vedlitz, A., & Shi, L. (2014). Examining the determinants of public environmental concern: Evidence from national public surveys. *Environmental Science & Policy*, *39*, 77–94. doi:10.1016/j.envsci.2014.02.006

Lu, L.-C., Chang, H.-H., & Chang, A. (2015). Consumer personality and green buying intention: The mediate role of consumer ethical beliefs. *Journal of Business Ethics*, *127*(1), 205–219. doi:10.100710551-013-2024-4

Majid, K. A., & Russell, C. A. (2015). Giving green a second thought: Modeling the value retention of green products in the secondary market. *Journal of Business Research*, *68*(5), 994–1002. doi:10.1016/j.jbusres.2014.10.001

McCarty, J., & Shrum, L. (1994). The recycling of solid wastes: Personal values, value orientations, and attitudes about recycling as antecedents of recycling behaviour. *Journal of Business Research*, *30*(1), 53–62. doi:10.1016/0148-2963(94)90068-X

McDonagh, P., & Prothero, A. (2014). Sustainability marketing research: Past, present, and future. *Journal of Marketing Management*, *30*(11-12), 1186–1219. doi:10.1080/0267257X.2014.943263

Mensah, J. (2019). Sustainable development: Meaning, history, principles, pillars, and implications for human action: Literature review. *Cogent Social Sciences*, *5*(1). DOI: .2019.1653531 doi:10.1080/23311886

Minton, E. A. (2013). Religion and religiosity's influence on sustainable consumption behaviors. In L. R. Kahle & E. Gurel-Atay (Eds.), *Communicating sustainability for the green economy* (pp. 73–87). M.E. Sharpe.

Minton, E. A., Kahle, L. R., & Kim, C.-H. (2015). Religion and motives for sustainable behaviors: A cross-cultural comparison and contrast. *Journal of Business Research*, *68*(9), 1937–1944. doi:10.1016/j.jbusres.2015.01.003

Minton, E. A., Spielmann, N., Kahle, L. R., & Kim, C.-H. (2018a). The subjective norms of sustainable consumption: A cross-cultural exploration. *Journal of Business Research*, *82*, 400–408. doi:10.1016/j.jbusres.2016.12.031

Minton, E. A., Xie, H. J., Gurel-Atay, E., & Kahle, L. R. (2018b). Greening up because of god: The relations among religion, sustainable consumption and subjective well-being. *International Journal of Consumer Studies*, *42*(6), 655–663. doi:10.1111/ijcs.12449

Newton, J. D., Tsarenko, Y., Ferraro, C., & Sands, S. (2015). Environmental concern and environmental purchase intentions: The mediating role of learning strategy. *Journal of Business Research*, *68*(9), 1974–1981. doi:10.1016/j.jbusres.2015.01.007

Papista, E., & Krystallis, A. (2013). Investigating the types of value and cost of green brands: Proposition of a conceptual framework. *Journal of Business Ethics*, *115*(1), 75–92. doi:10.100710551-012-1367-6

Peattie, K. (2001). Golden goose or wild goose? The hunt for the green consumer. *Business Strategy and the Environment*, *10*(4), 187–199. doi:10.1002/bse.292

Polonsky, M. J., Kilbourne, W., & Vocino, A. (2014). Relationship between the DSP, materialism, and environmental behaviors in four Asian economies. *European Journal of Marketing*, *48*(3-4), 522–551. doi:10.1108/EJM-07-2011-0351

Purvis, B., Mao, Y., & Robinson, D. (2019). Three pillars of sustainability: In search of conceptual origins. *Sustainability Science*, *14*(3), 681–695. doi:10.100711625-018-0627-5

Rettie, R., Burchell, K., & Barnham, C. (2014). Social normalisation: Using marketing to make green normal. *Journal of Consumer Behaviour*, *13*(1), 9–17. doi:10.1002/cb.1439

Rettie, R., Burchell, K., & Riley, D. (2012). Normalising green behaviours: A new approach to sustainability marketing. *Journal of Marketing Management*, *28*(3-4), 420–444. doi:10.1080/0267257X.2012.658840

Riskos, K., Paraskevi, D., Mylonas, N., & Tsourvakas, G. (2021). Ecolabels and the attitude-behavior relationship towards green product purchase: A multiple mediation model. *Sustainability (Basel)*, *13*(12), 6867. doi:10.3390u13126867

Robert, D. S., & James, A. R. (1999). Environmental segmentation alternatives: A look at green consumer behavior in the new millennium. *Journal of Consumer Marketing*, *16*(6), 558–575. doi:10.1108/07363769910297506

Roberts, J. A. (1995). Profiling levels of socially responsible behavior: A cluster analytic approach and its implications for marketing. *Journal of Marketing Theory and Practice*, *3*(4), 97–117. doi:10.1080/10696679.1995.11501709

Roberts, J. A. (1996). Green consumers in the 1990's: Profile and implications for advertising. *Journal of Business Research*, *36*(3), 217–231. doi:10.1016/0148-2963(95)00150-6

Roitner-Schobesberger, B., Darnhofer, I., Somsook, S., & Vogl, C. R. (2008). Consumer perceptions of organic foods in Bangkok. *Food Policy*, *33*(2), 112–121. doi:10.1016/j.foodpol.2007.09.004

Samdahl, D. M., & Robertson, R. (1989). Social determinants of environmental concern. *Environment and Behavior*, *21*(1), 57–81. doi:10.1177/0013916589211004

Schlegelmilch, B. B., Bohlen, G. M., & Diamantopoulos, A. (1996). The link between green purchasing decisions and measures of environmental consciousness. *European Journal of Marketing*, *30*(5), 35–55. doi:10.1108/03090569610118740

Schwartz, M. (2021). *How technology can provide a more sustainable future for the industrial sector.* Retrieved May 30, 2022, from https://forbes.com/2021/08/31

Schwepker, C. H. Jr, & Cornwell, T. B. (1991). An examination of ecologically concerned consumers and their intention to purchase ecologically packaged products. *Journal of Public Policy & Marketing*, *10*(2), 77–101. doi:10.1177/074391569101000205

Sharma, A. P. (2021). Consumers' purchase behavior and green marketing: A synthesis, review, and agenda. *International Journal of Consumer Studies*, *45*(6), 1217–1238. doi:10.1111/ijcs.12722

Shrum, L. J., McCarty, J. A., & Lowrey, T. M. (1995). Buyer characteristics of the green consumer and their implications for advertising strategy. *Journal of Advertising*, *24*(2), 71–82. doi:10.1080/00913367.1995.10673477

Singh, A., & Verma, P. (2017). Factors influencing Indian consumers' actual buying behavior towards organic food products. *Journal of Cleaner Production*, *167*, 473–483. doi:10.1016/j.jclepro.2017.08.106

Sreen, N., Purbey, S., & Sadarangani, P. (2018). Impact of culture, behavior and gender on green purchase intention. *Journal of Retailing and Consumer Services*, *41*, 177–189. doi:10.1016/j.jretconser.2017.12.002

Stadelmann, M., & Schubert, R. (2018). How do different designs of energy labels influence purchases of household appliances? A field study in Switzerland. *Ecological Economics*, *144*, 112–123. doi:10.1016/j.ecolecon.2017.07.031

Straughan, R. D., & Roberts, J. A. (1999). Environmental segmentation alternatives: A look at green consumer behavior in the new millennium. *Journal of Consumer Marketing*, *16*(6), 558–575. doi:10.1108/07363769910297506

Tanner, C., & Kast, S. W. (2003). Promoting sustainable consumption: Determinants of green purchases by Swiss consumers. *Psychology and Marketing*, *20*(10), 883–902. doi:10.1002/mar.10101

Testa, F., Iraldo, F., Vaccari, A., & Ferrari, E. (2015). Why eco-labels can be effective marketing tools: Evidence from a study on Italian consumers. *Business Strategy and the Environment*, *24*(4), 252–265. doi:10.1002/bse.1821

Testa, F., Pretner, G., Iovino, R., Bianchi, G., Tessitore, S., & Iraldo, F. (2021). Drivers to green consumption: A systematic review. *Environment, Development and Sustainability*, *23*(4), 4826-4880.

Tilikidou, I., & Delistavrou, A. (2012). The components of pro-environmental purchasing behavior. In *Proceedings of the International Conference on Contemporary Marketing Issues* (pp. 549-555). Academic Press.

Tung, J., Shih, C., Wei, S., & Chen, H. (2012). Attitudinal inconsistency toward organic food in relation to purchase intention and behavior: An illustration of Taiwan consumers. *British Food Journal*, *114*, 997–1015. doi:10.1108/00070701211241581

van Riper, C. J., Lum, C., Kyle, G. T., Wallen, K. E., Absher, J., & Landon, A. C. (2020). Values, motivations, and intentions to engage in pro-environmental behavior. *Environment and Behavior*, *52*(4), 437–462. doi:10.1177/0013916518807963

van Tonder, E., Fullerton, S., & de Beer, L. T. (2020). Cognitive and emotional factors contributing to green customer citizenship behaviours: A moderated mediation model. *Journal of Consumer Marketing*, *37*(6), 639–650. doi:10.1108/JCM-06-2019-3268

Vaughan, D. (2022). *What is the most widely practiced religion in the world?* Retrieved June 17, 2022, from https://www.britannica.com/story/what-is-the-most-widely-practiced-religion-in-the-world

Wang, H. (2022). Exploring the influencing factors of environmental deterioration: Evidence from China employing ARDL–VECM method with structural breaks. *International Journal of Climate Change Strategies and Management*. doi:10.1108/IJCCSM-10-2021-0114

Wang, Y. F., & Wang, C. J. (2016). Do psychological factors affect green food and beverage behavior? An application of the theory of planned behavior. *British Food Journal*, *118*(9), 2171–2199. doi:10.1108/BFJ-12-2015-0469

Watkins, L., Aitken, R., & Mather, D. (2016). Conscientious consumers: A relationship between moral foundations, political orientation and sustainable consumption. *Journal of Cleaner Production*, *134*, 137–146. doi:10.1016/j.jclepro.2015.06.009

WCED. (1987). Our common future. Oxford University Press.

Webster, F. E. Jr. (1975). Determining the characteristics of the socially conscious consumer. *The Journal of Consumer Research*, *2*(3), 188–196. doi:10.1086/208631

Wijekoon, R., & Sabri, M. F. (2021). Determinants that influence green product purchase intention and behavior: A literature review and guiding framework. *Sustainability (Basel)*, *13*(11), 6219. doi:10.3390u13116219

Witek, L., & Kuzniar, W. (2021). Green purchase behavior: The effectiveness of sociodemographic variables for explaining green purchases in an emerging market. *Sustainability (Basel)*, *13*(1), 209. doi:10.3390u13010209

Wolkomir, M. J., Futreal, M., Woodrum, E., & Hoban, T. (1997). Denominational subcultures of environmentalism. *Review of Religious Research*, *38*(4), 325–343. doi:10.2307/3512194

Wymer, W., & Polonsky, M. J. (2015). The limitations and potentialities of green marketing. *Journal of Nonprofit & Public Sector Marketing*, *27*(3), 239–262. doi: 10.1080/10495142.2015.1053341

ADDITIONAL READING

Bedard, S. A. N., & Tolmie, C. R. (2018). Millennials' green consumption behavior: Exploring the role of social media. *Corporate Social Responsibility and Environmental Management*, *25*(6), 1388–1396. doi:10.1002/csr.1654

Carson, R. (1962). *Silent Spring*. Houghton Mifflin Co.

Chabowski, B. R., Mena, J. A., & Gonzales-Padron, T. L. (2011). The structure of sustainability research in marketing, 1958–2008: A basis for future research opportunities. *Journal of the Academy of Marketing Science*, *39*(1), 55–70. doi:10.100711747-010-0212-7

Do Paço, A., Shiel, C., & Alves, H. (2019). A new model for testing green consumer behavior. *Journal of Cleaner Production*, *207*, 998–1006. doi:10.1016/j.jclepro.2018.10.105

Groening, C., Sarkis, J., & Zhu, Q. (2018). Green marketing consumer-level theory review: A compendium of applied theories and further research directions. *Journal of Cleaner Production*, *172*, 1848–1866. doi:10.1016/j.jclepro.2017.12.002

Otterbring, T., & Folwarczny, M. (2022). Firstborns buy better for the greater good: Birth order differences in green consumption values. *Personality and Individual Differences*, *186*, 111353. doi:10.1016/j.paid.2021.111353

Piligrimiene, Z., Zukauskaite, A., Korzilius, H., Banyte, J., & Dovaliene, A. (2020). Internal and external determinants of consumer engagement in sustainable consumption. *Sustainability (Basel)*, *12*(4), 1349. doi:10.3390u12041349

Saari, U. A., Damberg, S., Frömbling, L., & Ringle, C. M. (2021). Sustainable consumption behavior of Europeans: The influence of environmental knowledge and risk perception on environmental concern and behavioral intention. *Ecological Economics, 189*, 107155. doi:10.1016/j.ecolecon.2021.107155

Taufique, K. M. R., & Vaithianathan, S. (2018). A fresh look at understanding green consumer behavior among young urban Indian consumers through the lens of Theory of Planned Behavior. *Journal of Cleaner Production, 183*, 46–55. doi:10.1016/j.jclepro.2018.02.097

Yue, B., Sheng, G., She, S., & Xu, J. (2020). Impact of consumer environmental responsibility on green consumption behavior in China: The role of environmental concern and price sensitivity. *Sustainability (Basel), 12*(5), 2074. doi:10.3390u12052074

KEY TERMS AND DEFINITIONS

Contextual/Situational Factors: The specific circumstance and physical surroundings a consumer is confronted with while selecting a product/service to purchase.

Demographic Factors: Factors related to the socioeconomic characteristics of a population, such as gender, age, level of education, income, and occupation.

Environmental Sustainability: Acting responsibly towards the environment so as to preserve its quality both for the prevailing and coming generations.

Green Consumer: An individual who is considerate of the environment in which he/she lives.

Green Consumption Behavior: Selecting, purchasing, using, and disposing of products/services in such a way so as to give the least amount of harm to the environment.

Individual Factors: Factors that are related to internal influences (such as personality, attitudes, motivation, and perception) which affect individual consumers.

Pro-Environmental Behavior: All possible actions taken by individuals so as to protect the environment.

Social and Cultural Factors: Factors that exist in the social and cultural settings of consumers (such as values, ethical and moral considerations, and religion).

Sustainable Development: An approach to the process of growth and progress whereby societies meet their needs taking into consideration the coming generations, as well.

Chapter 11
Environmental Violations and Mandatory Disclosures by Oil and Gas Companies

Radiah Othman
Massey University, New Zealand

Rashid Ameer
IPU New Zealand, New Zealand

ABSTRACT

Environmental degradation will continue if the corporate continues to misbehave. This study offers evidence for whether environmental violators are more likely to disclose if they are in sensitive industries, faced higher penalties, and required to implement an environmental project. This chapter examines how environmental violations have shaped the mandatory and discretionary environmental disclosures of oil and gas companies in 10-k reports filed with the US Securities and Exchange Commission from 2011 to 2020. Using the Leximancer text mining tool, the violators' annual reports were examined before and after the violations. The results suggest that disclosure tended to be discretionary, and it is possible that selective information was traded off with other discretionary non-financial information for the sake of legitimacy. This chapter provides evidence of the effectiveness of regulators' efforts for higher mandated disclosures to encourage firms to deal with environmental matters proactively to prevent further environmental misbehaviour.

DOI: 10.4018/978-1-6684-6727-5.ch011

1. INTRODUCTION

Escalating environmental and climate change-related risks associated with extractive industry activities have intensified lobbying for increased environmental disclosure regulations by influential investor groups and non-governmental organisations. An analysis of shareholder voting trends from 2000 to 2018 was conducted by ISS Analytics, the research arm of the major proxy advisor Institutional Shareholders' Services (ISS). It concluded that the most significant change in investors' voting behaviour pertains to environmental and social issues (William and Nagy, 2021). The environmental and social justice rights of stakeholders are better safeguarded with more information about fossil fuel producers' potential contributions to environmental and climate change-related risks (Griffin and Jafee, 2018).

In the last decade or so, the two main regulators—the US Securities and Exchange Commission (SEC) and US Environmental Protection Agency (EPA)—have intensified their efforts towards higher mandated disclosures to encourage firms to deal with environmental matters proactively and prevent environmental misbehaviour (Peters and Romi, 2013). Anti-fraud provisions in federal laws require firms to disclose environmental violations imposed by the EPA in 10-k filings (EPA, 2001). In the settlement of civil cases, the EPA allows the violating firm to propose potential Supplementary Environmental Projects (SEP) to lessen the imposed penalty (Monti and Deason, 2012). Starting in 1998, the key principle of the SEP program was to implement environmentally beneficial projects that compensated the environment and the community where the alleged violation occurred beyond what firms were required to do by law (EPA, n.d). Undertaking SEPs is an initiative that promotes restorative justice (Lee and Xiao, 2020). If a firm agrees to implement a SEP as part of a settlement, then the amount of the EPA penalty does not exceed 80% of the estimated cost of the SEP implementation (EPA, 2015). Although the EPA's Clean Air Act and the Clean Water Act (CWA) mandate disclosure while the SEC's Regulation S-K (SEC, 2019, 2010) contains several provisions that directly address the disclosure of environmental liabilities, there is a lack of conclusive evidence that such disclosures help firms gain legitimacy in addressing environmental degradation.

The SEC's Regulation S-K Item 103 requires that SEC registrants disclose significant pending lawsuits or other legal proceedings known to be under consideration by a government authority to which they are a party if they arise under federal, state, or local provisions that have the primary purpose of protecting the environment (EPA, 2001). Environmental disclosures under Item 103 are initiated if at least one of the three conditions are met: (i) material proceedings; (ii) the relief sought amounts to over 10% of the registrants' current assets; or (iii) if government sanctions would amount to more than $100,000[1] (Pfund, 2004). The purpose of mandated SEC environmental disclosure is to reduce information asymmetry. If the size of the proposed penalty reaches the SEC's materiality threshold, the likelihood of the final penalty to be decided by the EPA tends to increase; thus, non-disclosure of the violations would only increase the probability of litigation by shareholders (Peters and Romi, 2013).

News of environmental violations by extractive industries automatically receive public attention due to controversial extraction processes for materials such as fossil fuels (Baudot et al., 2021). From a qualitative enquiry perspective, archival evidence is limited on how environmental violators from oil and gas industries choose accounting-related information in the disclosure mandated by the EPA in 10-k filings with the SEC. The theoretical rationale behind the sample selection is that the environmental violators have greater visibility that translates into public expectations and increases commitments for mandated disclosures and environmental improvements following the SEP announcement. Motivated by Peters and Romi's (2013) suggestions that US firms are more likely to provide sanction disclosures as per the SEC's Regulation S-K if they operate in environmentally sensitive industries, are subject to larger penalties, and voluntarily implement SEP, this chapter intends to validate these findings by choosing a sample of US oil and gas firms that have been the target of EPA enforcement actions between 2011 and 2020. A violating firm's adherence to the requirement reflects its attitude and commitment toward environmental communication (Peters and Romi, 2013). Consequently, our research question is: how do environmental violations shape mandatory environmental sanction disclosures for oil and gas companies? We identify and examine the main disclosure themes in different sections of the 10-k reports and pre- and post-environmental law violations among the top and bottom violators. We also examine whether the themes of violation and impact were prominent in violators' disclosures. A sentiment analysis was carried out to examine whether management's discretion translated into sentiment. As highlighted by Peters and Romi (2013), adherence to the requirement represents a potentially important indicator of a firm's attitude toward providing users of SEC filings with environmental disclosure. However, research that reports SEC environmental sanction disclosure is limited, and this chapter aims to fill this gap.

The chapter is structured as follows. The next section reviews the literature and is followed by the research method. The results are discussed in the findings and discussion sections. The final section concludes the chapter.

2. LITERATURE REVIEW

This literature builds on and complements the primarily quantitative research on US firms' environmental disclosure practices as they relate to US regulators' monitoring function of corporate behaviour (e.g., Alexander et al., 2011; Ettredge et al., 2001,

2011; Heitzman et al., 2010; Peters and Romi, 2013). An earlier study by Abdelrehim et al. (2015) reported how BP and Shell responded differently to crises in the 1950s. Other studies in the late 20th and early 21st centuries (e.g., Craig and Amernic, 2004; Guillamon-Saorin et al., 2012; Ogden and Clarke, 2005; Yang and Malone, 2008) did not focus on extractive industries nor did they examine disclosure responses after environmental violations. From the regulator's perspective, this information would be useful for better disclosure monitoring in the future (Alexander et al., 2011; Ettredge et al., 2001; Ettredge et al., 2011; Peters and Romi, 2013).

There is a call for research examining whether there is a strategic trade-off within the disclosure obligations (e.g., Fallan and Fallan, 2019; Heitzman et al., 2010). In the context of extractive industries, accounting information may be traded off with other discretionary non-financial information by firms to gain legitimacy. Accounting information such as that allocated in a budget for environmental remediation is a useful indication of commitment to mitigate the adverse effects of such violations (e.g., Cho et al., 2012). In previous research on environmental accounting information (EAI) in a non-US context, scholars have found that the type of information disclosed is limited and diverse across industry and jurisdiction (Cho and Patten, 2008; Criado-Jiménez et al., 2008; Larrinage et al., 2002; Senn and Giordano-Spring, 2020; Unerman et al., 2018). EAI provides both accounting and environmental information and as such is found to be value relevant to investors (Clarkson et al., 2004) as there is a demand for non-financial information (Senn and Giordano-Spring, 2020).

Past research has generally coded EAI into either category or assigned numerical values and it has thus lacked a measurement of the nature and extent of EAI being disclosed. Recently, Liu and Liu (2021) identified more information on the risks, regulations, and compliance disclosed by firms, but accounting information on environmental income, assets, rights, and liabilities was poorly disclosed, probably due to a lack of information in the accounting standards on what to disclose. Senn and Giordano-Spring (2020) suggested that one of the reasons for this poor disclosure is that management finds ambiguous definitions in regulations and an absence of accounting standards that impact on their decision to disclose EAI. Until accounting information is formalised into the environmental management system, the full cost of environmental violations cannot be determined. Cowan and Gadenne (2005) noted that in 10-k filings, management is selective about the sections in which they disclose the information. Their findings suggest that companies adopt different disclosure approaches when their environmental disclosures may be under increased scrutiny or surveillance. The disclosure may also influence the extent of EAI narratives. Heitzman et al. (2010) suggest that if a company's stock prices are sensitive to accounting information, management are more likely to disclose a given piece of sensitive environmental information.

The extant literature provides evidence of non-discretionary and discretionary disclosure for mandatory requirements (e.g., Alexander et al., 2011; Beyer et al., 2010; Ettredge et al., 2001; Hassan et al., 2008; Peters and Romi, 2013; Schwartz and Soo, 1996; Tsalavoutas et al., 2020; Wiedman and Wier, 1999). Discretionary disclosure theory suggests that managers act strategically in making disclosure decisions in both voluntary and mandatory settings (e.g., Alexander et al., 2011; Beyer et al., 2010; Blouin et al., 2010; Kothari et al., 2009; Wiedman and Wier, 1999). Corporate strategic disclosure is often driven by reputational concerns that weigh the costs and benefits of such disclosures and the extent of the information to be disclosed (Abdelrehim et al., 2015; Heitzman et al. 2010; Kothari et al. 2009; Leuz et al., 2004; Peters and Romi, 2013). Firms in environmentally sensitive industries are found to disclose more about their activities due to a higher propensity for environmental sanctions, implicit environmental liabilities, and scrutiny (Li et al., 1997; Nelson and Pritchard, 2016; Peters and Romi, 2013). According to Verrecchia (1983), firms in less competitive industries may see no costs associated with making public disclosures. In contrast, firms in highly competitive industries may regard public disclosure of any kind as potentially damaging if it reveals sensitive information to competitors.

Voluntary disclosure theorists (e.g., Clarkson et al., 2008) suggest that firms use voluntary rather than passive disclosure to proactively signal their societal concerns. However, when firms face mandatory reporting requirements because of environmental law violations, they tend to alter their disclosure practices because of managers' affirmative duty to disclose (Heitzman et al., 2010; Peters and Romi, 2013). Compliance is negatively associated with violations (Ettredge et al., 2011; Kothari et al., 2009; Nelson and Pritchard, 2016). Li et al. (1997) suggest that the magnitude of the penalty indicates the severity of the negative information to be disclosed, thus disincentivising management to disclose. Peters and Romi (2013) found that firms are more likely to provide sanction disclosures if they are subject to larger penalties and voluntarily participate in SEPs. It is therefore expected that environmental violations, pending litigation, and penalties announced by the EPA would be reported in more detail by firms in environmentally sensitive extractive industries such as oil and gas.

In sum, when firms' environmental violations are disclosed to the public, it increases public scrutiny, specifically from investors and creditors. EAI disclosures in the post-environmental violation period reveal organisational actions that may (or may not) meet expectations from regulators and society (Mobus, 2005). This chapter posits that when environmental violators undertake SEPs, they may be perceived as a monetary tool to compensate the environment (i.e., reparative justice). They may also convey a change in firm approach to what, when, where, and how EAI is disclosed discretionarily in the mandatory disclosures to the 10-k report users. Under the SEC's materiality threshold, accounting information disclosure might be shaped not only by mandated disclosure but also by public expectations.

3. MATERIALS AND METHOD

Employing a qualitative content analysis of 10-k reports, the study adopted the following sample-selection process. First, we identified the US oil and gas companies that violated the Clean Air Act from 2011 to 2020 using the EPA ECHO Database. Out of 45 firms, 15 firms agreed to SEPs in the EPA settlement, and 30 firms did not. Of the 15 firms, six were not registered as public companies. Next, the 10-k filings from the other nine companies were downloaded from the SEC website. Form 10-K is a detailed annual report that provides a comprehensive summary of a company's performance for the year. Item 1 requires a description of the business that is material to an understanding of the business's general development. Item 3 is the legal proceeding section in which the company is required to provide a brief description of material pending legal proceedings. Item 7 is the management discussion and analysis of financial conditions and operating results. We classified all 10-k reports into three categories: Item 1 (72 reports), Item 3 (73 reports), and Item 7 (58 reports). As illustrated in Table 1, these firms committed to SEPs that totalled more than $14 million.

Table 1. Sample firms by committed supplemental environmental project costs

	Companies Name	Committed SEP Amount ($)
1	Chevron	9,495,909.08 (67.5%)
2	Noble Energy	2,000,000 (14.2%)
3	Valero	1,125,000 (8.0%)
4	ConocoPhillips	400,000 (2.8%)
5	MPLX	385,000 (2.7%)
6	BreitBurn Energy	285,000 (2.0%)
7	Whiting Petroleum	204,280 (1.5%)
8	Berry Petroleum	114,000 (0.8%)
9	Phillips66	58,000 (0.4%)
	Total	14,067,189.08 (100%)

After this, the sampled firms were classified into two sub-groups: top and bottom violators, according to the amount of SEPs. The top violators incurred a higher cost for SEPs, representing at least 20% of the total fines imposed by the EPA for environmental violations. A previous study by Abdelrehim et al. (2015) suggested that oil companies reflected different concerns in times of crisis. Finally, the EPA ECHO Database was used to identify the year the violation occurred to segregate the documents into pre-and post-violations and determine whether firms reported similar concerns in these two events.

3.1 Automated Content Analysis Procedure

Text-mining software tool Leximancer was used for automated content analysis (ACA). Leximancer 5.0, a text analytics tool, was used for content analysis of textual documents and visualisation of extracted information (Leximancer, 2021). Leximancer generates objective and text-driven review documents with reproducible, reliable concept extractions and thematic clustering. Further, it minimises pre-conception bias (Biesenthal and Wilden, 2014; Cheng and Edwards, 2019; Smith and Humphreys, 2006). These capabilities are frequently used in qualitative research across many disciplines, such as communication (Rooney et al., 2010), tourism management (Tseng et al., 2015), and health research (Cretchley et al., 2010; Lemon and Hayes, 2020), but their use in accounting research is very limited. In the study described in this chapter, all concepts were auto-generated to discover the most common themes and concepts in the documents and identify those that were conspicuous by their absence, including the ability to auto-generate a sentiment analysis.

Leximancer extracts frequently occurring terms (concepts) in documents and tabulates their co-occurrence to identify inter-relationships (Guide, 2021). Underpinned by Bayesian theory, Leximancer quantifies texts by identifying the frequency of concepts and their relationships through an emergent iterative and unsupervised process (Cheng and Edwards, 2017; Smith and Humphreys, 2006). This text-mining tool read the 10-k reports and produced a map of key concepts, with their relationships indicated by their proximity on the map. Concepts in Leximancer are collections of words that travel together throughout the text and are weighted according to how frequently they occur in sentences containing the concept, compared to how frequently they occur elsewhere (Guide, 2021). Concept definitions are also used to determine the frequency of co-occurrence between the concepts. This co-occurrence measure was used to generate the concept map, which contains the main concepts within the text and is clustered into higher-level themes (Guide, 2021, p.20). Concepts that often appear together attract one another strongly and tend to settle near one another in the map space.

The text-mining tool also identifies important themes determined from the number of concepts present in each theme (Guide, 2021). The number of connections determines the size of the theme, and the count and relevance are calculated for each theme. Sentences are tagged as containing a concept if there was sufficient accumulated evidence. The themes are heat-mapped and colour-coded to indicate importance, with the red colour indicating the most important concept (Guide, 2021, Angus et al., 2013). There is consistency is how the texts are classified and how the relationships are identified among concepts, and the same result is produced regardless of the number of times data is coded and recoded (Guide, 2021; Fisk et al., 2014).

The total number of themes, relevance, and counts were analysed to identify the extent to which concepts were represented in 10-k reports. Leximancer provided a means for generating and recognising themes, including themes that might otherwise have been missed or overlooked. Smith and Humphreys (2006) also validated Leximancer according to the typology presented by Krippendorff (2004), which also includes predictive validity and correlative validity (including semantic and functional validity). Using the themes derived by the software, the researchers then returned to direct engagement with the text to explore and interpret its meanings further. Leximancer enables the themes to be traced from the documents; thus, excerpts from the documents are also included in the analysis.

4. RESULTS

4.1 Most Frequent and Important Themes and Concepts From 2011 to 2020

The study relied on the ACA capabilities of Leximancer to simultaneously search and analyse the text of 10-k reports in order to identify concepts and themes which were then compared to a manual text content analysis. As illustrated in Figure 1, the five dominant themes that appear most frequently in the 10-k reports from 2011 to 2020 are: operations, EPA, hazardous, violations, and remediation. These themes are consistent with the nature of the industry and the implicit risks associated with its operations (e.g., hazardous).

The themes were automatically heat-mapped and colour-coded to indicate their importance. 'Operations' was the dominant theme between 2011–2020 in mandated disclosure of legal proceedings by violating oil and gas companies (Item 103). Concepts that frequently appeared together attracted one another strongly and tended to settle near one another in the map space; thus, in the green circle, 'disposal' and 'substances' appeared near to 'hazardous'. This study also identified the relevance and likelihood of concept co-occurrence under themes when the concepts tab in

Leximancer was selected. The dimension 'costs' was 75% relevant to the discourse contained in the 10-k reports with 'financial' (52%) and 'liability' (30%). When the 'cost' concept was checked, the likelihood of it being related to 'future' was 48%, 'liability' 45%, and 'financial' at 40%. Overall, these findings suggest that both compliance with SEC environmental disclosure requirements and information about future financial liabilities associated with CWA and CAA violations has been at the core of EAI in the past decade.

Figure 1.

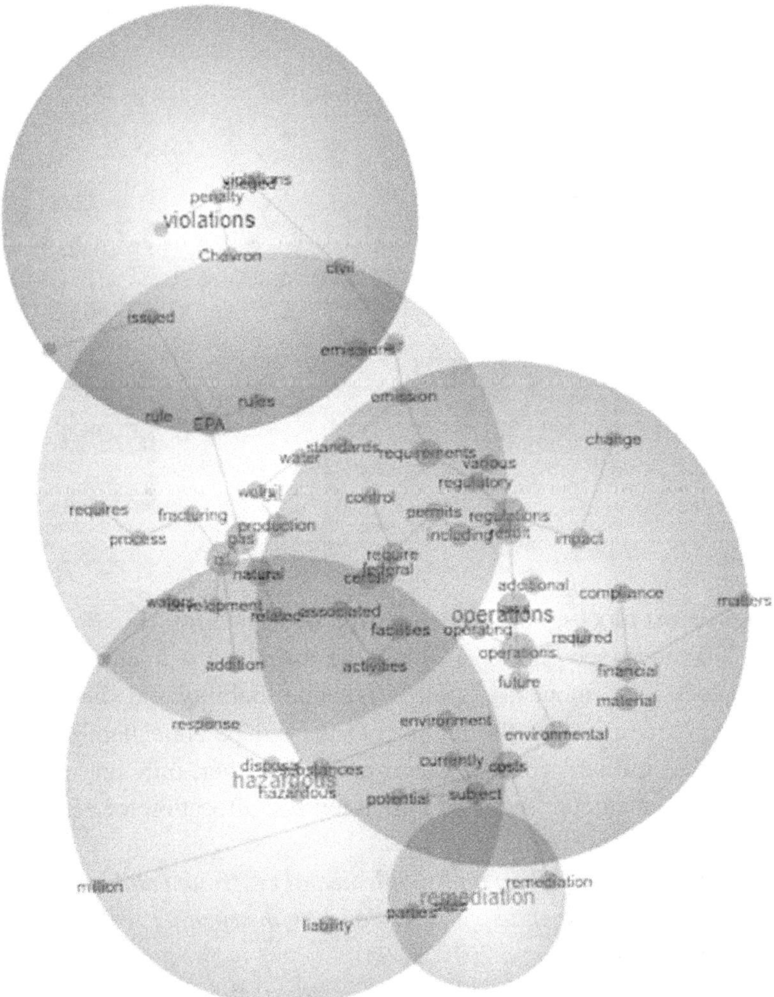

Figure 2.

	Co-Count	Likelihood
parties	168	62%
operating	182	55%
future	179	48%
result	228	45%
million	78	45%
liability	152	45%
sites	137	45%
remediation	173	45%
material	246	42%
compliance	150	42%
environment	129	41%
additional	119	41%
financial	230	40%

Over the context blocks, 'costs' co-occurred with 'liability' 152 times and with 'financial' 230 times (see Figure 2). When the document extracts were read, the disclosure tended to be monetary, such as potential liability and site remediation costs (Conoco, 2012) as well as the number of sites with potential liabilities (Phillip, 2012). When the 'million' concept was examined further, only one firm, Noble Energy, reported the amount spent for mitigation and allocation for SEPs:

Mitigation costs of USD4 million and supplemental environmental project costs of USD4 million are being expended in accordance with schedules established in the Consent Decree. (Noble Energy, 2019, p. 88)

Chevron and Phillips have disclosed expenses and capitalised environmental costs since 2011. Conoco reported the environmental costs accrued relative to the preceding year for remediation activities, and MPLX reported the yearly amount paid for incurred settlements.

4.1.1 Themes by Item in 10-k Filing

Under Item 1 of a 10-k filing, companies are required to provide a description of the general development of the registrant's business. The EPA has informed SEC's registrants that environmental information may need to be disclosed under this item (EPA, 2001). As illustrated in Table 2, 'oil', 'gas', 'operations', 'requirements', and 'regulations' were the top five themes under Item 1. The least discussed theme in this section was 'safety' (10%). Only a few companies, such as Conoco, reported its safety initiative in 2012 (p.22):

Assessments are conducted annually to capture progress and set new targets. We are also committed to continuously improving process safety and preventing releases of hazardous materials.

The top five themes under Items 1, 3, and 7 are illustrated in Table 2, which seemed to be consistent with the requirements of each section are: description of business, legal proceedings, and management discussion and analysis (MD&A). The 'operations' theme was in the top five in all sections, and the 'regulations' theme appeared in Item 1 and Item 7.

Table 2. Top-five themes by 10-k filing section

Rank	Item 1 Description of Business (Count; Relevance)	Item 3 Legal Proceedings (Count; Relevance)	Item 7 Management Discussion and Analysis (Count; Relevance)
	Themes		
1	Oil (680; 100%)	Adverse (222; 100%)	Environmental (393; 100%)
2	Gas (618; 91%)	Operations (219; 99%)	Costs (373; 95%)
3	Operations (560; 82%)	Violations (219; 99%)	Regulations (369; 94%)
4	Requirements (551; 81%)	Effect (188; 85%)	Laws (337; 86%)
5	Regulations (528; 78%)	Material (188; 85%)	Operations (301; 77%)

Under Item 3, companies must provide a brief description of any pending legal proceedings if the amount exceeds 10% of current consolidated assets, it is material, and potential monetary sanctions do not exceed $100,000. The proper noun that appeared most frequently in the texts was 'EPA' (this is an untabulated result). These results indicate that environmental sanctions and decisions by the EPA carried a higher weighting for environmental violators during this period. A disclosure on EPA sanctions can be found in Valero's 2020 (p.26) filing:

On December 11, 2020, the US EPA issued a Notice of Potential Violations and Opportunity to Confer related to a series of inspections conducted by the US EPA in 2019, arising out of the 2017 Pacific Gas and Electric Company power outage, and a 2019 emissions event.

In item 3, the top theme was 'adverse'. It occurred 222 times and was 100% relevant to the discussion of legal proceedings. This theme was clustered with 'operations', 'effect', 'material', 'position', 'financial', 'matters', 'proceedings', and 'outcomes'. For instance, in the MPLX's 2016 (p.72) filing:

In July 2015, representatives from the EPA and the United States Department of Justice conducted a raid on a MarkWest Liberty Midstream pipeline launcher/receiver site utilised for pipeline maintenance operations in Washington County, Pennsylvania pursuant to a search warrant issued by a magistrate of the United States District Court for the Western District of Pennsylvania. As part of this initiative, the US Attorney's Office for the Western District of Pennsylvania proceeded with an investigation of MarkWest Liberty Midstream's launcher/receiver, pipeline, and compressor station operations.

Leximancer identified this as 85% relevant to the discussion of 'legal proceedings'. The following excerpt was classified as material by Leximancer, even though the material concept was not mentioned, demonstrating that Leximancer was able to track the relevant concept and considered it to be material based on the context, not solely on its occurrence in the text:

In April 2015, we entered into a joint consent decree (Consent Decree) with the US Environmental Protection Agency, US Department of Justice, and State of Colorado to improve emission control systems at a number of our condensate storage tanks that are part of our upstream oil and natural gas operations within the Non-Attainment Area of the DJ Basin. The Consent Decree was entered by the Court on June 2, 2015. (Noble Energy, 2015, p.8)

For Item 7, the top-ranked theme was 'environmental', which was included in discussions and analyses by the firms' management. For instance, the MPLX discussion in 2020 (p.88) stated:

Our environmental capital expenditures are expected to approximate USD18 million in 2021. Actual expenditures may vary as the number and scope of environmental projects are revised as a result of improved technology or changes in regulatory requirements and could increase if additional projects are identified or additional requirements are imposed.

In the above excerpt, the monetary amount of capital expenditure was revealed in the discussion on the environment. Similarly, the theme of 'costs' appeared in the discussion section of Conoco in 2012 (p.60):

Expensed environmental costs were $575 million in 2012 and are expected to be about $478 million per year in 2013 and 2014. Capitalized environmental costs were $297 million in 2012 and are expected to be about $459 million per year in 2013 and 2014.

Our readings of the texts indicated that, historically, different sections of the filings could potentially be a deciding parameter for monetary accounting-related information to be disclosed. When environmental accounting information such as environmental expenditures were incurred or occurred, it was not always clear whether the costs were devoted to environmental remediation or mitigation of future violations through new investment and technology. It is interesting to note that the 'environmental' theme ranked first in Item 7, while 'adverse', 'operations', and 'violations' ranked first in Item 3, considering that the sampled companies have violated the Clean Air Act. This finding is supported by legitimacy theory, which implies that environmental violators have greater public expectations placed on them to increase their commitment to environmental protection.

4.1.2 Themes in 10-k Report Pre- and Post-Violations

In this section, the theme in the 10-k report was identified by dividing the period into pre- and post-violation. The year of environmental violation was identified for each firm from the EPA violation database to identify whether the firms' disclosure emphasised similar themes in the pre- and post-EPA enforcement action events. Only the top three ranked themes were discussed for comparison purposes.

Table 3. Top-three themes pre- and post-violations

	Pre-Violation (Ranked Theme; Relevance)	Post-Violation (Ranked Theme; Relevance)
Item 1		
Description of business EPA relevancy	Operations (100%)	Oil (100%)
	Requirements (89%)	Gas (92%)
	Oil (81%)	Regulations (71%)
	54%	49%
Item 3		
Legal proceeding EPA relevancy	Civil (100%)	Violations (100%)
	Alleged (80%)	Operations (95%)
	Result (59%)	Results (89%)
	32%	54%
Item 7		
Management discussion and analysis on performance EPA relevancy	Environmental (100%)	Environmental (100%)
	Costs (89%)	Regulations (98%)
	Regulations (76%)	Laws (95%)
	0%	35%

Table 3 illustrates that in the pre-violation years, the sample firms' disclosure was related to normal business operations in the discussion of 'oil', 'gas', and 'operations' in Item 1, which is expected as companies discussed their business. When the texts were tracked, our readings confirmed that the companies described their business operations and requirements as follows:

Air Emissions. The Clean Air Act, and comparable state laws, regulate emissions of various air pollutants through air emissions permitting programs and the imposition of other requirements. (Breitburn, 2012, p.15)

In the post-violation years, 'operations' was replaced by 'operations in the three most frequently discussed themes. This text referred to current and future regulations, as disclosed by Breitburn in 2016 (p.18).

The EPA has developed, and continues to develop, stringent regulations governing emissions of air pollutants at specified sources. For example, in October 2015, the EPA lowered the National Ambient Air Quality Standard ("NAAQS") for ozone from 75 to 70 parts per billion.

The likelihood of 'requirements' and 'regulations' being mentioned together in Item 1 was 34% (pre-violation) and 41% (post-violation), indicating that more than one-third of the texts simultaneously discussed 'requirements' and 'regulations'. Interestingly, the discussion on 'EPA' was reduced to 49% in the post-violation period compared to 54% in the pre-violation period. However, in Item 3, where management explained its involvement with legal proceedings, the 'EPA' became more relevant both in this section and in Item 7 post-violation, which demonstrated a big jump from no mention during the pre-period to 35% post-violation. 'Violations' became the most frequent theme in the Item 3 discussions compared to 'civil' (penalty) in the pre-violation period. The discussion on 'violations' includes pending allegations, resolution, and potential pay-outs. For example, in the 2015 filing, Phillip disclosed that:

Our US refineries are implementing two separate consent decrees, regarding alleged violations of the Federal Clean Air Act, with the U.S. Environmental Protection Agency (EPA), six states and one local air pollution agency. Some of the requirements and limitations contained in the decrees provide for stipulated penalties for violations.

The results remained in the top three themes in both the pre- and post-violation periods. Discussions related to the results of 'litigation cases' (e.g., Whiting, 2019), 'settlement agreements' (e.g., Whiting, 2020), and 'proceedings' (e.g., Conoco, 2016). All of this would affect the results of the operation as described by Berry in its 2012 (p.85) filing:

In the opinion of management, the resolution of these matters will not have a material adverse effect on its business, financial condition, or results of operations; however, cash flow could be significantly impacted in the reporting periods in which such matters are resolved.

We found that only Whiting disclosed accounting-related information to describe the impact of violations and litigations on its accounts. For instance, in the 2019 (p.42) filing, it stated:

We are involved in litigation related to a payment arrangement with a third party, which currently claims damages up to $41 million, as well as court costs and interest, that is scheduled to go to trial in May 2020. Certain amounts have been accrued in accrued liabilities and other in the consolidated balance sheet as of December 31, 2019 and general and administrative expenses in the consolidated statement of operations for the year ended December 31, 2019 based on the determination that it is probable that a loss has been incurred and can be reasonably estimated.

In item 7 (MD&A), both 'environmental' and 'regulations' were listed in the top three most common themes in the pre-and post-violation periods. There was no significant difference in how the companies reported environmental pre- and post-violations. For example, Chevron in 2015 (p. FS—17) (pre-violation) reported the following:

Included in these expenditures were approximately $0.9 billion of environmental capital expenditures and $1.8 billion of costs associated with the prevention, control, abatement or elimination of hazardous substances and pollutants from operating, closed or divested sites, and the abandonment and restoration of sites...For 2016, total worldwide environmental capital expenditures are estimated at $0.6 billion.

In the post-violation period, in 2018, Chevron disclosed similar information. Pre-violation, the emphasis on costs, post-violations, and laws was relevant at 95%, but these concepts were general. An example of how costs were discussed during the pre-violation period:

Future expenditures may be required to comply with the Clean Air Act and other federal, state and local requirements for our various sites, including our pipelines and storage assets. The impact of these legislative and regulatory developments, if enacted or adopted, could result in increased compliance costs and additional operating restrictions on our business, each of which could have an adverse impact on our financial position, results of operations and liquidity. (MPLX, 2014, p.20)

Overall, these findings demonstrate that the most persistent themes in the pre- and post-violation periods were the industry (oil), the regulations that the companies were subjected to, the implicit environmental risk from their operation, and their results and performance. There seemed to be an increase in these themes post violations compared to the pre-violation period. These findings are supported by disclosure and legitimacy theories.

4.1.3 Concepts and Themes by Top and Bottom Violators

In Table 4, the sampled firms were classified into top and bottom violator categories based on the number of SEPs (refer to Table 1).

The nature of the industries—oil and gas—were emphasised by both top and bottom violators in Item 1 when describing their businesses. One of the top violators, Noble Energy, reported its mitigation initiative in 2015:

We have undertaken emission reduction projects such as our US Vapor Recovery Unit (VRU) program, where we have installed VRUs to capture natural gas that would otherwise be flared on a substantial number of our tank batteries.

Table 4. Top concepts and themes by top and bottom violators

	Top Violators (Ranked Concept; Relevance)	Bottom Violators (Ranked Concept; Relevance)
Item 1		
Description of business EPA relevancy	Gas (100%)	Oil (100%)
	Oil (96%)	Gas (85%)
	Requirements (70%)	Laws (80%)
	30%	63%
Item 3		
Legal proceeding EPA relevancy	Violations (100%)	Violations (100%)
	Penalty (84%)	Resolve (98%)
	Alleged (83%)	Effect (86%)
	50%	70%
Item 7		
Management discussion and analysis on performance EPA relevancy	Environmental (100%)	Costs (100%)
	Regulations (99%)	Laws (87%)
	Operations (93%)	Regulations (82%)
	19%	49%

The bottom violators emphasised the discussion on laws under Item 1, which, as per the EPA, required firms to give a general description of the laws and regulations affecting their operations. Phillips 2019 (p.10), for instance, reported the following:

We believe our facilities are in substantial compliance with applicable environmental laws and regulations. However, these laws and regulations are subject to changes, or to changes in their interpretation, by regulatory authorities, and continued and future compliance with such laws and regulations may require us to incur significant expenditures.

Both the top and bottom violators emphasised the discussion surrounding violations in Item 3. The following is an example:

The New Mexico Environment Department has issued a Notice of Violation (NOV) to ConocoPhillips alleging failure to comply with two air emission monitoring requirements at the East Vacuum Liquid Recovery/CO2 Plant in southeastern New Mexico. The Plant has corrected these issues and has resolved this NOV by paying a penalty of USD34,003. (Conoco, 2014, p. 27)

Other than violations, the top violators emphasised penalties and allegations, while the bottom violators focused on reports about resolution and effect. For example, one of the top violators, Chevron, in 2018 (p.21–22), stated:

As previously disclosed, on August 6, 2012, a piping failure and fire occurred at the Chevron refinery in Richmond, California. The United States Environmental Protection Agency (EPA) issued alleged findings of violation related to the incident on December 17, 2013, pursuant to its authority under the Clean Air Act Risk Management Plan program (RMP).

When the results in Leximancer were examined, the 'resolve' concept also focused on the companies' progress in resolving the violations with the EPA or other related parties involved. For example, Whiting in 2017 (p.15) reported the following:

We are currently engaged in discussions with the Colorado Department of Public Health and Environment (the "CDPHE") concerning certain equipment used in our Retail facilities and our compliance with various air permits and applicable federal and state air quality laws and regulations over the control of air pollutant emissions from those facilities. We and the CDPHE are negotiating the terms of a settlement agreement to resolve this matter.

For Item 7, the management's narrative analysis should explain the reasons for material changes in the number of revenue and expenses items between the most recent fiscal year presented and the immediately preceding fiscal year. 'Regulations' appeared in the top three for both top and bottom violators. However, 'environmental' was the most frequently mentioned concept for the top violators, and 'costs' seemed to be the main focus of discussion for the bottom violators. Phillips (2019, p.59) disclosed the following:

As of December 31, 2018, we reported that we had been notified of potential liability under CERCLA and comparable state laws at 27 sites within the United States. During 2019, there were no new sites for which we received notice of potential liability, nor were any existing sites deemed resolved and closed, leaving 27 unresolved sites with potential liability at December 31, 2019.

It can be inferred from Item 7 that accounting-related information, such as 'costs', did not receive similar emphasis by the top violators who focused on the environmental concept. However, when the excerpts generated by Leximancer were checked, it was found that 'environmental' and 'costs' were linked. For example, an excerpt on 'environmental' by Conoco (top violator) (2012, p.60) revealed the following:

At December 31, 2012, our balance sheet included total accrued environmental costs of $364 million, and we expect to incur a substantial amount of these expenditures within the next 30 years. On December 31, 2011, accrued environmental costs were $922 million, of which $542 million related to the Downstream business.

Similar information was reported by Phillips in 2012 (p.45) on 'costs':

At December 31, 2012, our balance sheet included total accrued environmental costs of $530 million, compared with $542 million at December 31, 2011, and $554 million at December 31, 2010. We expect to incur a substantial amount of these expenditures within the next 30 years.

Table 4 indicates that both the top and bottom violators' narratives were quite similar in the extent of the discussion of their industry, 'oil and gas', and 'regulations'. 'Violations' were at the top of their agenda, but surprisingly, 'EPA' was more relevant in the narratives of the bottom violators than the top violators. Similarly, accounting-related information, such as 'costs', was one of the three most commonly mentioned concepts by the bottom violators.

4.2 Additional Analysis

4.2.1 Environmental Violations and Related Concepts in Item 3 Legal Proceedings

This section examines and approximates the conditional probability of other concepts being discussed when the 'violation' concept is mentioned. This study analysed Item 3 to identify the concepts most connected with violations. When the 'violation' concept is scrutinised, the strength of the relationship (co-occurrence) is displayed in relation to the other concepts. The likelihood of violations associated with the EPA was 58%.

Table 5. Co-occurrence and likelihood of violations

	Violations	Co-Count	Likelihood
1	Groundwater	21	100%
2	Air	60	83%
3	Permit	62	81%
4	Facilities	3	9%
5	Environmental	5	9%

Co-count is the total number of text context blocks across the data when a concept co-occurs with the 'violation' concept. Table 5 illustrates that 'permit', 'air', and 'groundwater' had the highest number of narrated co-counts, but 'violations' was most likely to be discussed related to 'groundwater'. In 2017, Chevron disclosed:

As initially disclosed in the Quarterly Report on Form 10-Q for the quarter ended March 31, 2017, on November 18, 2016, Chevron received an Administrative Order (AO) from the EPA alleging noncompliance with the water permit that governed conveyances of captured groundwater and spring water from the former Questa mine located in New Mexico to its associated tailing facility. Chevron is concluding its negotiations with EPA regarding this matter. (Chevron, 2017, p. 23)

As the sampled companies violated the Clean Air Act, they were also in breach of other statutes. Air violations are expected, but historically, these were reported depending on the stages, such as alleging potential noncompliance, negotiating the terms of consent, and resolving notice of violations. When the locations in the documents were tracked, between 2011 to 2013 Conoco disclosed the proposed penalty of $2,000 per well drilled from its Bakken Pool facility, which would result in a possible civil penalty exceeding $100,000. In 2013 and 2014, Valero reported a penalty assessment of $180,911 for alleged air emission and reporting violations, as well as $166,000 for unauthorised emissions. In 2012, Chevron disclosed the possibility of a civil penalty of $100,000 or more.

There were only three disclosures on the penalty paid for 'violations'. First, Conoco paid $34,000 in 2014 for its failure to comply with two air emission monitoring requirements. Second, Chevron mentioned two settlements in 2012, one related to the Richmond Refinery which resulted in the payment of a $192,500 civil penalty, and another related to the San Jose and Sacramento terminals and which resulted in the payment of a $205,000 civil penalty.

4.2.2 Impact in Item 7 Management Discussion and Analysis

The management discussions and analyses were also examined to identify concepts and themes most connected with impact.

Table 6. Co-occurrence and likelihood of impact on financial, environment, employee, community, and health and safety

	Impact	Co-Count	Likelihood
1	Financial	81	49%
2	Operations	81	27%
3	Emissions	20	11%
4	Environmental	43	11%
5	Remediation	11	4%

In 2012, Conoco discussed the financial impact of environmental laws and regulations, global climate change initiatives, and the impact on their operations:

The ultimate financial impact arising from environmental laws and regulations is neither clearly known nor easily determinable as new standards, such as air emission standards, water quality standards, and stricter fuel regulations, continue to evolve. However, environmental laws and regulations, including those that may arise to address concerns about global climate change, are expected to continue to have an increasing impact on our operations in the United States and in other countries in which we operate. (p.59)

In the 2011 (p.57) filing, Noble Energy, while admitting the industry as the source of certain greenhouse gases (GHGs), indicated the possibility of its restriction in the future:

During 2011, the United Nations-sponsored Intergovernmental Panel on Climate Change, a scientific body which provides an assessment of the risk of climate change, issued its Special Report on Managing the Risks of Extreme Events and Disasters to Advance Climate Change Adaptation, in which it concluded that it is likely that climate change is fuelling extreme weather and predicted that there will be an escalation of impacts on people and economies. The oil and natural gas exploration and production industry is a source of certain GHGs, namely carbon dioxide and methane, and future restrictions on the combustion of fossil fuels or the venting of natural gas could have a significant impact on our future operations.

However, in 2015 (p.65), Noble Energy shifted criticism of the international effort for its failure and the financial impact it had on its business:

And to the extent that international efforts are not successful in preventing climate change, any resulting increase in severity or frequency of storms, rise in sea levels, extreme temperatures or other extreme environmental effects may have an adverse impact on our financial condition, results of operations and cash flows.

Noble Energy also discussed the possibility of future restrictions on GHG emissions or related measures to encourage the use of renewable energy as having a significant impact on their future operations and a reduction in demand for their products (Noble Energy, 2015). Phillips (2015, p. 57) took a different approach. While discussing the impact of future restrictions on its operation, it also reported on its initiatives to meet future requirements:

We have met the increasingly stringent requirements to date while establishing implementation, operating and capital strategies, along with advanced technology development, to address projected future requirements. It is uncertain how various future requirements contained in EISA, and the regulations promulgated thereunder, may be implemented and what their full impact may be on our operations.

Only MPLX disclosed how it accounted for environmental remediation. For instance, in 2017 (p.90), it reported that:

However, the specific impact on each competitor may vary depending on a number of factors, including, but not limited to, the age and location of its operating facilities. Our environmental expenditures for each of the past three years were: We accrue for environmental remediation activities when the responsibility to remediate is probable, and the amount of associated costs can be reasonably estimated.

Historically, different companies had different approaches to discussing their performance. However, they were reporting on uncertainties regarding future restrictions and initiatives on emission restrictions, environmental laws, and regulations that might affect operations, products, and financial conditions.

Table 7. Sentiment analysis of top 5 ranked concepts

Ranked Concept/ Item	Item 1 Description of Business	Favourable F); Unfavourable (UF) % Likelihood	Item 3 Legal Proceedings	Favourable (F); Unfavourable (UF) % Likelihood	Item 7 Management Discussion and Analysis	Favourable F); Unfavourable (UF) % Likelihood
1	Oil	UF:40%	Adverse	UF:7%	Environmental	F: 58% UF: 34%
2	Gas	UF: 32%	Operations	UF:7%	Costs	F: 19% UF: 71%
3	Operations	F: 67% UF: 53%	Violations	UF:28%	Regulations	F: 98% UF: 34%
4	Requirements	UF: 63%	Effect	UF:7%	Laws	F: 77% UF: 14%
5	Regulations	UF: 49%	Material	UF:7%	Operations	F: 68% UF: 24%

1 All instances of the use of $ imply USD.

4.3 Automatic sentiment analysis of top five ranked concepts

The sentiment lens automatically generates insight into positive and negative sentiments in the text and will only apply those sentiment terms identified as relevant and as used consistently within the document set during processing (Guide, 2021, pp.80–81). A list of sentiment thesaurus terms which are irrelevant or inconsistent in the text will be grey. In this study, only favourable and unfavourable terms were deemed relevant. To illustrate, Table 3 above was reproduced to identify the sentiment of the top five ranked themes for Items 1, 3, and 7 for the 10-k filings.

Table 7 illustrates that Leximancer identified the ranked themes in Item 1 as unfavourable with the exception of 'operations', which was classified as both favourable and unfavourable. An example of text that represents an unfavourable narrative of 'operations' is from Noble Energy in 2016 (p.30):

These court decisions have the effect of giving local communities in Pennsylvania more authority to regulate oil and gas operations, which could make it more difficult to develop our Marcellus Shale acreage in some municipalities. Furthermore, the state has finalized new rules for surface operations at oil and gas sites that, among other things, would increase public participation in the permitting process, increase mitigation obligations and require surveys for abandoned wells.

As expected, for all ranked concepts in Item 3, 'legal proceedings' were marked unfavourably at a 7% likelihood, except for 'violations' at 28%. It is expected that disclosures on violations were made in this section. Based on Leximancer, the sentiment lens was more prominent in Item 7, where management discussed and analysed their performance. The most counterintuitive was how 'environment' was depicted as favourable and unfavourable. The favourable text on 'environmental' was traced to Valero's filing in 2014 (p.49), which referred to the company's positive initiatives:

Accruals for environmental liabilities are based on best estimates of probable undiscounted future costs over a 20-year time period using currently available technology and applying current regulations, as well as our own internal environmental policies.

However, in the same year, Valero (2014, p.46) created the following environmental concept, which was categorised as unfavourable by Leximancer:

Our operations are subject to extensive environmental regulations by governmental authorities relating to the discharge of materials into the environment, waste management, pollution prevention measures, greenhouse gas emissions, and characteristics and composition of gasoline and distillates. Because environmental laws and regulations are becoming more complex and stringent and new environmental laws and regulations are continuously being enacted or proposed, the level of future expenditures required for environmental matters could increase in the future, as previously discussed above in "OUTLOOK."

The study also examined how 'costs' and 'regulations' were depicted as favourable because they were scored at 98%. It was found that texts that discussed the regulations and reduction of greenhouse emissions were coded as favourable. For example, Phillips in 2015 (p.57) provided these statements in Item 7:

European Union Trading Directive resulting in the European Union Emissions Trading Scheme (EU ETS), which uses a market-based mechanism to incentivize the reduction of greenhouse gas emissions. These laws and their implementing regulations set limits on emissions and, in the case of discharges to water, establish water quality limits.

In 2017 (p.61), Conoco made a similar disclosure:

These laws and their implementing regulations set limits on emissions and, in the case of discharges to water, establish water quality limits. They also, in most cases, require permits in association with new or modified operations.

However, when this study tracked down the excerpt that Leximancer provided to support the 'costs' concept as favourable, all discussions focused on how costs would be determined or how uncertainties related to the magnitude of costs to be incurred. The actual costs, if stated, were classified as unfavourable.

5. DISCUSSION

Disclosure aids in resolving agency problems by facilitating the monitoring of managers by shareholders and regulators (Vashishtha, 2014). Based on a decade of selected sections of 10-k filings for corporate violators, it was found that accounting-related information was poorly emphasised, which supports early disclosure observations (e.g., Cho and Patten, 2008; Criado-Jiménez et al., 2008; Larrinaga et al., 2002). A recent study by Senn and Giordano-Spring (2020) also found that environmental accounting information is still in its infancy. Although environmental expenditures were disclosed, they were not always accompanied by monetary values. The nature and extent of violations and impacts were reported as required, but disclosure on remediation costs was minimal, even for corporate violators who committed to undertake SEPs. It was at management's discretion to portray disclosure compliance, excluding voluntary commentary on how the violations would be mitigated in the future and the type and progress of SEPs undertaken. Similar to Laine et al. (2017), it was observed that the provision of accounting-related information, such as environmental expenditure and liabilities, was mainly arbitrary and through subjective allocations.

While it was appreciated that environmental liabilities and remediations involve estimates and possibly evolve as the environmental regulation evolves, remediation of sites where violations occurred should receive wider coverage than the violators currently report. The silence regarding the required disclosures limits the likelihood that a violating firm will disclose its violations. General information may be directed to investors simply to update them of events (Abdelrehim et al., 2015). As noted by Peters and Romi (2013), regulators do not have adequate resources to fully complete their monitoring duties. It is believed therefore that minimal compliance would be the preferred choice for management. After all, full disclosure demands additional costs and might be deemed unnecessary if low detection and enforcement are anticipated. Schwartz and Soo (1996) found the cost attributed to mandatory disclosure to be one of the main reasons for noncompliance with mandatory disclosure. Consequently, it was found that the sampled companies did not always produce homogeneity in disclosing behaviours among companies. Bebbington et al. (2012) commented that formal legislation alone might not be sufficient to create a norm, and the absence of specific requirements in the law did not help either. Narver (1971) argued that such requirements were irrelevant without effective enforcement and willing compliance.

Consistent with Cowan and Gadenne (2005), our comparative analysis of information disclosed under legal proceedings and other sections indicates that management has a preference for what information to present and where to emphasise it. Schwartz and Soo (1996) asserted that companies are more sensitive to potential penalties and future scrutiny and thus disclose promptly after a violation event. Nevertheless, accounting-related information was not an obvious choice when describing their business legal proceedings or discussing and analysing their performance.

Skinner (1994) argues that managers make pre-emptive disclosures to reduce their expected litigation costs. Our analysis of pre- and post-violation periods suggests that similar industry-dominated themes were emphasised, but the 'laws' concept emerged post-violation. As suggested by Karpoff et al. (2005), firms that violate environmental laws suffer statistically significant losses of similar magnitudes to the legal penalties imposed. Discussion of the laws was thus not unexpected. Other changes in firms' operating environments, such as a decline in performance, increased litigation concerns, and increased uncertainty about future outcomes, may also explain the change in focus (see Vashishtha, 2014). The literature suggests that following violations, environmental violators often increase environmental disclosures and report favourable environmental information (Deegan and Rankin, 1996; Ettredge et al., 2001; Patten, 1992). In our study, the emphasis changed pre- and post-violation. The choice of detailed or general argument suggests that each company considered its specific needs (Abdelrehim et al., 2015) at their discretion.

As indicated by the ranked concepts, our analysis demonstrated that bottom violators emphasised 'EPA' and 'costs' more often in their narratives. This is consistent with Cho et al. (2010), who demonstrated that the worst performers (in this chapter, top violators) exhibit significantly more optimism and less certainty than their counterparts. As stated by Meng et al. (2014), poor performers disclosed more soft information and avoided disclosing negative environmental information. These communication strategies may have been to implicitly deflect responsibility (Cho, 2009). Both the top and bottom violators ranked discussion on 'violations' most frequently. Peters and Romi (2013) suggested that firms are less likely to disclose sanctions involving judicial proceedings.

Leximancer auto generated the sentiments expressed in different sections of the filings, and the sentiments were assigned differently for each section based on the text blocks in the documents. Notably, the legal proceedings were all classified as unfavourable, and management discussions and analyses contained favourable and unfavourable narratives. As management discretion was present, favourable narratives were prevalent. Consistent with previous findings, management seemed to use biased language and verbal tone as a tool to influence society's perception of their operations and ongoing existence (Cho et al., 2010; Deegan et al., 2000).

6. CONCLUSION

Extractive industries have a significant impact on the natural environment. This chapter uses a qualitative research method and disclosure and legitimacy theory to examine how environmental violations have shaped oil and gas companies' mandatory and discretionary environmental disclosures in the different sections of the 10-k reports filed with SEC between 2011 to 2020. Using a qualitative interpretive lens on environmental violators in the US oil and gas industries, 10-k disclosures allowed us to identify themes and demonstrate the environmental concerns for these violators pre- and post-violations in the past decade.

Overall, to a certain extent our findings suggest that disclosures made in compliance with SEC environmental sanctions reference accounting disclosure of financial liabilities associated with CWA and CAA. In particular, 'environmental' ranked first in Item 7 disclosures, while themes of 'adverse', 'operations', and 'violations' ranked higher in Item 3 disclosures. The sentiment analysis demonstrated that text referencing 'legal proceedings' was classified as unfavourable, as expected. Consistent with discretionary disclosure theory, the text of the MD&A sections was favourable. Management discretion seemed evident through their commitment to SEPs progress. The impact of such environmental protection expenditures or investments was not, however, updated over the years. Despite accounting measures used by regulators to determine the materiality threshold for environmental disclosures, it was found that to some extent such measures shaped discretionary environmental disclosures for firms involved in SEP. Accounting information was not widely disclosed, such as the amount spent on environmental remediation, including the use of new technology and investment for mitigation. It is possible that accounting-related information was traded for capital market benefits instead (e.g., Leuz et al., 2004).

The conclusion which can be drawn is that mandating disclosures would not be effective if management were allowed to exercise their discretion. This has resulted in non- and minimal disclosure of accounting information. This information could be useful for regulators for future improvement of legislating and monitoring services (see Leuz et al., 2016). More transparency is required to break down silos between the traditionally discrete domains of financial reporting and sustainability reporting (Unerman et al., 2018). There is also a need for a strong commitment to mitigate the adverse effects of such violations (see Cho et al., 2012). Scholars have emphasised the need for accounting practices to be pushed and pulled by various actors to diffuse and become norms over time (Brunnee and Toope, 1997; Finnmore and Sikkink, 1998).

It was found that the Leximancer software provides accounting researchers with the capability to use qualitative research methods or mixed-methods research in accounting disclosure. Leximancer has been shown to be an effective tool for performing comprehensive analyses and a sub-category ACA, including a sentiment analysis. Nonetheless, it is acknowledged that such text mining tools do not in any way replace a human researcher's competence and skills in data interpretation and the need to engage with the data.

This study had some limitations. First, in 2019, the SEC proposed the Modernisation of Regulation S-K Items 101, 103, and 105. This caused Item 3's materiality threshold to increase from $100,000 to $300,000, effective from November 2020 (SEC, 2019). It is expected this had minimal impact on our findings, as it only impacts the last two months of our disclosures. Future research comparing old and new regimes or examining any differences in the disclosures would be beneficial. Another limitation of this study stems from the methodological choices and the setting analysed. Although the researchers removed possible biases by resorting to the full auto-generated processes by Leximancer, the results might be different if specific concepts were seeded into the software. This does not, however, eliminate interpretation bias.

REFERENCES

Abdelrehim, N., Maltby, J., & Toms, S. (2015). Narrative reporting and crises: British Petroleum and Shell 1950–1958. *Accounting History*, *20*(2), 138–157. doi:10.1177/1032373214563323

Alexander, R., Ettredge, M., Stone, M., & Sun, L. (2011). Are mandatory disclosure decisions made strategically? The case of SAB 74 estimates preceding adoption of FIN 48. *Research in Accounting Regulation*, *23*(2), 160–166. doi:10.1016/j.racreg.2011.06.001

Angus, D., Rintel, S., & Wiles, J. (2013). Making sense of the big text: A visual-first approach for analysing text data using Leximancer and Discursis. *International Journal of Social Research Methodology*, *16*(3), 261–267. doi:10.1080/13645579.2013.774186

Baudot, L., Huang, Z., & Wallace, D. (2021). Stakeholder perceptions of risk in mandatory corporate responsibility disclosure. *Journal of Business Ethics*, *172*(1), 151–174. doi:10.100710551-020-04476-7

Bebbington, J., Kirk, E. A., & Larrinaga, C. (2012). The production of normativity: A comparison of reporting regimes in Spain and the UK. *Accounting, Organizations and Society*, *37*(2), 78–94. doi:10.1016/j.aos.2012.01.001

Beyer, A., Cohen, D. A., Lys, T. Z., & Walther, B. R. (2010). The financial reporting environment: Review of recent literature. *Journal of Accounting and Economics*, *50*(2–3), 296–343. doi:10.1016/j.jacceco.2010.10.003

Biesenthal, C., & Wilden, R. (2014). Multi-level project governance: Trends and opportunities. *International Journal of Project Management*, *32*(8), 1291–1308. doi:10.1016/j.ijproman.2014.06.005

Blouin, J., Gleason, C., Mills, L., & Sikes, S. (2010). Pre-empting disclosure? Firms' decisions prior to FIN 48. *The Accounting Review*, *85*(3), 791–815. doi:10.2308/accr.2010.85.3.791

Brunnée, J., & Toope, S. J. (1997). Environmental security and freshwater resources: Ecosystem regime building. *The American Journal of International Law*, *91*(1), 26–59. doi:10.2307/2954139

Cheng, M., & Edwards, D. (2019). A comparative automated content analysis approach on the review of the sharing economy discourse in tourism and hospitality. *Current Issues in Tourism*, *22*(1), 35–49. doi:10.1080/13683500.2017.1361908

Cho, C. H. (2009). Legitimation strategies used in response to environmental disaster: A French case study of Total SA's Erika and AZF incidents. *European Accounting Review*, *18*(1), 33–62. doi:10.1080/09638180802579616

Cho, C. H., Freedman, M., & Patten, D. M. (2012). Corporate disclosure of environmental capital expenditures: A test of alternative theories. *Accounting, Auditing & Accountability Journal*, *25*(3), 486–507. doi:10.1108/09513571211209617

Cho, C. H., & Patten, D. M. (2008). Did the GAO get it right? Another look at corporate environmental disclosure. *Social and Environmental Accountability Journal*, *28*(1), 21–32. doi:10.1080/0969160X.2008.9651788

Cho, C. H., Roberts, R. W., & Patten, D. M. (2010). The language of US corporate environmental disclosure. *Accounting, Organizations and Society*, *35*(4), 431–443. doi:10.1016/j.aos.2009.10.002

Clarkson, P. M., Li, Y., & Richardson, G. D. (2004). The market valuation of environmental Capital expenditures by pulp and paper companies. *The Accounting Review*, *79*(2), 329–353. doi:10.2308/accr.2004.79.2.329

Clarkson, P. M., Li, Y., Richardson, G. D., & Vasvari, F. P. (2008). Revisiting the relation between environmental performance and environmental disclosure: An empirical analysis. *Accounting, Organizations and Society*, *33*(4/5), 303–327. doi:10.1016/j.aos.2007.05.003

Cowan, S., & Gadenne, D. (2005). Australian corporate environmental reporting: A comparative analysis of disclosure practices across voluntary and mandatory disclosure systems. *Journal of Accounting & Organizational Change*, *1*(2), 165–179. doi:10.1108/18325910510635344

Craig, J., & Amernic, H. (2004). Enron discourse: The rhetoric of a resilient capitalism. *Critical Perspectives on Accounting*, *15*(6/7), 813–851. doi:10.1016/j.cpa.2002.12.001

Cretchley, J., Gallois, C., Chenery, H., & Smith, A. (2010). Conversations between carers and people with Schizophrenia: A qualitative analysis using Leximancer. *Qualitative Health Research*, *20*(12), 1611–1628. doi:10.1177/1049732310378297 PMID:20675536

Criado-Jiménez, I., Fernández-Chulián, M., Larrinaga-González, C., & Husillos-Carqués, F. J. (2008). Compliance with mandatory environmental reporting in financial statements: The case of Spain (2001–2003). *Journal of Business Ethics*, *79*(3), 245–262. doi:10.100710551-007-9375-7

Deegan, C., & Rankin, M. (1996). Do Australian companies report environmental news objectively? An analysis of environmental disclosures by firms prosecuted successfully by the environmental protection authority. *Accounting, Auditing & Accountability Journal*, *9*(2), 50–67. doi:10.1108/09513579610116358

Deegan, C., Rankin, M., & Voght, P. (2000). Firms' disclosure reactions to major social incidents: Australian evidence. *Accounting Forum*, *24*(1), 101–130. doi:10.1111/1467-6303.00031

Environmental Protection Agency (EPA) (2001). *U.S. EPA Notifying Defendants of Securities and Exchange Commission's Environmental Disclosure Requirements Enforcement Alert 4(3)*. EPA.

Environmental Protection Agency (EPA). (2015) U.*S. Environmental Protection Agency supplemental environmental projects*. https://www.epa.gov/sites/default/files/2015-04/documents/sepupdatedpolicy15.pdf

Environmental Protection Agency (EPA) (n.d.). *Supplemental Environmental Projects (SEPs)*. EPA.

Ettredge, M., Johnstone, K., Stone, M., & Wang, Q. (2011). The effects of firm size corporate governance quality and bad news on disclosure compliance. *Review of Accounting Studies*, *16*(4), 866–889. doi:10.100711142-011-9153-8

Ettredge, M. L., Smith, D. B., & Stone, M. S. (2001). How effective is joint public and private monitoring? The case of the AICPA auditor change notification letter. *Auditing*, *20*(1), 29–43. doi:10.2308/aud.2001.20.1.29

Fallan, E., & Fallan, L. (2019). Corporate tax behaviour and environmental disclosure: Strategic trade-offs across elements of CSR? *Scandinavian Journal of Management*, *35*(3), 101042. Advance online publication. doi:10.1016/j.scaman.2019.02.001

Finnemore, M., & Sikkink, K. (1998). International norm dynamics and political change. *International Organization*, *52*(4), 887–917. doi:10.1162/002081898550789

Griffin, P. A., & Jaffe, A. M. (2018). Are fossil fuel firms informing investors well enough about the risks of climate change? *Journal of Energy & Natural Resources Law*, *36*(4), 381–410. doi:10.1080/02646811.2018.1502240

Guide. (2021). *Leximancer User Guide Release 5.0*. Leximancer Pvt. Ltd.

Guillamon-Saorin, E., Osma, G. B., & Jones, M. J. (2012). Opportunistic disclosure in press release headlines. *Accounting and Business Research*, *42*(2), 143–168. doi:10.1080/00014788.2012.632575

Hasan, T., Karim, W., & Quayes, S. (2008). Regulatory change and the quality of compliance to mandatory disclosure requirements. *Research in Accounting Regulation*, *20*, 193–203. doi:10.1016/S1052-0457(07)00211-1

Heitzman, S., Wasley, C., & Zimmerman, J. (2010). The joint effects of materiality thresholds and Voluntary disclosure incentives on firms' disclosure decisions. *Journal of Accounting and Economics*, *49*(1-2), 109–132. doi:10.1016/j.jacceco.2009.10.002

Karpoff, J. M., Lott, J. R. Jr, & Wehrly, E. W. (2005). The Reputational Penalties for Environmental Violations: Empirical Evidence. *The Journal of Law & Economics*, *48*(2), 653–675. doi:10.1086/430806

Kothari, S. P., Shu, S., & Wysocki, P. D. (2009). Do managers withhold bad news? *Journal of Accounting Research*, *47*(1), 241–276. doi:10.1111/j.1475-679X.2008.00318.x

Krippendorff, K. (2004). Measuring the Reliability of Qualitative Text Analysis. *Quality & Quantity*, *38*(6), 787–800. doi:10.100711135-004-8107-7

Laine, M., Järvinen, J. T., Hyvönen, T., & Kantola, H. (2017). Ambiguity of financial environmental information: A case study of a Finnish energy company. *Accounting, Auditing & Accountability Journal*, *30*(3), 593–619. doi:10.1108/AAAJ-02-2015-1961

Larrinaga, C., Carrasco, F., Correa, C., Llena, F., & Moneva, J. M. (2002). Accountability and accounting regulation: The case of the Spanish environmental disclosure standard. *European Accounting Review*, *11*(4), 723–740. doi:10.1080/0963818022000001000

Lee, G., & Xiao, X. (2020). Voluntary Engagement in Environmental Projects: Evidence from Environmental Violators. *Journal of Business Ethics*, *164*(2), 325–348. doi:10.100710551-018-4074-0

Lemon, L. L., & Hayes, J. (2020). Enhancing trustworthiness of qualitative findings: Using Leximancer for qualitative data analysis triangulation. *Qualitative Report*, *25*(3), 604–614. doi:10.46743/2160-3715/2020.4222

Leuz, C., Pfaff, D., & Hopwood, A. (2004). *The Economics and Politics of Accounting: International Perspectives on Trends Policy and Practice*. Oxford University Press. doi:10.1093/0199260621.001.0001

Leximancer. (2021). https://info.leximancer.com

Li, Y., Richardson, G. D., & Thornton, D. B. (1997). Corporate disclosure of environmental liability information: Theory and evidence. *Contemporary Accounting Research, 14*(3), 435–474. doi:10.1111/j.1911-3846.1997.tb00535.x

Liu, Z., & Liu, M. (2021). Quality evaluation of enterprise environmental accounting information disclosure based on projection pursuit model. *Journal of Cleaner Production, 279*, 123679. Advance online publication. doi:10.1016/j.jclepro.2020.123679

Meng, Zeng, Shi, Qi, & Zhang. (2014). The relationship between corporate environmental performance and environmental disclosure: An empirical study. *China Journal of Environmental Management, 145*, 357-367.

Mobus, J. L. (2005). Mandatory environmental disclosures in a legitimacy theory context. *Accounting, Auditing & Accountability Journal, 18*(4), 492–517. doi:10.1108/09513570510609333

Monti, G., & Deason, J. P. (2012). Applying an ecosystem services approach to supplemental environmental projects. *Journal of Environmental Planning and Management, 55*(8), 1047–1057. doi:10.1080/09640568.2011.636221

Narver, J. C. (1971). Rational management responses to external effects. *Academy of Management Journal, 14*(1), 99–115. doi:10.2307/254714

Nelson, K. K., & Pritchard, A. C. (2016). Carrot or Stick? The Shift from untary to Mandatory Disclosure of Risk Factors. *Journal of Empirical Legal Studies, 13*(2), 266–297. doi:10.1111/jels.12115

Ogden, S., & Clarke, J. (2005). Customer disclosures impression management and the construction of legitimacy: Corporate reports in the UK privatised water industry. *Accounting, Auditing & Accountability Journal, 18*(3), 313–345. doi:10.1108/09513570510600729

Patten, D. M. (1992). Intra-industry environmental disclosures in response to the Alaskan oil spill: A note on legitimacy theory. *Accounting, Organizations and Society, 17*(5), 471–475. doi:10.1016/0361-3682(92)90042-Q

Patten, D. M. (2005). The accuracy of financial report projections of future environmental capital expenditures: A research note. *Accounting, Organizations and Society, 30*(5), 457–468. doi:10.1016/j.aos.2004.06.001

Peters, G. F., & Romi, A. M. (2013). Discretionary compliance with mandatory environmental disclosures: Evidence from SEC filings. *Journal of Accounting and Public Policy*, *32*(4), 213–236. doi:10.1016/j.jaccpubpol.2013.04.004

Pfund, T. (2004). Corporate Environmental Accountability: Expanding SEC Disclosures to Promote Market Based Environmentalism. *Monthly Environmental Law & Policy Review*, *11*(2), 118–137.

Rooney, D., Paulsen, N., Callan, V. J., Brabant, M., Gallois, C., & Jones, E. (2010). A new role for place identity in managing organizational change. *Management Communication Quarterly, 24*(1), 44-73.

Schneider, T., Maier, M., & Michelon, G. (2017). Environmental liabilities and diversity in practice under international financial reporting standards. *Accounting, Auditing & Accountability Journal*, *30*(2), 378–403. doi:10.1108/AAAJ-01-2014-1585

Schwartz, K. B., & Soo, B. S. (1996). Evidence of regulatory noncompliance with SEC disclosure rules on auditor changes. *The Accounting Review*, *71*(4), 555–572.

Securities and Exchange Commission (SEC). (2010). *Commission guidance regarding disclosure related to climate change.* Release Nos. 33-9106; 34-61469; FR-82 February 2 2010, Securities and Exchange Commission.

Securities and Exchange Commission (SEC). (2019). Modernization of Regulation S-K Items 101 103 and 105. Release Nos. 33-10668; 34-86614; File S7-11-19; RIN 3235-AL78; 17 CFR 229 239 and 240. August 8 2019 Securities and Exchange Commission.

Senn, J., & Giordano-Spring, S. (2020). The limits of environmental accounting disclosure: Enforcement of regulations standards and interpretative strategies. *Accounting, Auditing & Accountability Journal*, *33*(6), 1367–1393. doi:10.1108/AAAJ-04-2018-3461

Skinner, D. (1994). Why firms untarily disclose bad news. *Journal of Accounting Research*, *32*(1), 38–60. doi:10.2307/2491386

Smith, A. E., & Humphreys, M. S. (2006). Evaluation of unsupervised semantic mapping of natural language with Leximancer concept mapping. *Behavior Research Methods*, *38*(2), 262–279. doi:10.3758/BF03192778 PMID:16956103

Tsalavoutas, I., Tsoligkas, F., & Evans, L. (2020). Compliance with IFRS Mandatory Disclosure Requirements: A Structured Literature Review. *Journal of International Accounting, Auditing & Taxation, 40*, 100338. Advance online publication. doi:10.1016/j.intaccaudtax.2020.100338

Tseng, C., Wu, B., Morrison, A. M., Zhang, J., & Chen, Y. C. (2015). Travel blogs on China as a destination image formation agent: A qualitative analysis using Leximancer. *Tourism Management, 46*, 347–358. doi:10.1016/j.tourman.2014.07.012

Unerman, J., Bebbington, J., & O'Dwyer, B. (2018). Corporate reporting and accounting for externalities. *Accounting and Business Research, 48*(5), 497–522. doi:10.1080/00014788.2018.1470155

Vashishtha, R. (2014). The role of bank monitoring in borrowers- discretionary disclosure: Evidence from covenant violations. *Journal of Accounting and Economics, 57*(2-3), 176–195. doi:10.1016/j.jacceco.2014.04.002

Verrecchia, R. E. (1983). Discretionary Disclosure. *Journal of Accounting and Economics, 5*, 179–194. doi:10.1016/0165-4101(83)90011-3

Wiedman, C., & Wier, H. (1999). Management of note disclosures: The case of unconsolidated subsidiaries prior to FAS 94. *Journal of Accounting, Auditing & Finance, 14*(1), 73–94. doi:10.1177/0148558X9901400104

Williams, C. A., & Nagy, D. M. (2021). ESG and climate change blind spots: Turning the corner on SEC disclosure. *Texas Law Review, 99*(7), 1453–1485.

Yang, J., & Malone, E. (2008). Working to shape what society's expectations of us should be: Philip Morris's societal alignment strategy. *Tobacco Control, 17*(6), 391–398. doi:10.1136/tc.2008.026476 PMID:18845623

Compilation of References

Abdelrehim, N., Maltby, J., & Toms, S. (2015). Narrative reporting and crises: British Petroleum and Shell 1950–1958. *Accounting History*, *20*(2), 138–157. doi:10.1177/1032373214563323

Adedoyin, F. F., Alola, U. V., & Bekun, F. V. (2022). On the nexus between globalization, tourism, economic growth, and biocapacity: Evidence from top tourism destinations. *Environmental Science and Pollution Research International*, *29*(17), 24995–25005. doi:10.100711356-021-17651-8 PMID:34837107

Adhikari, D., & Chen, Y. (2014). Energy productivity convergence in Asian countries: A spatial panel data approach. *International Journal of Economics and Finance*, *6*(7), 94–107. doi:10.5539/ijef.v6n7p94

Adom, P. K., Agradi, M., & Vezzulli, A. (2021). Energy efficiency- economic growth nexus: What is the role of income inequality? *Journal of Cleaner Production*, *310*, 1–20. doi:10.1016/j.jclepro.2021.127382

ahad, S., Adnan, M., & Saud, S. (Eds.). (2022). *Improvement of Plant Production in the Era of Climate Change*. CRC Press. doi:10.1201/9781003286417

Ahluwalia, M. S. (1976). Inequality, poverty and development. *Journal of Development Economics*, *3*(4), 307–342. doi:10.1016/0304-3878(76)90027-4 PMID:12339111

Ahmed, K. (2020). Environmental policy stringency, related technological change and emissions inventory in 20 OECD countries. *Journal of Environmental Management*, *274*, 111209. doi:10.1016/j.jenvman.2020.111209 PMID:32798847

Ahmed, Z., Asghar, M. M., Malik, M. N., & Nawaz, K. (2020). Moving towards a sustainable environment: The dynamic linkage between natural resources, human capital, urbanization, economic growth, and ecological footprint in China. *Resources Policy*, *67*, 101677. doi:10.1016/j.resourpol.2020.101677

Aigner, D. J., & Chu, S. (1968). On Estimating the Industry Production Function. *The American Economic Review*, *58*, 826–839.

Aigner, D., Lovell, C. A. K., & Schmidt, P. (1977). Formulation and estimation of stochastic frontier production function models. *Journal of Econometrics*, *6*(1), 21–37. doi:10.1016/0304-4076(77)90052-5

Akadiri, S. S., Lasisi, T. T., Uzuner, G., & Akadiri, A. C. (2019). Examining the impact of globalization in the environmental Kuznets curve hypothesis: The case of tourist destination states. *Environmental Science and Pollution Research International*, *26*(12), 12605–12615. doi:10.100711356-019-04722-0 PMID:30852745

Alam, M. J., Begum, I. A., Buysse, J., Rahman, S., & Van Huylenbroeck, G. (2011). Dynamic modeling of the causal relationship between energy consumption, CO2 emissions and economic growth in India. *Renewable & Sustainable Energy Reviews*, *15*(6), 3243–3251. doi:10.1016/j.rser.2011.04.029

Alexander, R., Ettredge, M., Stone, M., & Sun, L. (2011). Are mandatory disclosure decisions made strategically? The case of SAB 74 estimates preceding adoption of FIN 48. *Research in Accounting Regulation*, *23*(2), 160–166. doi:10.1016/j.racreg.2011.06.001

Algan, N., & Akdoğan, M. (2001). Veri Zarflama Analiz Tekniği Kullanılarak Karadeniz Bölgesi'nde Yer Alan İllerin Etkinlik Değerlendirilmesi. In Ulusal Bölge Bilimi, Bölge Planlama Kongresi Bildiriler Kitabı (pp. 241-251). Trabzon.

Allingham, M. G. (1972). The measurement of inequality. *Journal of Economic Theory*, *5*(1), 163–169. doi:10.1016/0022-0531(72)90124-X

Alola, A. A., Eluwole, K. K., Lasisi, T. T., & Alola, U. V. (2021). Perspectives of globalization and tourism as drivers of ecological footprint in top 10 destination economies. *Environmental Science and Pollution Research International*, *28*(24), 31607–31617. doi:10.100711356-021-12871-4 PMID:33609247

Alper, A., & Oguz, O. (2016). The role of renewable energy consumption in economic growth: Evidence from asymmetric causality. *Renewable & Sustainable Energy Reviews*, *60*, 953–959. doi:10.1016/j.rser.2016.01.123

Altinay, G., & Karagol, E. (2004). Structural break, unit root, and the causality between energy consumption and GDP in Turkey. *Energy Economics*, *26*(6), 985–994. doi:10.1016/j.eneco.2004.07.001

Angus, D., Rintel, S., & Wiles, J. (2013). Making sense of the big text: A visual-first approach for analysing text data using Leximancer and Discursis. *International Journal of Social Research Methodology*, *16*(3), 261–267. doi:10.1080/13645579.2013.774186

Anser, M. K., Yousaf, Z., Nassani, A. A., Abro, M. M. Q., Zaman, K., & Kabbani, A. (2020). Evaluating ecological footprints through inbound tourism, population density, and global trade. *Polish Journal of Environmental Studies*, *30*(1), 555–560. doi:10.15244/pjoes/122445

Apergis, N., & Christou, C. (2016). Energy productivity convergence: New evidence from club converging. *Applied Economics Letters*, *23*(2), 142–145. doi:10.1080/13504851.2015.1058899

Apergis, N., & Payne, J. E. (2011). The renewable energy consumption–growth nexus in Central America. *Applied Energy*, *88*(1), 343–347. doi:10.1016/j.apenergy.2010.07.013

Apergis, N., & Payne, J. E. (2012). Renewable and non-renewable energy consumption-growth nexus: Evidence from a panel error correction model. *Energy Economics*, *34*(3), 733–738. doi:10.1016/j.eneco.2011.04.007

Arellano, M. (1987). Computing robust standart errors for within group estimators. *Oxford Bulletin of Economics and Statistics*, *49*(4), 431–434. doi:10.1111/j.1468-0084.1987.mp49004006.x

Arias-Ortiz, A., Masqué, P., Glass, L., Benson, L., Kennedy, H., Duarte, C. M., Garcia-Orellana, J., Benitez-Nelson, C. R., Humphries, M. S., Ratefinjanahary, I., Ravelonjatovo, J., & Lovelock, C. E. (2021). Losses of soil organic carbon with deforestation in mangroves of Madagascar. *Ecosystems (New York, N.Y.)*, *24*(1), 1–19. doi:10.100710021-020-00500-z

Armiento, M. (2018). The sustainable welfare index: Towards a threshold effect for Italy. *Ecological Economics*, *152*, 296–309. doi:10.1016/j.ecolecon.2018.06.014

Arvola, A., Vassallo, M., Dean, M., Lampila, P., Saba, A., Lahteenmaki, R., & Shepherd, R. (2008). Predicting intentions to purchase organic food: The role of affective and moral attitudes in the theory of planned behavior. *Appetite*, *50*(2-3), 443–454. doi:10.1016/j.appet.2007.09.010 PMID:18036702

Asamoah, O., Kuittinen, S., Abrefa Danquah, J., Quartey, E. T., Bamwesigye, D., Mario Boateng, C., & Pappinen, A. (2020). Assessing wood waste by timber industry as a contributing factor to deforestation in Ghana. *Forests*, *11*(9), 939. doi:10.3390/f11090939

Aslan, M. (2005). Political Economic Analysis of Turkish Economy: Structural Problems and the Role of the State. *Eskişehir Osmangazi Üniversitesi Sosyal Bilimler Dergisi*, *6*(2), 15-27. Retrieved from https://dergipark.org.tr/tr/pub/ogusbd/issue/10987/131485

Atan, M., Özgür, E., & Güler, H. (2004). Çok Değişkenli İstatistiksel Analizler ve VZA ile İllerin Gelişmişlik Düzeylerinin Karşılaştırılması. *Gazi Üniversitesi İktisadi ve İdari Bilimler Fakültesi Dergisi*, *6*(2), 25-42. Retrieved from https://dergipark.org.tr/en/pub/gaziuiibfd/issue/28341/301243

Atkinson, A. B. (1970). On the measurement of inequality. *Journal of Economic Theory*, *2*(3), 244–263. doi:10.1016/0022-0531(70)90039-6

AtlasW. (2013). https://www.worldatlas.com/aatlas/infopage/oceans.htm

Avtar, R., Tripathi, S., Aggarwal, A. K., & Kumar, P. (2019). Population–urbanization–energy nexus: A review. *Resources*, *8*(3), 136. doi:10.3390/resources8030136

Aydemir, Z. C. (2002). *Bölgesel Rekabet Edebilirlik Kapsamında İllerin Kaynak Kullanım Görece Verimlilikleri: Veri Zarflama Analizi Uygulaması* [Master's Thesis]. DPT Yayınları.

Aydın, N. (2012). Türkiye'de İstihdam Politikalarının Bulanık Veri Zarflama Analizi Yöntemiyle Etkinlik Analizi. *Finans Politik & Ekonomik Yorumlar Dergisi*, *49*, 15–28.

Aydoğan, E. (2017). Foreign Direct Investment in Turkey: A Comparison with the Central and Eastern European Countries. *Finans Politik ve Ekonomik Yorumlar Dergisi*, 626.

Ay, S. (2012). Türkiye'de İşsizliğin Nedenleri: İstihdam Politikaları Üzerine Bir Değerlendirme. *Yönetim ve Ekonomi Dergisi, 19*(2), 321–341.

Azami, S., & Almasi, S. (2020). Energy consumption and sustainable economic welfare: New evidence of organization of petroleum exporting countries. *International Journal of Energy Economics and Policy, 10*(5), 31–40. doi:10.32479/ijeep.9435

Bagader, A., El-Sabbagh, A., Al-Glayand, M., & Samarrai, M. (2011). *Environmental protection in Islam: A general introduction.* Retrieved June 17, 2022, from https://www.islamreligion.com/articles/307/viewall/environmental-protection-in-islam/

Bahmani-Oskooee, M., Chang, T., & Wu, T. (2014). Revisiting purchasing power parity in African countries: Panel stationary test with sharp and smooth breaks. *Applied Financial Economics, 24*(22), 1429–1438. doi:10.1080/09603107.2014.925068

Bakırcı, F., Ekinci, E. D., & Şahinoğlu, T. (2014). Bölgesel Kalkınma Politikalarının Etkinliği: Türkiye Alt Bölgeler Bazında Bir Uygulama. *Atatürk Üniversitesi Sosyal Bilimler Enstitüsü Dergisi, 18*(2), 281–298.

Baloch, M. A., Zhang, J., Iqbal, K., & Iqbal, Z. (2019). The effect of financial development on ecological footprint in bri countries: Evidence from panel data estimation. *Environmental Science and Pollution Research International, 26*(6), 6199–6208. doi:10.100711356-018-3992-9 PMID:30617883

Balsalobre-Lorente, D., Driha, O. M., Shahbaz, M., & Sinha, A. (2020). The effects of tourism and globalization over environmental degradation in developed countries. *Environmental Science and Pollution Research International, 27*(7), 7130–7144. doi:10.100711356-019-07372-4 PMID:31879881

Bamidele, R., Ozturk, I., Gyamfi, B. A., & Bekun, F. V. (2021). Tourism-induced pollution emission amidst energy mix: Evidence from Nigeria. *Environmental Science and Pollution Research International, 29*(13), 19752–19761. doi:10.100711356-021-17233-8 PMID:34718978

Banerjee, B., & McKeage, K. (1994). How green is my value: Exploring the relationship between environmentalism and materialism. *Advances in Consumer Research. Association for Consumer Research (U. S.), 21*, 147–152.

Banker, R. D. (1993). Maximum-likelihood, consistency and data envelopment analysis - A statistical foundation. *Management Science, 39*(10), 1265–1273. doi:10.1287/mnsc.39.10.1265

Banker, R. D., & Cooper, C. A. (1984). WW. Some models for estimating technical and scale inefficiencies in data envelopment analysis. *Management Science, 30*(9), 1078–1092. doi:10.1287/mnsc.30.9.1078

Barnwell, G. C., Stroud, L., & Watson, M. (2020). "Nothing green can grow without being on the land": Mine-affected communities' psychological experiences of ecological degradation and resistance in Rustenburg, South Africa. *Community Psychology in Global Perspective, 6*(2), 87-109.

Basit, I., Faizi, F., Mahmood, K., Bilgili, M. S., Yildirim, Y., & Mushtaq, F. (2022). Geospatial alternatives for quantification of bio-thermal influence zone in the vicinity of a solid waste dump. *Waste Management & Research.*

Baudot, L., Huang, Z., & Wallace, D. (2021). Stakeholder perceptions of risk in mandatory corporate responsibility disclosure. *Journal of Business Ethics, 172*(1), 151–174. doi:10.100710551-020-04476-7

Baumeister, C., & Kilian, L. (2016). Forty years of oil price fluctuations: Why the price of oil may still surprise us. *Journal of Economic Perspectives, 30*(1), 139-60.

BBC. (2022). *Bitesize / Environment.* Retrieved February 25, 2022, from http://www.bbc.co.uk >revision

Bebbington, J., Kirk, E. A., & Larrinaga, C. (2012). The production of normativity: A comparison of reporting regimes in Spain and the UK. *Accounting, Organizations and Society, 37*(2), 78–94. doi:10.1016/j.aos.2012.01.001

Beça, P., & Santos, R. (2010). Measuring sustainable welfare: A new approach to the ISEW. *Ecological Economics, 69*(4), 810–819. doi:10.1016/j.ecolecon.2009.11.031

Becker, R., Enders, W., & Lee, J. (2006). A stationarity test in the presence of an unknown number of smooth breaks. *Journal of Time Series Analysis, 27*(3), 381–409. doi:10.1111/j.1467-9892.2006.00478.x

Ben Jebli, M., Ben Youssef, S., & Apergis, N. (2015). The dynamic interaction between combustible renewables and waste consumption and international tourism: The case of Tunisia. *Environmental Science and Pollution Research International, 22*(16), 12050–12061. doi:10.100711356-015-4483-x PMID:25874437

Berndt, E. R., & Morrisson, C. J. (1979). Income redistribution and employment effects of rising energy prices. *Resources and Energy, 2*(2-3), 131–150. doi:10.1016/0165-0572(79)90026-4

Berry, R. S., Salamon, P., & Heal, G. (1978). On a relation between economic and thermodynamic optima. *Resources and Energy, 1*(2), 125–137. doi:10.1016/0165-0572(78)90002-6

Beyer, A., Cohen, D. A., Lys, T. Z., & Walther, B. R. (2010). The financial reporting environment: Review of recent literature. *Journal of Accounting and Economics, 50*(2–3), 296–343. doi:10.1016/j.jacceco.2010.10.003

Bhattacharya, M., Inekwe, J. N., & Sadorsky, P. (2020). Convergence of energy productivity in Australian states and territories: Determinants and forecasts. *Energy Economics, 85*, 104538. doi:10.1016/j.eneco.2019.104538

Bhattacharya, M., Inekwe, J. N., Sadorsky, P., & Saha, A. (2018). Convergence of energy productivity across Indian states and territories. *Energy Economics*, *74*, 427–440. doi:10.1016/j.eneco.2018.07.002

Bhattacharya, M., Paramati, S. R., Ozturk, I., & Bhattacharya, S. (2016). The effect of renewable energy consumption on economic growth: Evidence from top 38 countries. *Applied Energy*, *162*, 733–741. doi:10.1016/j.apenergy.2015.10.104

Biçerli, K. (2009). *Çalışma Ekonomisi* (5th ed.). Beta Yayınları.

Biesenthal, C., & Wilden, R. (2014). Multi-level project governance: Trends and opportunities. *International Journal of Project Management*, *32*(8), 1291–1308. doi:10.1016/j.ijproman.2014.06.005

Bisht, A., Kamboj, N., Kamboj, V., & Bisht, A. (2020). A review on the role of emerging anthropogenic activities in environmental degradation and emphasis on their mitigation. *Archives of Agriculture and Environmental Science*, *5*(3), 419–425. doi:10.26832/24566632.2020.0503025

Bliadze, M., Chanturia, G., & Kereselidze, D. (2021). *World Geography*. Bakur Sulakauri Edition.

Blouin, J., Gleason, C., Mills, L., & Sikes, S. (2010). Pre-empting disclosure? Firms' decisions prior to FIN 48. *The Accounting Review*, *85*(3), 791–815. doi:10.2308/accr.2010.85.3.791

Bojanic, D. C., & Warnick, R. B. (2020). The relationship between a country's level of tourism and environmental performance. *Journal of Travel Research*, *59*(2), 220–230. doi:10.1177/0047287519827394

Bokil, Y., & Sinha, D. K. (2021). Green Marketing and Mindful Consumption for Sustainable Development. *Central Asian Journal of Innovations on Tourism Management and Finance*, *2*(9), 4–7.

Borjas, G. J. (2013). Measuring Labor Supply. In *Labor Economics* (pp. 22–23). McGraw-Hill Education.

Bowden, N., & Payne, J. E. (2010). Sectoral analysis of the causal relationship between renewable and non-renewable energy consumption and real output in the US. *Energy Sources. Part B, Economics, Planning, and Policy*, *5*(4), 400–408. doi:10.1080/15567240802534250

Boztepe, Y. (2007). *Türkiye'de İşsizlik Kavramı ve İşsizliğin Ortadan Kaldırılması ile İlgili Bir Model Oluşturulması* [Master's Thesis]. Yıldız Teknik Üniversitesi Sosyal Bilimler Enstitüsü, İstanbul.

Breitung, J., & Candelon, B. (2006). Testing for short- and long-run causality: A frequency-domain approach. *Journal of Econometrics*, *132*(2), 363–378. doi:10.1016/j.jeconom.2005.02.004

British Petroleum. (2020). *Statistical Review of World Energy 2020*. Author.

Brunnée, J., & Toope, S. J. (1997). Environmental security and freshwater resources: Ecosystem regime building. *The American Journal of International Law*, *91*(1), 26–59. doi:10.2307/2954139

Bulut, U., & Durusu-Ciftci, D. (2018). Revisiting energy intensity convergence: New evidence from OECD countries. *Environmental Science and Pollution Research International*, *25*(13), 12391–12397. doi:10.100711356-018-1551-z PMID:29460242

Bulut, U., & Muratoglu, G. (2018). Renewable energy in Turkey: Great potential, low but increasing utilization, and an empirical analysis on renewable energy-growth nexus. *Energy Policy*, *123*, 240–250. doi:10.1016/j.enpol.2018.08.057

Burnett, J. W., & Madariaga, J. (2016). The convergence of US state-level energy intensity. *Energy Economics*, *62*, 357–370. doi:10.1016/j.eneco.2016.03.029

Butnariu, A., & Luca, F. A. (2019). An index of sustainable economic welfare for Romania. *New Trends and Issues Proceedings on Humanities and Social Sciences*, *6*(8), 1–10. doi:10.18844/prosoc.v6i8.4543

Cam, E., & Atan, M. (2018). Türkiye'de İl Bazında İstihdam Politikalarının Etkinliği. *Bilgi Sosyal Bilimler Dergisi*, (1), 102–123.

Canel, C., Guris, S., Guris, B., Öktem, B., & Oktem, R. (2017). Convergence of energy intensity in OECD countries. *Modern Economy*, *8*(7), 946–958. doi:10.4236/me.2017.87066

Carrion-i-Silvestre, J. L., Del Barrio-Castro, T., & Lopez-Bazo, E. (2005). Breaking the panels: An application to the GDP per capita. *The Econometrics Journal*, *8*(2), 159–175. doi:10.1111/j.1368-423X.2005.00158.x

Castaneda, B. E. (1999). An index of sustainable economic welfare (ISEW) for Chile. *Ecological Economics*, *28*(2), 231–244. doi:10.1016/S0921-8009(98)00037-8

Chamorro, R., Rubio, S., & Miranda, F. J. (2009). Characteristics of research on green marketing. *Business Strategy and the Environment*, *18*(4), 223–239. doi:10.1002/bse.571

Chang, J., Fang, W., Liang, J., Zhang, P., Zhang, G., Zhang, H., Zhang, Y., & Wang, Q. (2022). A critical review on interaction of microplastics with organic contaminants in soil and their ecological risks on soil organisms. *Chemosphere*, *306*, 135573. doi:10.1016/j.chemosphere.2022.135573 PMID:35797912

Chan, R. Y. K. (1999). Environmental attitudes and behavior of consumers in China. *Journal of International Consumer Marketing*, *11*(4), 25–52. doi:10.1300/J046v11n04_03

Chan, R. Y. K. (2001). Determinants of Chinese consumers' green purchase behavior. *Psychology and Marketing*, *18*(4), 389–413. doi:10.1002/mar.1013

Chan, T.-S. (1996). Concerns for environmental issues and consumer purchase preferences: A two-country study. *Journal of International Consumer Marketing*, *9*(1), 43–55. doi:10.1300/J046v09n01_04

Charnes, A., Cooper, W. W., & Rhodes, E. (1978). Measuring the efficiency of decision making units. *European Journal of Operational Research*, *2*(6), 429–444. doi:10.1016/0377-2217(78)90138-8

Chekima, B., Wafa, S. A., Igau, O. A., Chekima, S., & Sondoh, S. L. (2016). Examining green consumerism motivational drivers: Does premium price and demographics matter to green purchasing? *Journal of Cleaner Production*, *112*, 3436–3450. doi:10.1016/j.jclepro.2015.09.102

Cheng, M., & Edwards, D. (2019). A comparative automated content analysis approach on the review of the sharing economy discourse in tourism and hospitality. *Current Issues in Tourism*, *22*(1), 35–49. doi:10.1080/13683500.2017.1361908

Cheng, Y., Awan, U., Ahmad, S., & Tan, Z. (2021). How do technological innovation and fiscal decentralization affect the environment? A story of the fourth industrial revolution and sustainable growth. *Technological Forecasting and Social Change*, *162*, 120398. doi:10.1016/j.techfore.2020.120398

Chen, X., Peterson, M. N., Hull, V., Lu, C., Hong, D., & Liu, J. (2013). How perceived exposure to environmental harm influences environmental behavior in urban China. *Ambio*, *42*(1), 52–60. doi:10.100713280-012-0335-9 PMID:22821144

Chitadze, N. (2004). *Caspian oil projects and transit role of Georgia*. Tbilisi State University.

Chitadze, N. (2017). *World Geography. Political, Economic and Demographic Dimensions*. Scholars Press.

Cho, C. H. (2009). Legitimation strategies used in response to environmental disaster: A French case study of Total SA's Erika and AZF incidents. *European Accounting Review*, *18*(1), 33–62. doi:10.1080/09638180802579616

Cho, C. H., Freedman, M., & Patten, D. M. (2012). Corporate disclosure of environmental capital expenditures: A test of alternative theories. *Accounting, Auditing & Accountability Journal*, *25*(3), 486–507. doi:10.1108/09513571211209617

Cho, C. H., & Patten, D. M. (2008). Did the GAO get it right? Another look at corporate environmental disclosure. *Social and Environmental Accountability Journal*, *28*(1), 21–32. doi:10.1080/0969160X.2008.9651788

Cho, C. H., Roberts, R. W., & Patten, D. M. (2010). The language of US corporate environmental disclosure. *Accounting, Organizations and Society*, *35*(4), 431–443. doi:10.1016/j.aos.2009.10.002

Christiansen, L. E., Lin, H., Pereira, J., Topalova, P. B., & Turk, R. (2016). Unlocking Female Employment Potential in Europe: Drivers and Benefits. *International Monetary Fund, Publication Services*. Retrieved June 5, 2022, from https://www.imf.org/external/pubs/ft/dp/2016/eur1601.pdf

Christie, A., & Brathwaite, R. (2011). *Mineral Commodity Report 14 — Gold*. Institute of Geological and Nuclear Sciences Ltd.

Chudik, A., & Pesaran, M. H. (2015a). Large panel data models with cross-sectional dependence: A survey. In B. H. Baltagi (Ed.), *The oxford handbook of panel data* (pp. 3–45). Oxford University Press. doi:10.1093/oxfordhb/9780199940042.013.0001

Chudik, A., & Pesaran, M. H. (2015b). Common correlated effects estimation of heterogeneous dynamic panel data models with weakly exogenous regressors. *Journal of Econometrics*, *188*(2), 393–420. doi:10.1016/j.jeconom.2015.03.007

CIA. (2022). *Country Comparisons - Current Account Balance*. The World Factbook. Retrieved July 31, 2022, from https://www.cia.gov/the-world-factbook/field/current-account-balance/country-comparison

Clark, C. F., Kotchen, M. J., & Moore, M. R. (2003). Internal and external influences on pro-environmental behavior: Participation in a green electricity program. *Journal of Environmental Psychology*, *23*(3), 237–246. doi:10.1016/S0272-4944(02)00105-6

Clarke, M. (2004). Widening development prescriptions: Policy implications of an Index of Sustainable Economic Welfare (ISEW) for Thailand. *International Journal of Environment and Sustainable Development*, *3*(3/4), 262–275. doi:10.1504/IJESD.2004.005075

Clarke, M., & Islam, S. M. (2005). Diminishing and negative welfare returns of economic growth: An index of sustainable economic welfare (ISEW) for Thailand. *Ecological Economics*, *54*(1), 81–93. doi:10.1016/j.ecolecon.2004.10.003

Clarkson, P. M., Li, Y., & Richardson, G. D. (2004). The market valuation of environmental Capital expenditures by pulp and paper companies. *The Accounting Review*, *79*(2), 329–353. doi:10.2308/accr.2004.79.2.329

Clarkson, P. M., Li, Y., Richardson, G. D., & Vasvari, F. P. (2008). Revisiting the relation between environmental performance and environmental disclosure: An empirical analysis. *Accounting, Organizations and Society*, *33*(4/5), 303–327. doi:10.1016/j.aos.2007.05.003

Clements, J. M., McCright, A. M., & Xiao, C. (2014). Green Christians? An empirical examination of environmental concern within the U.S. general public. *Organization & Environment*, *27*(1), 85–102. doi:10.1177/1086026613495475

Cobb, C. W., & Douglas, P. H. (1928). A Theory of Production. *The American Economic Review*, *18*(Supplement), 139–165.

Cowan, S., & Gadenne, D. (2005). Australian corporate environmental reporting: A comparative analysis of disclosure practices across voluntary and mandatory disclosure systems. *Journal of Accounting & Organizational Change*, *1*(2), 165–179. doi:10.1108/18325910510635344

Craig, J., & Amernic, H. (2004). Enron discourse: The rhetoric of a resilient capitalism. *Critical Perspectives on Accounting*, *15*(6/7), 813–851. doi:10.1016/j.cpa.2002.12.001

Cretchley, J., Gallois, C., Chenery, H., & Smith, A. (2010). Conversations between carers and people with Schizophrenia: A qualitative analysis using Leximancer. *Qualitative Health Research*, *20*(12), 1611–1628. doi:10.1177/1049732310378297 PMID:20675536

Criado-Jiménez, I., Fernández-Chulián, M., Larrinaga-González, C., & Husillos-Carqués, F. J. (2008). Compliance with mandatory environmental reporting in financial statements: The case of Spain (2001–2003). *Journal of Business Ethics*, *79*(3), 245–262. doi:10.100710551-007-9375-7

Croux, C., & Reusens, P. (2013). Do stock prices contain predictive power for the future economic activity? A Granger causality analysis in the frequency domain. *Journal of Macroeconomics*, *35*, 93–103. doi:10.1016/j.jmacro.2012.10.001

Daly, L., & Posner, S. (2012). *Beyond GDP new measures for a new economy*. Dēmos.

Dangelico, R. M., Nonino, F., & Pompei, A. (2021). Which are the determinants of green purchase behaviour? A study of Italian consumers. *Business Strategy and the Environment*, *30*(5), 2600–2620. doi:10.1002/bse.2766

Danish & Ulucak, R. (2020). How do environmental technologies affect green growth? Evidence from BRICS economies. *Science of the Total Environment*, *712*, 136504. doi:10.1016/j.scitotenv.2020.136504

Danish, & Wang, Z. (2018). Dynamic relationship between tourism, economic growth, and environmental quality. *Journal of Sustainable Tourism*, *26*(11), 1928-1943.

Das, A., & Basu, T. (2020). Assessment of peri-urban wetland ecological degradation through importance-performance analysis (IPA): A study on Chatra Wetland, India. *Ecological Indicators*, *114*, 106274. doi:10.1016/j.ecolind.2020.106274

Dayıoğlu, M., & Ercan, H. (2010). *Developing More Effective Labour Market Policies and Institutions in Emerging Economies: Turkey. International Labour Organization Project, (76). Employment Analysis and Research Unit*. EMP/ANALYSIS.

De Pelsmacker, P., Driesen, L., & Rayp, G. (2005). Do consumers care about ethics? Willingness to pay for fair-trade coffee. *The Journal of Consumer Affairs*, *39*(2), 363–385. doi:10.1111/j.1745-6606.2005.00019.x

Dean, M., Raats, M. M., & Shepherd, R. (2008). Moral concerns and consumer choice of fresh and processed organic foods. *Journal of Applied Social Psychology*, *38*(8), 2088–2107. doi:10.1111/j.1559-1816.2008.00382.x

Deegan, C., & Rankin, M. (1996). Do Australian companies report environmental news objectively? An analysis of environmental disclosures by firms prosecuted successfully by the environmental protection authority. *Accounting, Auditing & Accountability Journal*, *9*(2), 50–67. doi:10.1108/09513579610116358

Deegan, C., Rankin, M., & Voght, P. (2000). Firms' disclosure reactions to major social incidents: Australian evidence. *Accounting Forum*, *24*(1), 101–130. doi:10.1111/1467-6303.00031

DeFries, R. S., Edenhofer, O., Halliday, A. N., Heal, G. M., Lenton, T., Puma, M., ... Ward, B. (2019). *The missing economic risks in assessments of climate change impacts*. Academic Press.

Deletic, A., & Wang, H. (2019). Water pollution control for sustainable development. *Engineering (Beijing)*, *5*(5), 839–840. doi:10.1016/j.eng.2019.07.013

Dell, E. (1987). Mercantilism. In *The Politics of Economic Interdependence* (pp. 168–185). Palgrave Macmillan. doi:10.1007/978-1-349-18874-1_11

Demir, F., & Erdem, N. (2010). Labor Market Performance after Structural Adjustment in Developing Countries: The Interesting But not so Unique Case of Turkey. In L. K. Valencia & B. J. Hahn (Eds.), *Employment and Labor Issues: Unemployment, Youth Employment and Child Labor* (pp. 1–55). Nova Science Publishers.

Dergiades, T., Martinopoulos, G., & Tsoulfidis, L. (2013). Energy consumption and economic growth: Parametric and non-parametric causality testing for the case of Greece. *Energy Economics*, *36*, 686–697. doi:10.1016/j.eneco.2012.11.017

Destek, M. A., & Aslan, A. (2017). Renewable and non-renewable energy consumption and economic growth in emerging economies: Evidence from bootstrap panel causality. *Renewable Energy*, *111*, 757–763. doi:10.1016/j.renene.2017.05.008

Diamantopoulos, A., Schlegelmilch, B. B., Sinkovics, R. R., & Bohlen, G. M. (2003). Can socio-demographics still play a role in profiling green consumers? A review of the evidence and an empirical investigation. *Journal of Business Research*, *56*(6), 465–480. doi:10.1016/S0148-2963(01)00241-7

Dickey, D. A., & Fuller, W. A. (1979). Distribution of the estimators for autoregressive time series with a unit root. *Journal of the American Statistical Association*, *74*, 427–431.

Dickey, D. A., & Fuller, W. A. (1981). Likelihood ratio statistics for autoregressive time series with a unit root. *Econometrica*, *49*(4), 1057. doi:10.2307/1912517

Dictionary, F. (2022). *Recreation*. Retrieved from: https://www.thefreedictionary.com/recreation

Dimitri, C., & Dettmann, R. (2012). Organic food consumers: What do we really know about them? *British Food Journal*, *114*(8), 1157–1183. doi:10.1108/00070701211252101

Ditzen, J. (2018). Estimating dynamic common-correlated effects in stata. *The Stata Journal*, *18*(3), 585–617. doi:10.1177/1536867X1801800306

Do Paço, A. M. F., Alves, H., Shiel, C., & Filho, W. L. (2013). Development of a green consumer behavior model. *International Journal of Consumer Studies*, *37*(4), 414–421. doi:10.1111/ijcs.12009

Do Paço, A. M. F., & Raposo, M. L. B. (2009). Green segmentation: An application to the Portuguese consumer market. *Marketing Intelligence & Planning*, *27*(3), 364–379. doi:10.1108/02634500910955245

Do Paço, A. M. F., & Raposo, M. L. B. (2010). Green consumer market segmentation: Empirical findings from Portugal. *International Journal of Consumer Studies*, *34*(4), 429–436. doi:10.1111/j.1470-6431.2010.00869.x

Do Paço, A. M. F., Raposo, M. L. B., & Filho, W. L. (2009). Identifying the green consumer: A segmentation study. *Journal of Targeting. Measurement and Analysis for Marketing*, *17*(1), 17–25. doi:10.1057/jt.2008.28

Duguma, L. A., Atela, J., Minang, P. A., Ayana, A. N., Gizachew, B., Nzyoka, J. M., & Bernard, F. (2019). Deforestation and forest degradation as an environmental behavior: Unpacking realities shaping community actions. *Land (Basel)*, *8*(2), 26. doi:10.3390/land8020026

Du, K. (2017). Econometric convergence test and club clustering using Stata. *The Stata Journal*, *17*(4), 882–900. doi:10.1177/1536867X1801700407

Dunlap, A. (2022). The Self-Reinforcing Cycle of Ecological Degradation and Repression: Revealing the Ecological Cost of Policing and Militarization. In Enforcing Ecocide (pp. 153-176). Palgrave Macmillan.

Dvoskin, D., & Heady, E. O. (1976). Farming practices, environmental quality, and the energy crises. *Agriculture and Environment*, *3*(1), 1–13. doi:10.1016/0304-1131(76)90002-3

Ekin, N. (2003). Türkiye'de İşsizlik: İş Aramayan İşsizler – Kırsal Yoksullar – Kentsel Kayıtdışı Yapay İstihdamdakiler – Açık İşsizler. *Kamu-İş Dergisi*, (7), 2–21.

El Menyari, Y. (2021). The effects of international tourism, electricity consumption, and economic growth on CO_2 emissions in North Africa. *Environmental Science and Pollution Research International*, *28*(32), 44028–44038. doi:10.100711356-021-13818-5 PMID:33844140

Elder, S., & Smith, A. (2010). *Women in labour markets: Measuring progress and identifying challenges*. International Labour Office.

ElHaffar, G., Durif, F., & Dube, L. (2020). Towards closing the attitude-intention-behavior gap in green consumption: A narrative review of the literature and an overview of future research directions. *Journal of Cleaner Production*, *275*(122556), 1–20. doi:10.1016/j.jclepro.2020.122556

Elizbarashvili, N., & Davitashvili, Z. (2012). *Global Geography*. Tbilisi State University.

Emirmahmutoglu, F., & Kose, N. (2011). Testing for Granger causality in heterogeneous mixed panels. *Economic Modelling*, *28*(3), 870–876. doi:10.1016/j.econmod.2010.10.018

Enders, W., & Lee, J. (2012). A unit root test using a Fourier series to approximate smooth breaks. *Oxford Bulletin of Economics and Statistics*, *74*(4), 574–599. doi:10.1111/j.1468-0084.2011.00662.x

Energy, G. (2010). *World Natural Resources*. Retrieved from: http://greenenergyhomedesign.tk/tag/green-energy/page/237/

Engle, R. F., & Granger, C. W. (1987). Co-integration and error correction: Representation, estimation, and testing. *Econometrica*, *55*(2), 251–276. doi:10.2307/1913236

Environmental Protection Agency (EPA) (2001). *U.S. EPA Notifying Defendants of Securities and Exchange Commission's Environmental Disclosure Requirements Enforcement Alert 4(3)*. EPA.

Environmental Protection Agency (EPA) (n.d.). *Supplemental Environmental Projects (SEPs)*. EPA.

Environmental Protection Agency (EPA). (2015) U.S. *Environmental Protection Agency supplemental environmental projects*. https://www.epa.gov/sites/default/files/2015-04/documents/sepupdatedpolicy15.pdf

Ercan, H. (2007a). Demographic Trends and Rural-to-Urban Migration in Turkey: Implications for the Labour Market and Education. In World Bank project on Investing in Turkey's Next Generation: Making Youth Transition from School to Work Contribute to Turkey's Development (pp. 1-7). World Bank Publications - Reports, The World Bank Group.

Ercan, H. (1998). Non-Wage Labor Costs in the Turkish Manufacturing Industry: An International Comparison. In T. Bulutay (Ed.), *Impositions and Policies Related to the Turkish Labour Market* (pp. 113–130). Ankara State Institute of Statistics Publication.

Ercan, H. (2007b). *Youth Employment in Turkey*. International Labour Organization Publications.

Ercan, H. (2010a). The Impact of the Global Financial Crisis on Employment in Turkey. In *Crisis and Turkey: Impact Analysis of Crisis Response Measures* (pp. 73–97). International Labour Office Publications.

Ercan, H. (2010b). *EEO Review: Youth Employment Measures 2010*. European Employment Observatory.

Erdoğan, S., Gedikli, A., & Genç, S. Y. (2018). An overview of Turkey's national energy policies. *Politico-Economic Evaluation of Current Issues*, 143.

Erdoğan, S., Yıldırım, S., Yıldırım, D. Ç., & Gedikli, A. (2020). The effects of innovation on sectoral carbon emissions: Evidence from G20 countries. *Journal of Environmental Management*, *267*, 110637. doi:10.1016/j.jenvman.2020.110637 PMID:32349957

Eroğlu, İ., & Bursal, M. (2020). Türkiye'nin Cari Açık Dinamiklerinde Enerji Açığının Rolü ve Finansman Kalitesi. *MANAS Sosyal Araştırmalar Dergisi*, *9*(4), 2231–2247. doi:10.33206/mjss.689842

Erol, I., Cakar, N., Erel, D., & Sari, R. (2009). Sustainability in the Turkish retailing industry. *Sustainable Development (Bradford)*, *17*(1), 49–67. doi:10.1002d.369

Ertz, M., Karakaş, F., & Sarıgöllü, E. (2016). Exploring pro-environmental behaviors of consumers: An analysis of contextual factors, attitudes, and behaviors. *Journal of Business Research*, *69*(10), 3971–3980. doi:10.1016/j.jbusres.2016.06.010

Ettredge, M. L., Smith, D. B., & Stone, M. S. (2001). How effective is joint public and private monitoring? The case of the AICPA auditor change notification letter. *Auditing*, *20*(1), 29–43. doi:10.2308/aud.2001.20.1.29

Ettredge, M., Johnstone, K., Stone, M., & Wang, Q. (2011). The effects of firm size corporate governance quality and bad news on disclosure compliance. *Review of Accounting Studies*, *16*(4), 866–889. doi:10.100711142-011-9153-8

European Commission. (2021a). *Paris Agreement.* https://ec.europa.eu/clima/policies/international/negotiations/paris_en

European Commission. (2021b). *2030 Climate & Energy Framework.* https://ec.europa.eu/clima/policies/strategies/2030_en

Eurostat. (2022a, October 27). *Energy efficiency.* https://ec.europa.eu/eurostat/databrowser/view/nrg_ind_eff/default/table?lang=en

Eurostat. (2022b, October 27). *Energy imports dependency,* https://ec.europa.eu/eurostat/databrowser/view/nrg_ind_id/default/table?lang=en

Eurostat. (2022c, October 27). *Energy productivity.* https://ec.europa.eu/eurostat/databrowser/view/t2020_rd310/default/table?lang=en

Eurostat. (2022d, October 27). *Inequality of income distribution.* ec.europa.eu/eurostat/databrowser/view/tespm151/default/table?lang=en

Ezcurra, R. (2007). Distribution dynamics of energy intensities: A cross-country analysis. *Energy Policy, 35*(10), 5254–5259. doi:10.1016/j.enpol.2007.05.006

Fallan, E., & Fallan, L. (2019). Corporate tax behaviour and environmental disclosure: Strategic trade-offs across elements of CSR? *Scandinavian Journal of Management, 35*(3), 101042. Advance online publication. doi:10.1016/j.scaman.2019.02.001

FAO. (2022). *Annual Report.* Retrieved from: https://www.fao.org/common-pages/search/en/?q=Annual%20Report%202022

Farrell, M. J. (1957). The measurement of productive efficiency. *Journal of the Royal Statistical Society. Series A (General), 120*(3), 253–281. doi:10.2307/2343100

Felix, R., & Braunsberger, K. (2016). I believe therefore I care. The relationship between religiosity, environmental attitudes, and green product purchase in Mexico. *International Marketing Review, 33*(1), 137–155. doi:10.1108/IMR-07-2014-0216

Finnemore, M., & Sikkink, K. (1998). International norm dynamics and political change. *International Organization, 52*(4), 887–917. doi:10.1162/002081898550789

Foteinis, S. (2020). How small daily choices play a huge role in climate change: The disposable paper cup environmental bane. *Journal of Cleaner Production, 255,* 120294. doi:10.1016/j.jclepro.2020.120294

Frankena, F. (1978). Regional socioeconomic impacts of declining net energy. *Urban Ecology, 3*(2), 101–110. doi:10.1016/0304-4009(78)90001-3

Fred. (2022a). *Average Hours of Work Per Week, Total, Household Survey for United States (2022).* Retrieved April 22, 2019, from https://fred.stlouisfed.org/series/M08354USM310NNBR

Froot, K. A. (1989). Consistent covariance matrix estimation with cross-sectional dependence and heteroscedasticity in financial data. *Journal of Financial and Quantitative Analysis*, *24*(3), 333–355. doi:10.2307/2330815

Fuss, M. A. (1977). The demand for the energy in Canadian manufacturing: An example of the estimation of production structures with many inputs. *Journal of Econometrics*, *5*(1), 89–116. doi:10.1016/0304-4076(77)90036-7

Gaddis, I., & Klasen, S. (2014). Economic development, structural change, and women's labor force participation: A reexamination of the feminization U hypothesis. *Journal of Population Economics*, *27*(3), 639–681. doi:10.100700148-013-0488-2

Garnier, M., & Hout, M. (1976). Inequality of educational opportunity in France and the United States. *Social Science Research*, *5*(3), 225–246. doi:10.1016/0049-089X(76)90001-6

Gastwirth, J. L. (1975). The estimation of a family of measures of economic inequality. *Journal of Econometrics*, *3*(1), 61–70. doi:10.1016/0304-4076(75)90067-6

GDRC (Global Development Research Center). (2021). *Sustainable development*. Retrieved April 29, 2021, from http://www.gdrc.org/sustdev/definitions

Gedamu, M. T. (2020). Soil Degradation and its Management Options in Ethiopia: A Review. *International Journal of Research and Innovations in Earth Science*, *7*(5), 59–76.

Ghazali, E. M., Mutum, D. S., & Ariswibowo, N. (2018). Impact of religious values and habit on an extended green purchase behaviour model. *International Journal of Consumer Studies*, *42*(6), 639–654. doi:10.1111/ijcs.12472

Gilpatrick, E. (1966). On The Classification of Unemployment: A View of The Structural-Inadequate Demand Debate. *Industrial & Labor Relations Review*, *19*(2), 201–212. doi:10.1177/001979396601900203

Gleim, M. R., Smith, J. S., Andrews, D., & Cronin, J. J. Jr. (2013). Against the green: A multi-method examination of the barriers to green consumption. *Journal of Retailing*, *89*(1), 44–61. doi:10.1016/j.jretai.2012.10.001

Global Change. (2021). *Current Lectures on Land Degradation*. Retrieved from: http://www.globalchange.umich.edu/globalchange2/current/lectures/land_deg/land_deg.html

Global Footprint Network. (2022). *National footprint accounts*. Retrieved May 10, 2022, from https://data.footprintnetwork.org/#/

Goldin, C. (1995). The U-Shaped Female Labor Force Function in Economic Development and Economic History. In *Schultz TP Investment in Women's Human Capital and Economic Development* (pp. 61–90). University of Chicago Press.

Gonçalves, H. M., Lourenço, T. F., & Silva, G. M. (2016). Green buying behavior and the theory of consumption values: A fuzzy-set approach. *Journal of Business Research*, *69*(4), 1484–1491. doi:10.1016/j.jbusres.2015.10.129

González, E. M., Felix, R., Carrete, L., Centeno, E., & Castaño, R. (2015). Green shades: A segmentation approach based on ecological consumer behavior in an emerging economy. *Journal of Marketing Theory and Practice*, *23*(3), 287–302. doi:10.1080/10696679.2015.1032395

Gorrie, M. (2020). Regulating our Consumer Culture: What Role Can the Law Play in Addressing Excessive Consumption? In *Sustainability and Law* (pp. 119-134). Springer.

Gorus, M. S., & Karagol, E. T. (2022). Reactions of energy intensity, energy efficiency, and activity indexes to income and energy price changes: The panel data evidence from OECD countries. *Energy*, *254*, 124281. doi:10.1016/j.energy.2022.124281

Greening, L. A., Greene, D. L., & Difiglio, C. (2000). Energy efficiency and consumption—the rebound effect—A survey. *Energy Policy*, *28*(6-7), 389–401. doi:10.1016/S0301-4215(00)00021-5

Greenman, J. V. (1979). An energy analysis of the investment problem. *Energy Economics*, *1*(4), 219–223. doi:10.1016/0140-9883(79)90004-5

Griffin, K. (1974). The international transmission of inequality. *World Development*, *2*(3), 3–15. doi:10.1016/0305-750X(74)90091-6

Griffin, P. A., & Jaffe, A. M. (2018). Are fossil fuel firms informing investors well enough about the risks of climate change? *Journal of Energy & Natural Resources Law*, *36*(4), 381–410. doi:10.1080/02646811.2018.1502240

Grossman, G. M., & Krueger, A. B. (1991). Environmental impacts of a north American free trade agreement. *NBER Working Papers Series*, 1–39.

Gu, D., Gao, S., Wang, R., Jiang, J., & Xu, Y. (2020). The negative associations between materialism and pro-environmental attitudes and behaviors: Individual and regional evidence from China. *Environment and Behavior*, *52*(6), 611–638. doi:10.1177/0013916518811902

Guerrero, V. T., & Guevara, F. C. (2021). Influence of Deforestation and Environmental Degradation on Agriculture in Mexico. *Nongxue Xuebao*, *5*(1).

Guide. (2021). *Leximancer User Guide Release 5.0*. Leximancer Pvt. Ltd.

Guide, C. I. M. O. (2008). *Measurement of sunshine duration*. World Meteorological Organization.

Guillamon-Saorin, E., Osma, G. B., & Jones, M. J. (2012). Opportunistic disclosure in press release headlines. *Accounting and Business Research*, *42*(2), 143–168. doi:10.1080/00014788.2012.632575

Gupta, R., Gil-Alana, L. A., & Yaya, O. S. (2015). Do sunspot numbers cause global temperatures? Evidence from a frequency domain causality test. *Applied Economics*, *47*(8), 798–808. doi:10.1080/00036846.2014.980575

Gürsel, S., & Atasoy, V. (2000). Genç İşsiz, Kadın İşsiz. *Forum TOBB Dergisi*, *7*(6), 18–22.

Gu, W., Zhao, X., Yan, X., Wang, C., & Li, Q. (2019). Energy technological progress, energy consumption, and CO_2 emissions: Empirical evidence from China. *Journal of Cleaner Production*, *236*, 117666. doi:10.1016/j.jclepro.2019.117666

Haberstroh, J. R. (1978). The case of Hungary: Liberal socialism under stress. *Journal of Comparative Economics*, *2*(2), 111–125. doi:10.1016/0147-5967(78)90043-4

Haller, M., & Hoellinger, F. (1994). Female Employment and the Change of Gender Roles: The Conflictual Relationship between Participation and Attitudes in International Comparison. *International Sociology*, *9*(1), 87–112. doi:10.1177/026858094009001006

Hansen, B. E. (1995). Rethinking the univariate approach to unit root testing: Using covariates to increase power. *Econometric Theory*, *11*(5), 1148–1171. doi:10.1017/S0266466600009993

Hansen, C., & Skinner, R. (2005). *Nuclear Power and Renewables: Strange Bedfellows?* Oxford Institute for Energy Studies.

Hanson, D. L., & Pledger, G. (1976). Consistency in concave regression. *Annals of Statistics*, *4*(6), 1038–1050. doi:10.1214/aos/1176343640

Harper, A. B., Powell, T., Cox, P. M., House, J., Huntingford, C., Lenton, T. M., Sitch, S., Burke, E., Chadburn, S. E., Collins, W. J., Comyn-Platt, E., Daioglou, V., Doelman, J. C., Hayman, G., Robertson, E., van Vuuren, D., Wiltshire, A., Webber, C. P., Bastos, A., ... Shu, S. (2018). Land-use emissions play a critical role in land-based mitigation for Paris climate targets. *Nature Communications*, *9*(1), 1–13. doi:10.103841467-018-05340-z PMID:30087330

Hasan, T., Karim, W., & Quayes, S. (2008). Regulatory change and the quality of compliance to mandatory disclosure requirements. *Research in Accounting Regulation*, *20*, 193–203. doi:10.1016/S1052-0457(07)00211-1

Hassan, S. T., Batool, B., Zhu, B., & Khan, I. (2022). Environmental complexity of globalization, education, and income inequalities: New insights of energy poverty. *Journal of Cleaner Production*, *340*, 1–12. doi:10.1016/j.jclepro.2022.130735

Hayes, A. (2002). What is Unemployment. *Investopia*. Retrieved June 6, 2001, from https://www.investopedia.com/terms/u/unemployment.asp

Heitzman, S., Wasley, C., & Zimmerman, J. (2010). The joint effects of materiality thresholds and Voluntary disclosure incentives on firms' disclosure decisions. *Journal of Accounting and Economics*, *49*(1-2), 109–132. doi:10.1016/j.jacceco.2009.10.002

Herrerias, M. J. (2012). World energy intensity convergence revisited: A weighted distribution dynamics approach. *Energy Policy*, *49*, 383–399. doi:10.1016/j.enpol.2012.06.044

Hildreth, C. (1954). Point Estimates of Ordinates of Concave Functions. *Journal of the American Statistical Association*, *49*(267), 598–619. doi:10.1080/01621459.1954.10483523

Honkanen, P., Verplanken, B., & Olsen, S. O. (2006). Ethical values and motives driving organic food choice. *Journal of Consumer Behaviour*, *5*(5), 420–430. doi:10.1002/cb.190

Hussain, M., Arshad, Z., & Bashir, A. (2022). Do economic policy uncertainty and environment-related technologies help in limiting ecological footprint? *Environmental Science and Pollution Research International, 29*(31), 1–8. doi:10.100711356-022-19000-9 PMID:35171421

Hussain, M., & Dogan, E. (2021). The role of institutional quality and environment-related technologies in environmental degradation for BRICS. *Journal of Cleaner Production, 304*, 127059. doi:10.1016/j.jclepro.2021.127059

Hussain, M., & Khan, J. A. (2021). The nexus of environment-related technologies and consumption-based carbon emissions in top five emitters: Empirical analysis through dynamic common correlated effects estimator. *Environmental Science and Pollution Research International, 30*(10), 1–10. doi:10.100711356-021-15333-z PMID:34254244

Hussain, M., Mir, G. M., Usman, M., Ye, C., & Mansoor, S. (2022). Analysing the role of environment-related technologies and carbon emissions in emerging economies: A step towards sustainable development. *Environmental Technology, 43*(3), 367–375. doi:10.1080/09593330.2020.1788171 PMID:32579488

İlkkaracan, İ., & Selim, R. (2003). The Role of Unemployment in Wage Determination: Further Evidence on the Wage Curve from Turkey. *Applied Economics, 35*(14), 1589–1598. doi:10.1080/0003684022000040939

İlkkaracan, İ., & Yörükoğlu, A. (2004). Ücretlerin Belirlenme Süreçleri. In *Türkiye'de İşgücü Piyasasının Kurumsal Yapısı ve İşsizlik* (pp. 95–145). TÜSİAD.

Illich, I. (1974). *Energy and equity*. Marion Boyars.

Im, K. S., Lee, J., & Tieslau, M. A. (2014). More powerful unit root tests with non-normal errors. *Festschrift in honor of Peter Schmidt* (pp. 315–342). Springer.

Im, K. S., Pesaran, M. H., & Shin, Y. (2003). Testing for unit roots in heterogeneous panels. *Journal of Econometrics, 115*(1), 53–74. doi:10.1016/S0304-4076(03)00092-7

Im, K. S., & Schmidt, P. (2008). More efficient estimation under non-normality when higher moments do not depend on the regressors, using residual augmented least squares. *Journal of Econometrics, 144*(1), 219–233. doi:10.1016/j.jeconom.2008.01.003

Incze, J. S., & Merriam, J. Jr. (1973). The energy crisis. *Lancet, 302*(7834), 918. doi:10.1016/S0140-6736(73)92057-6 PMID:4126964

International Labour Organization. (2019a). *Employment to population ratio, 15+, female (%) (modeled ILO estimate) - Country Ranking*. ILOSTAT database. Retrieved September 19, 2019, from https://www.indexmundi.com/facts/indicators/SL.EMP.TOTL.SP.FE.ZS/rankings

International Labour Organization. (2019b). *Employment to population ratio, 15+, male (%) (modeled ILO estimate) - Country Ranking*. ILOSTAT database. Retrieved September 19, 2019, from https://www.indexmundi.com/facts/indicators/SL.EMP.TOTL.SP.MA.ZS/rankings

International Labour Organization. (2019c). *Employment to population ratio, 15+, total (%) (modeled ILO estimate) - Country Ranking*. ILOSTAT database. Retrieved September 19, 2019, from https://www.indexmundi.com/facts/indicators/SL.EMP.TOTL.SP.ZS/rankings

Işık, Ş. (2005). Türkiye'de Kentleşme ve Kentleşme Modelleri. *Ege Coğrafya Dergisi, 14*(1-2), 57-71. Retrieved June 7, 2022, from https://dergipark.org.tr/tr/pub/ecd/issue/4878/66933

Işik, C., Kasımatı, E., & Ongan, S. (2017). Analyzing the causalities between economic growth, financial development, international trade, tourism expenditure and/on the CO_2 emissions in Greece. *Energy Sources. Part B, Economics, Planning, and Policy, 12*(7), 665–673. doi:10.1080/15567249.2016.1263251

Jaini, A., Quoquab, F., Mohammad, J., & Hussin, N. (2020). Antecedents of green purchase behavior of cosmetics products: An empirical investigation among Malaysian consumers. *International Journal of Ethics and Systems, 36*(2), 185–203. doi:10.1108/IJOES-11-2018-0170

Jain, S. K., & Kaur, G. (2006). Role of sociodemographics in segmenting and profiling green consumers: An exploratory study of consumers in India. *Journal of International Consumer Marketing, 18*(3), 107–117. doi:10.1300/J046v18n03_06

Jaiswal, D., & Kant, R. (2018). Green purchasing behavior: A conceptual framework and empirical investigation of Indian consumers. *Journal of Retailing and Consumer Services, 41*, 60–69. doi:10.1016/j.jretconser.2017.11.008

Jansson, J., Marell, A., & Nordlund, A. (2010). Green consumer behavior: Determinants of curtailment and eco-innovation adoption. *Journal of Consumer Marketing, 27*(4), 358–370. doi:10.1108/07363761011052396

Johnstone, L., & Lindh, C. (2018). The sustainability-age dilemma: A theory of (un) planned behavior via influencers. *Journal of Consumer Behaviour, 17*(1), 127–139. doi:10.1002/cb.1693

Johnstone, M.-L., & Tan, L. P. (2015). Exploring the gap between consumers' green rhetoric and purchasing behavior. *Journal of Business Ethics, 132*(2), 311–328. doi:10.100710551-014-2316-3

Joseph, N., Ryu, D., Malano, H. M., George, B., & Sudheer, K. P. (2020). A review of the assessment of sustainable water use at continental-to-global scale. *Sustainable Water Resources Management, 6*(2), 1–20. doi:10.100740899-020-00379-7

Joshi, Y., & Rahman, Z. (2015). Factors affecting green purchase behavior and future research directions. *International Strategic Management Review, 3*(1-2), 128–143. doi:10.1016/j.ism.2015.04.001

Kafandaris, S. (1978). Restructuring images of inequality. *Habitat International, 3*(1), 83–87. doi:10.1016/0197-3975(78)90040-1

Kalamas, M., Cleveland, M., & Laroche, M. (2014). Pro-environmental behaviors for thee but not for me: Green giants, green Gods, and external environmental locus of control. *Journal of Business Research, 67*(2), 12–22. doi:10.1016/j.jbusres.2013.03.007

Kaplan, B., & Iyer, E. S. (2021). Motivating sustainable behaviors: The role of religiosity in a cross-cultural context. *The Journal of Consumer Affairs*, *55*(3), 792–820. doi:10.1111/joca.12376

Karabıyık, İ. (2012). Türkiye'de Çalışma Hayatında Kadın İstihdamı. *Marmara Üniversitesi İktisadi ve İdari Bilimler Dergisi*, *32*(1), 231–260.

Karabulut, A. (2007). *Türkiye'de İşsizliği Önlemede Aktif İstihdam Politikalarının Rolü ve Etkinliği* [Master's Thesis]. Çalışma ve Sosyal Güvenlik Bakanlığı Türkiye İş Kurumu Genel Müdürlüğü.

Karanassou, M., Sala, H., & Salvador, P. (2008). Capital Accumulation and Unemployment: New Insights on the Nordic Experience. *Cambridge Journal of Economics*, *32*(6), 977–1001. doi:10.1093/cje/ben022

Karimu, A., Brännlund, R., Lundgren, T., & Söderholm, P. (2017). Energy intensity and convergence in Swedish industry: A combined econometric and decomposition analysis. *Energy Economics*, *62*, 347–356. doi:10.1016/j.eneco.2016.07.017

Karpoff, J. M., Lott, J. R. Jr, & Wehrly, E. W. (2005). The Reputational Penalties for Environmental Violations: Empirical Evidence. *The Journal of Law & Economics*, *48*(2), 653–675. doi:10.1086/430806

Karumidze, A. (2021). *International Organizations* (7th ed.). Universali.

Kassarjian, H. H. (1971). Incorporating ecology into marketing strategy: The case of air pollution. *Journal of Marketing*, *35*(3), 61–65. doi:10.1177/002224297103500310

Katircioğlu, S. T. (2014). Testing the tourism-induced EKC hypothesis: The case of Singapore. *Economic Modelling*, *41*, 383–391. doi:10.1016/j.econmod.2014.05.028

Katircioglu, S. T., Feridun, M., & Kilinc, C. (2014). Estimating tourism-induced energy consumption and CO_2 emissions: The case of Cyprus. *Renewable & Sustainable Energy Reviews*, *29*, 634–640. doi:10.1016/j.rser.2013.09.004

Katircioglu, S., Gokmenoglu, K. K., & Eren, B. M. (2018). Testing the role of tourism development in ecological footprint quality: Evidence from top 10 tourist destinations. *Environmental Science and Pollution Research International*, *25*(33), 33611–33619. doi:10.100711356-018-3324-0 PMID:30276681

Khan, A., Bibi, S., Ardito, L., Lyu, J., Hayat, H., & Arif, A. M. (2020). Revisiting the dynamics of tourism, economic growth, and environmental pollutants in the emerging economies–sustainable tourism policy implications. *Sustainability (Basel)*, *12*(6), 2533. doi:10.3390u12062533

Khan, I., Hou, F., Le, H. P., & Ali, S. A. (2021). Do natural resources, urbanization, and value-adding manufacturing affect environmental quality? Evidence from the top ten manufacturing countries. *Resources Policy*, *72*, 102109. doi:10.1016/j.resourpol.2021.102109

Kılavuz, E., Oralhan, B., Sarıgül, S. S., & Uluğ, E. E. (2021). The validity of the tourism-induced EKC hypothesis: The case of Turkey. *International Journal of Business and Economic Studies*, *3*(2), 124–138. doi:10.54821/uiecd.1033651

Kilbourne, W. E., & Beckmann, S. C. (1998). Review and critical assessment of research on marketing and the environment. *Journal of Marketing Management*, *14*(6), 513–532. doi:10.1362/026725798784867716

Kilbourne, W. E., & Pickett, G. (2008). How materialism affects environmental beliefs, concern, and environmentally responsible behavior. *Journal of Business Research*, *61*(9), 885–893. doi:10.1016/j.jbusres.2007.09.016

Kilian, L. (2008). Exogenous oil supply shocks: How big are they and how much do they matter for the US economy? *The Review of Economics and Statistics*, *90*(2), 216–240. doi:10.1162/rest.90.2.216

Kilian, L. (2015). Energy Price Shocks. In S. N. Durlauf & L. E. Blume (Eds.), *The New Palgrave Dictionary of Economics*. Palgrave Macmillan. doi:10.1057/978-1-349-95121-5_3008-1

Kim, Y., & Choi, S. M. (2005). Antecedents of green purchase behaviour: An examination of collectivism, environmental concern and PCE. *Advances in Consumer Research. Association for Consumer Research (U. S.)*, *32*, 592–59.

Kinnear, T. C., Taylor, J. R., & Ahmed, S. A. (1974). Ecologically concerned consumers: Who are they? *Journal of Marketing*, *38*, 20–24.

Kıran, B. (2008). *Kalkınmada Öncelikli İllerin Ekonomik Etkinlikleri-nin Veri Zarflama Analizi Yöntemi ile Değerlendirilmesi* [Unpublished Master's Thesis]. Çukurova Üniversitesi Sosyal Bilimler Enstitüsü, Adana.

Kiran, B. (2013). Energy intensity convergence in OECD countries. *Energy Exploration & Exploitation*, *31*(2), 237–247. doi:10.1260/0144-5987.31.2.237

Koçak, E., Ulucak, R., & Ulucak, Z. Ş. (2020). The impact of tourism developments on CO_2 emissions: An advanced panel data estimation. *Tourism Management Perspectives*, *33*(100611), 1–10.

Kollmuss, A., & Agyeman, J. (2002). Mind the gap: Why do people act environmentally and what are the barriers to pro-environmental behavior. *Environmental Education Research*, *8*(3), 239–260. doi:10.1080/13504620220145401

Köne, A. Ç. (2014). Türkiye'de Bölgesel Eko-Verimliliğin Ölçülmesi: Veri Zarflama Analizi ile Bir Değerlendirme. *TİSK Akademi Dergisi*, *9*(16), 138–155.

Kopittke, P. M., Menzies, N. W., Wang, P., McKenna, B. A., & Lombi, E. (2019). Soil and the intensification of agriculture for global food security. *Environment International*, *132*, 105078. doi:10.1016/j.envint.2019.105078 PMID:31400601

Korkmaz, D., & Korkut, A. (2012). Türkiye'de Kadın İşgücüne Katılımın Belirleyicileri. *Süleyman Demirel Üniversitesi İktisadi ve İdari Bilimler Fakültesi Dergisi*, *17*(2), 41–65.

Kothari, S. P., Shu, S., & Wysocki, P. D. (2009). Do managers withhold bad news? *Journal of Accounting Research*, *47*(1), 241–276. doi:10.1111/j.1475-679X.2008.00318.x

Kraft, J., & Kraft, A. (1978). On the relationship between energy and GNP. *The Journal of Energy and Development*, 401-403.

Krippendorff, K. (2004). Measuring the Reliability of Qualitative Text Analysis. *Quality & Quantity*, *38*(6), 787–800. doi:10.100711135-004-8107-7

Kumar, V., Singh, J., & Kumar, P. (Eds.). (2020). Environmental Degradation: Causes and Remediation Strategies (Vol. 1). Agro Environ Media, Publication Cell of AESA, Agriculture and Environmental Science Academy.

Kumbhakar, S., & Lovell, C. A. K. (2000). *Stochastic frontier analysis*. Cambridge University Press. doi:10.1017/CBO9781139174411

Kuosmanen, T. (2008). *Data Envelopment Analysis as Least-Squares Regression*. Working Paper. Retrieved July 30, 2022, from https://tinyurl.com/DEA-Least-Squares-Regression

Kuosmanen, T., & Johnson, A. (2010). Data Envelopment Analysis as Nonparametric Least-Squares Regression. *Operations Research*, *58*(1), 149–160. doi:10.1287/opre.1090.0722

Kuosmanen, T., & Kortelainen, M. (2012). Stochastic non-smooth envelopment of data: Semiparametric frontier estimation subject to shape constraints. *Journal of Productivity Analysis*, *38*(1), 11–28. doi:10.100711123-010-0201-3

Kuznets, S. (1955). Economic growth and income inequality. *The American Economic Review*, *45*(1), 1–28.

Kwon, D. S., Cho, J. H., & Sohn, S. Y. (2017). Comparison of technology efficiency for CO2 emissions reduction among European countries based on DEA with decomposed factors. *Journal of Cleaner Production*, *151*, 109–120. doi:10.1016/j.jclepro.2017.03.065

Laine, M., Järvinen, J. T., Hyvönen, T., & Kantola, H. (2017). Ambiguity of financial environmental information: A case study of a Finnish energy company. *Accounting, Auditing & Accountability Journal*, *30*(3), 593–619. doi:10.1108/AAAJ-02-2015-1961

Lamb, W. F., Wiedmann, T., Pongratz, J., Andrew, R., Crippa, M., Olivier, J. G., Wiedenhofer, D., Mattioli, G., Khourdajie, A. A., House, J., Pachauri, S., Figueroa, M., Saheb, Y., Slade, R., Hubacek, K., Sun, L., Ribeiro, S. K., Khennas, S., de la Rue du Can, S., ... Minx, J. (2021). A review of trends and drivers of greenhouse gas emissions by sector from 1990 to 2018. *Environmental Research Letters*, *16*(7), 073005. doi:10.1088/1748-9326/abee4e

Laroche, M., Bergeron, J., & Barbaro-Forleo, G. (2001). Targeting consumers who are willing to pay more for environmentally friendly products. *Journal of Consumer Marketing*, *18*(6), 503–520. doi:10.1108/EUM0000000006155

Laroche, M., Tomiuk, M.-A., Bergeron, J., & Barbaro-Forleo, G. (2002). Cultural differences in environmental knowledge, attitudes, and behaviors of Canadian consumers. *Canadian Journal of Administrative Sciences*, *19*(3), 267–283. doi:10.1111/j.1936-4490.2002.tb00272.x

Larrinaga, C., Carrasco, F., Correa, C., Llena, F., & Moneva, J. M. (2002). Accountability and accounting regulation: The case of the Spanish environmental disclosure standard. *European Accounting Review*, *11*(4), 723–740. doi:10.1080/0963818022000001000

Law, K. L., Starr, N., Siegler, T. R., Jambeck, J. R., Mallos, N. J., & Leonard, G. H. (2020). The United States' contribution of plastic waste to land and ocean. *Science Advances*, *6*(44), eabd0288. doi:10.1126ciadv.abd0288 PMID:33127684

Le Pen, Y., & Sévi, B. (2010). On the non-convergence of energy intensities: Evidence from a pair-wise econometric approach. *Ecological Economics*, *69*(3), 641–650. doi:10.1016/j.ecolecon.2009.10.001

Leary, R. B., Minton, E. A., & Mittelstaedt, J. D. (2016). Thou shall not? The influence of religion on beliefs of stewardship and dominion, sustainable behaviors, and marketing systems. *Journal of Macromarketing*, *36*(4), 457–470. doi:10.1177/0276146715626219

Leckie, G. (1993). Global oil, and gas fields, sizes tallied and analyzed. Oil and Gas Journal.

Lee, C. C., Chang, C. P., & Chen, P. F. (2008). Energy-income causality in OECD countries revisited: The key role of capital stock. *Energy Economics*, *30*(5), 2359–2373. doi:10.1016/j.eneco.2008.01.005

Lee, C. C., & Chen, M. P. (2021). Ecological footprint, tourism development, and country risk: International evidence. *Journal of Cleaner Production*, *279*(123671), 1–24. doi:10.1016/j.jclepro.2020.123671

Lee, C. C., Xing, W., & Lee, C. C. (2022). The impact of energy security on income inequality: The key role of economic development. *Energy*, *248*, 1–14. doi:10.1016/j.energy.2022.123564

Lee, G., & Xiao, X. (2020). Voluntary Engagement in Environmental Projects: Evidence from Environmental Violators. *Journal of Business Ethics*, *164*(2), 325–348. doi:10.100710551-018-4074-0

Lee, H., Lee, J., & Im, K. (2015). More powerful cointegration tests with non-normal errors. *Studies in Nonlinear Dynamics and Econometrics*, *19*(4), 397–413. doi:10.1515nde-2013-0060

Lee, J. W., & Brahmasrene, T. (2013). Investigating the influence of tourism on economic growth and carbon emissions: Evidence from panel analysis of the European union. *Tourism Management*, *38*, 69–76. doi:10.1016/j.tourman.2013.02.016

Lee, K. (2008). Opportunities for green marketing: Young consumers. *Marketing Intelligence & Planning*, *26*(6), 573–586. doi:10.1108/02634500810902839

Lemmens, A., Croux, C., & Dekimpe, M. G. (2008). Measuring and testing Granger causality over the spectrum: An application to European production expectation surveys. *International Journal of Forecasting*, *24*(3), 414–431. doi:10.1016/j.ijforecast.2008.03.004

Lemon, L. L., & Hayes, J. (2020). Enhancing trustworthiness of qualitative findings: Using Leximancer for qualitative data analysis triangulation. *Qualitative Report*, *25*(3), 604–614. doi:10.46743/2160-3715/2020.4222

Leonidou, C. N., & Leonidou, L. C. (2011). Research into environmental marketing/management: A bibliographic analysis. *European Journal of Marketing*, *45*(1/2), 68–103. doi:10.1108/03090561111095603

Leonidou, C. N., Leonidou, L. C., & Kvasova, O. (2010). Antecedents and outcomes of consumer environmentally friendly attitudes and behavior. *Journal of Marketing Management*, *26*(13-14), 1319–1344. doi:10.1080/0267257X.2010.523710

Leuz, C., Pfaff, D., & Hopwood, A. (2004). *The Economics and Politics of Accounting: International Perspectives on Trends Policy and Practice*. Oxford University Press. doi:10.1093/0199260621.001.0001

Leximancer. (2021). https://info.leximancer.com

Leybourne, S., Newbold, P., & Vougas, D. (1998). Unit roots and smooth transitions. *Journal of Time Series Analysis*, *19*(1), 83–97. doi:10.1111/1467-9892.00078

Liddle, B. (2010). Revisiting world energy intensity convergence for regional differences. *Applied Energy*, *87*(10), 3218–3225. doi:10.1016/j.apenergy.2010.03.030

Liew, C. K. (1980). The impact of higher energy prices on growth and inflation in an industrializing economy: The Korean experience. *Journal of Policy Modeling*, *2*(3), 389–408. doi:10.1016/0161-8938(80)90030-7

Lilien, D. M. (1982). Sectoral Shifts and Cyclical Unemployment. *Journal of Political Economy*, *90*(4), 777–793. doi:10.1086/261088

Lin, B., & Wu, W. (2018). Why people want to buy electric vehicle: An empirical study in first-tier cities of China. *Energy Policy*, *112*, 233–241. doi:10.1016/j.enpol.2017.10.026

Lin, S. T., & Niu, H. J. (2018). Green consumption: Environmental knowledge, environmental consciousness, social norms, and purchasing behavior. *Business Strategy and the Environment*, *27*(8), 1679–1688. doi:10.1002/bse.2233

Li, Q., Long, R., & Chen, H. (2017). Empirical study of the willingness of consumers to purchase low-carbon products by considering carbon labels: A case study. *Journal of Cleaner Production*, *161*, 1237–1250. doi:10.1016/j.jclepro.2017.04.154

Liu, X., Vedlitz, A., & Shi, L. (2014). Examining the determinants of public environmental concern: Evidence from national public surveys. *Environmental Science & Policy*, *39*, 77–94. doi:10.1016/j.envsci.2014.02.006

Liu, Y., Zhu, X., & Wang, Y. (2022). Revisiting and evaluation of the index of sustainable economic welfare based on artificial intelligence: Data from 30 Chinese provinces from 2003 to 2019. *Environment, Development and Sustainability*, 1–30. PMID:36618554

Liu, Z., & Liu, M. (2021). Quality evaluation of enterprise environmental accounting information disclosure based on projection pursuit model. *Journal of Cleaner Production*, *279*, 123679. Advance online publication. doi:10.1016/j.jclepro.2020.123679

Li, Y., Richardson, G. D., & Thornton, D. B. (1997). Corporate disclosure of environmental liability information: Theory and evidence. *Contemporary Accounting Research*, *14*(3), 435–474. doi:10.1111/j.1911-3846.1997.tb00535.x

Longwe, B. (2022). A Comparative Study of Ecological Consumer Behavior Amongst Consumers In India and Pakistan. *Educational Research*, *4*(1), 296–307.

Lubowiecki-Vikuk, A., Dąbrowska, A., & Machnik, A. (2021). Responsible consumer and lifestyle: Sustainability insights. *Sustainable Production and Consumption*, *25*, 91–101. doi:10.1016/j.spc.2020.08.007 PMID:32864405

Lu, L.-C., Chang, H.-H., & Chang, A. (2015). Consumer personality and green buying intention: The mediate role of consumer ethical beliefs. *Journal of Business Ethics*, *127*(1), 205–219. doi:10.100710551-013-2024-4

Lumsdaine, R. L., & Papell, D. H. (1997). Multiple trend breaks and the unit-root hypothesis. *The Review of Economics and Statistics*, *79*(2), 212–218. doi:10.1162/003465397556791

Magnani, E. (2001). The Environmental Kuznets curve: Development path or policy result? *Environmental Modelling & Software*, *16*(2), 157–165. doi:10.1016/S1364-8152(00)00079-7

Majid, K. A., & Russell, C. A. (2015). Giving green a second thought: Modeling the value retention of green products in the secondary market. *Journal of Business Research*, *68*(5), 994–1002. doi:10.1016/j.jbusres.2014.10.001

Maksakovsky, V. (2009). World Social and Economic Geography. Aspects Press.

Malley, J., & Moutos, T. (2001). Capital Accumulation and Unemployment: A Tale of Two "Continents". *The Scandinavian Journal of Economics*, *130*(1), 79–99. doi:10.1111/1467-9442.00231

Manowska, A., & Nowrot, A. (2019). The importance of heat emission caused by global energy production in terms of climate impact. *Energies*, *12*(16), 3069. doi:10.3390/en12163069

Map of World. (2018). *Forests for the World Map*. Retrieved from: https://www.mapsofworld.com/world-natural-forest.htm

Marinebio. (2001). *Ocean Resources*. Retrieved from: https://marinebio.org/oceans/ocean-resources

Markandya, A., Pedroso-Galinato, S., & Streimikiene, D. (2006). Energy intensity in transition economies: Is there convergence towards the EU average? *Energy Economics*, *28*(1), 121–145. doi:10.1016/j.eneco.2005.10.005

Mayer, L. S. (1978). Estimating the effects of the onset of the energy crisis on residential energy demand. *Resources and Energy*, *1*(1), 57–92. doi:10.1016/0165-0572(78)90014-2

McCarty, J., & Shrum, L. (1994). The recycling of solid wastes: Personal values, value orientations, and attitudes about recycling as antecedents of recycling behaviour. *Journal of Business Research*, *30*(1), 53–62. doi:10.1016/0148-2963(94)90068-X

McDonagh, P., & Prothero, A. (2014). Sustainability marketing research: Past, present, and future. *Journal of Marketing Management*, *30*(11-12), 1186–1219. doi:10.1080/0267257X.2014.943263

Meeusen, W., & Vandenbroeck, J. (1977). Efficiency estimation from cobb-douglas production functions with composed error. *International Economic Review*, *18*(2), 435–445. doi:10.2307/2525757

Menegaki, A. (2018). *The basic, the solid, the site-specific and the full or total index of Sustainable Economic Welfare (ISEW) for Turkey*. Academic Press.

Menegaki, A. N., & Tsagarakis, K. P. (2015). More indebted than we know? Informing fiscal policy with an index of sustainable welfare for Greece. *Ecological Indicators*, *57*, 159–163. doi:10.1016/j.ecolind.2015.04.037

Menegaki, A. N., & Tugcu, C. T. (2018). Two versions of the Index of Sustainable Economic Welfare (ISEW) in the energy-growth nexus for selected Asian countries. *Sustainable Production and Consumption*, *14*, 21–35. doi:10.1016/j.spc.2017.12.005

Menegaki, A., & Tiwari, A. K. (2019). Energy Efficiency in Europe; Stochastic-Convergent and Non-Convergent Countries. In M. Shahbaz & D. Balsalobre (Eds.), *Energy and Environmental Strategies in the Era of Globalization* (pp. 305–333). Springer. doi:10.1007/978-3-030-06001-5_12

Meng, Zeng, Shi, Qi, & Zhang. (2014). The relationship between corporate environmental performance and environmental disclosure: An empirical study. *China Journal of Environmental Management, 145*, 357-367.

Mensah, J. (2019). Sustainable development: Meaning, history, principles, pillars, and implications for human action: Literature review. *Cogent Social Sciences*, *5*(1). DOI: .2019.1653531 doi:10.1080/23311886

Mensah, C. N., Long, X., Dauda, L., Boamah, K. B., Salman, M., Appiah-Twum, F., & Tachie, A. K. (2019). Technological innovation and green growth in the Organization for Economic Cooperation and Development economies. *Journal of Cleaner Production*, *240*, 118204. doi:10.1016/j.jclepro.2019.118204

Metcalf, G. E. (2008). An empirical analysis of energy intensity and its determinants at the state level. *The Energy Journal (Cambridge, Mass.)*, *29*(3), 1–26. doi:10.5547/ISSN0195-6574-EJ-Vol29-No3-1

Mielnik, O., & Goldemberg, J. (2000). Converging to a common pattern of energy use in developing and industrialized countries. *Energy Policy*, *28*(8), 503–508. doi:10.1016/S0301-4215(00)00015-X

Mikayilov, J. I., Mukhtarov, S., Mammadov, J., & Azizov, M. (2019). Re-evaluating the environmental impacts of tourism: Does EKC exist? *Environmental Science and Pollution Research International*, *26*(19), 19389–19402. doi:10.100711356-019-05269-w PMID:31073834

Minton, E. A. (2013). Religion and religiosity's influence on sustainable consumption behaviors. In L. R. Kahle & E. Gurel-Atay (Eds.), *Communicating sustainability for the green economy* (pp. 73–87). M.E. Sharpe.

Minton, E. A., Kahle, L. R., & Kim, C.-H. (2015). Religion and motives for sustainable behaviors: A cross-cultural comparison and contrast. *Journal of Business Research*, *68*(9), 1937–1944. doi:10.1016/j.jbusres.2015.01.003

Minton, E. A., Spielmann, N., Kahle, L. R., & Kim, C.-H. (2018a). The subjective norms of sustainable consumption: A cross-cultural exploration. *Journal of Business Research*, *82*, 400–408. doi:10.1016/j.jbusres.2016.12.031

Minton, E. A., Xie, H. J., Gurel-Atay, E., & Kahle, L. R. (2018b). Greening up because of god: The relations among religion, sustainable consumption and subjective well-being. *International Journal of Consumer Studies*, *42*(6), 655–663. doi:10.1111/ijcs.12449

Mobus, J. L. (2005). Mandatory environmental disclosures in a legitimacy theory context. *Accounting, Auditing & Accountability Journal*, *18*(4), 492–517. doi:10.1108/09513570510609333

Mohsin, M., Abbas, Q., Zhang, J., Ikram, M., & Iqbal, N. (2019). Integrated effect of energy consumption, economic development, and population growth on CO2 based environmental degradation: A case of transport sector. *Environmental Science and Pollution Research International*, *26*(32), 32824–32835. doi:10.100711356-019-06372-8 PMID:31502046

Mondal, T., & Tripathy, B. (2020). Riverbank Erosion and Environmental degradation. *Int J Res*, *7*(7).

Monti, G., & Deason, J. P. (2012). Applying an ecosystem services approach to supplemental environmental projects. *Journal of Environmental Planning and Management*, *55*(8), 1047–1057. doi:10.1080/09640568.2011.636221

Mourdoukoutas, P. (1988). Seasonal Employment, Seasonal Unemployment and Unemployment Compensation: The Case of the Tourist Industry of the Greek Island. *American Journal of Economics and Sociology*, *47*(3), 315–329. doi:10.1111/j.1536-7150.1988.tb02044.x

Mulder, P., & De Groot, H. L. (2012). Structural change and convergence of energy intensity across OECD countries, 1970–2005. *Energy Economics*, *34*(6), 1910–1921. doi:10.1016/j.eneco.2012.07.023

Narayan, P. K., & Prasad, A. (2008). Electricity consumption–real GDP causality nexus: Evidence from a bootstrapped causality test for 30 OECD countries. *Energy Policy*, *36*(2), 910–918. doi:10.1016/j.enpol.2007.10.017

Narver, J. C. (1971). Rational management responses to external effects. *Academy of Management Journal*, *14*(1), 99–115. doi:10.2307/254714

Nathaniel, S. P., Barua, S., & Ahmed, Z. (2021). What drives ecological footprint in top ten tourist destinations? Evidence from advanced panel techniques. *Environmental Science and Pollution Research International*, *28*(28), 38322–38331. doi:10.100711356-021-13389-5 PMID:33728610

Nazlioglu, S., Kayhan, S., & Adiguzel, U. (2014). Electricity consumption and economic growth in Turkey: Cointegration, linear and nonlinear Granger causality. *Energy Sources. Part B, Economics, Planning, and Policy*, *9*(4), 315–324. doi:10.1080/15567249.2010.495970

Neidze, V. (2004). *World Economic and Social Geography*. Institute of Geography of Georgia.

Nellemann, C., & Corcoran, E. (2010). *Dead Planet, Living Planet - Biodiversity and Ecosystem Restoration for Sustainable Development: A Rapid Response Assessment. United Nations Environment Program*. GRID-Arendal.

Nelson, K. K., & Pritchard, A. C. (2016). Carrot or Stick? The Shift from untary to Mandatory Disclosure of Risk Factors. *Journal of Empirical Legal Studies*, *13*(2), 266–297. doi:10.1111/jels.12115

Nepal, R., Al Irsyad, M. I., & Nepal, S. K. (2019). Tourist arrivals, energy consumption and pollutant emissions in a developing economy–implications for sustainable tourism. *Tourism Management*, *72*, 145–154. doi:10.1016/j.tourman.2018.08.025

Neumayer, E. (1999). The ISEW: Not an index of sustainable economic welfare. *Social Indicators Research*, *48*(1), 77–101. doi:10.1023/A:1006914023227

Newton, J. D., Tsarenko, Y., Ferraro, C., & Sands, S. (2015). Environmental concern and environmental purchase intentions: The mediating role of learning strategy. *Journal of Business Research*, *68*(9), 1974–1981. doi:10.1016/j.jbusres.2015.01.007

Nguyen, K. H., & Kakinaka, M. (2019). Renewable energy consumption, carbon emissions, and development stages: Some evidence from panel cointegration analysis. *Renewable Energy*, *132*, 1049–1057. doi:10.1016/j.renene.2018.08.069

Nillesen, H. L. P., Haffner, C. G. R., & Ozbugday, F. C. (2013). A Global Perspective on the Long-Term Impact of Increased Energy Efficiency. In F. P. Sioshansi (Ed.), *Energy Efficiency Towards the End of Demand Growth* (pp. 87–110). Academic Press. doi:10.1016/B978-0-12-397879-0.00003-7

Nilsson, L. J. (1993). Energy intensity trends in 31 industrial and developing countries 1950–1988. *Energy*, *18*(4), 309–322. doi:10.1016/0360-5442(93)90066-M

Norris, P. (2001). Breaking the Barriers: Positive Discrimination Policies for Women. In J. Klausen & C. S. Maier (Eds.), *Has Liberalism Failed Women?*

O'Mahony, T. (2021). Spain's Index of Sustainable Economic Welfare. Are welfare and sustainability compatible? Cuadernos económicos de ICE, 159-183.

Ocean resources Defense Council. (2014). *Oceans*. Retrieved from: https://www.nrdc.org/oceans/

Ochoa-Moreno, W. S., Quito, B., Enríquez, D. E., & Álvarez-García, J. (2022). Evaluation of the environmental Kuznets curve hypothesis in a tourism development context: Evidence for 15 Latin American countries. *Business Strategy and the Environment, 31*(5), 1–13. doi:10.1002/bse.3012

Odume, O. N., Onyima, B. N., Nnadozie, C. F., Omovoh, G. O., Mmachaka, T., Omovoh, B. O., Uku, J. E., Akamagwuna, F. C., & Arimoro, F. O. (2022). Governance and Institutional Drivers of Ecological Degradation in Urban River Ecosystems: Insights from Case Studies in African Cities. *Sustainability (Basel), 14*(21), 14147. doi:10.3390u142114147

OECD. (2012). *Linking Renewable Energy to Rural Development, OECD Greeen Growth Studies.* OECD Publishing.

OECD. (2021). *FDI in Figures October 2021: Global FDI flows rebound to exceed pre-pandemic levels.* Foreign Direct Investment Statistics: Data, Analysis and Forecasts. Retrieved July 30, 2022, from https://www.oecd.org/investment/investment-policy/FDI-in-Figures-October-2021.xlsx

OECD. (2022a). Employment rate (indicator). https://data.oecd.org/emp/employment-rate.htm

OECD. (2022b). Labour force participation rate (indicator). https://data.oecd.org/emp/labour-force-participation-rate.htm

OECD. (2022c). *Labour force participation rate by sex, 15+,15-64 and 15-24 years old.* Retrieved July 30, 2022, from https://www.oecd.org/gender/data/labour-force-participation-by-sex-age.htm

OECD. (2022d). *Current Account Balance.* OECD Quarterly International Trade Statistics. Retrieved August 31, 2022, from https://data.oecd.org/trade/current-account-balance.htm

Ogden, S., & Clarke, J. (2005). Customer disclosures impression management and the construction of legitimacy: Corporate reports in the UK privatised water industry. *Accounting, Auditing & Accountability Journal, 18*(3), 313–345. doi:10.1108/09513570510600729

Oğuz, F., & Oğuz, S. (2016). Kamuda Girişimcilik, Rekabet ve İstihdam Politikalarının Etkinliği. *Maliye Dergisi. Temmuz-Aralık, 2016*(171), 34–54.

Onaran, Ö., & Aydıner-Avşar, N. (2006). The controversy over employment policy: Low labor costs and openness, or demand policy? A sectoral analysis for Turkey. *Department of Economics Working Paper Series,* (97), 24.

Onaran, Ö. (2002). Measuring Wage Flexibility: The Case of Turkey Before and After Structural Adjustment. *Applied Economics, 34*(6), 767–781. doi:10.1080/00036840110056934

Onaran, Ö. (2007). İşsizlik ve Alternatif Politikalar. In *Türkiye'de İşsizliğin Çözüm Yolları ve Sendikalar Çalıştayı* (p. 13). DİSK.

Öncel, A. (2010). *Türkiye'de Bölgelerarası Gelişmişlik Farklılıklarını Gidermeye Yönelik Politika Uygulamaları ve Etkin Kaynak Kullanımı Üzerine Bir Analiz* [Unpublished Doctoral Dissertation]. Sakarya Üniversitesi Sosyal Bilimler Enstitüsü, Adapazarı.

Organization for Economic Cooperation and Development. (2020). *Environment at a glance indicators*. Retrieved July, 2022, from https://www.oecd.org/environment/environment-at-a-glance/Climate-Change-Archive-June-2020.pdf

Özsoy, F. (2021). Investigation of the relationship between tourism sector and environmental pollution in the framework of the environmental Kuznets curve hypothesis. *Journal of Abant Social Sciences, 21*(1), 1–19.

Ozturk, I. (2010). A literature survey on energy–growth nexus. *Energy Policy, 38*(1), 340–349. doi:10.1016/j.enpol.2009.09.024

Ozturk, İ., Al-Mulali, U., & Saboori, B. (2016). Investigating the environmental Kuznets curve hypothesis: The role of tourism and ecological footprint. *Environmental Science and Pollution Research International, 23*(2), 1916–1928. doi:10.100711356-015-5447-x PMID:26408117

Ozturk, I., Aslan, A., & Kalyoncu, H. (2010). Energy consumption and economic growth relationship: Evidence from panel data for low and middle-income countries. *Energy Policy, 38*(8), 4422–4428. doi:10.1016/j.enpol.2010.03.071

Paço, A., Leal Filho, W., Ávila, L. V., & Dennis, K. (2021). Fostering sustainable consumer behavior regarding clothing: Assessing trends on purchases, recycling and disposal. *Textile Research Journal, 91*(3-4), 373–384. doi:10.1177/0040517520944524

Pala, A. (2016). Which energy-growth hypothesis is valid in OECD countries? Evidence from panel Granger causality. *International Journal of Energy Economics and Policy, 6*(1), 28–34.

Papista, E., & Krystallis, A. (2013). Investigating the types of value and cost of green brands: Proposition of a conceptual framework. *Journal of Business Ethics, 115*(1), 75–92. doi:10.100710551-012-1367-6

Paramati, S. R., Alam, M. S., & Chen, C. F. (2017). The Effects of tourism on economic growth and CO_2 emissions: A comparison between developed and developing economies. *Journal of Travel Research, 56*(6), 712–724. doi:10.1177/0047287516667848

Parlaktuna, İ. (2010). Türkiye'de Cinsiyete Dayalı Mesleki Ayrımcılığın Analizi. *Ege Academic Review, 10*(4), 1217–1230. doi:10.21121/eab.2010419607

Pata, U. K., Yilanci, V., Hussain, B., & Naqvi, S. A. A. (2022). Analyzing the role of income inequality and political stability in environmental degradation: Evidence from South Asia. *Gondwana Research, 107*, 13–29. doi:10.1016/j.gr.2022.02.009

Patten, D. M. (1992). Intra-industry environmental disclosures in response to the Alaskan oil spill: A note on legitimacy theory. *Accounting, Organizations and Society, 17*(5), 471–475. doi:10.1016/0361-3682(92)90042-Q

Patten, D. M. (2005). The accuracy of financial report projections of future environmental capital expenditures: A research note. *Accounting, Organizations and Society, 30*(5), 457–468. doi:10.1016/j.aos.2004.06.001

Paul, S., & Bhattacharya, R. N. (2004). Causality between energy consumption and economic growth in India: A note on conflicting results. *Energy Economics*, *26*(6), 977–983. doi:10.1016/j.eneco.2004.07.002

Peattie, K. (2001). Golden goose or wild goose? The hunt for the green consumer. *Business Strategy and the Environment*, *10*(4), 187–199. doi:10.1002/bse.292

Pesaran, M. H. (2004). General diagnostic tests for cross section dependence in panels (IZA Discussion Paper No. 1240). *Institute for the Study of Labor (IZA)*, 1-39.

Pesaran, M. H. (2006). Estimation and inference in large heterogeneous panels with a multifactor error structure. *Econometrica*, *74*(4), 967–1012. doi:10.1111/j.1468-0262.2006.00692.x

Pesaran, M. H., Ullah, A., & Yamagata, T. (2008). A bias-adjusted LM test of error cross-section independence. *The Econometrics Journal*, *11*(1), 105–127. doi:10.1111/j.1368-423X.2007.00227.x

Pesaran, M. H., & Yamagata, T. (2008). Testing slope homogeneity in large panels. *Journal of Econometrics*, *142*(1), 50–93. doi:10.1016/j.jeconom.2007.05.010

Peters, G. F., & Romi, A. M. (2013). Discretionary compliance with mandatory environmental disclosures: Evidence from SEC filings. *Journal of Accounting and Public Policy*, *32*(4), 213–236. doi:10.1016/j.jaccpubpol.2013.04.004

Pfund, T. (2004). Corporate Environmental Accountability: Expanding SEC Disclosures to Promote Market Based Environmentalism. *Monthly Environmental Law & Policy Review*, *11*(2), 118–137.

Phillips, P. C., & Sul, D. (2007). Transition modeling and econometric convergence tests. *Econometrica*, *75*(6), 1771–1855. doi:10.1111/j.1468-0262.2007.00811.x

Phillips, P. C., & Sul, D. (2009). Economic transition and growth. *Journal of Applied Econometrics*, *24*(7), 1153–1185. doi:10.1002/jae.1080

Pigram, J. J. (1980). Environmental implications of tourism development. *Annals of Tourism Research*, *7*(4), 554–583. doi:10.1016/0160-7383(80)90049-3

Pinet, P. R. (1996). *Invitation to Oceanography* (3rd ed.). West Publishing Co.

Politkovskaya, I., Khvichiya, D., & Artamonova, L. (2020). Economic Support of the Policy of Transition to Sustainable Development of the Territories of the Russian North. In *E3S Web of Conferences* (Vol. 174, p. 04047). EDP Sciences. doi:10.1051/e3sconf/202017404047

Polonsky, M. J., Kilbourne, W., & Vocino, A. (2014). Relationship between the DSP, materialism, and environmental behaviors in four Asian economies. *European Journal of Marketing*, *48*(3-4), 522–551. doi:10.1108/EJM-07-2011-0351

Priesmann, J., Spiegelburg, S., Madlener, R., & Praktiknjo, A. (2022). Does renewable electricity hurt the poor? Exploring levy programs to reduce income inequality and energy poverty across German households. *Energy Research & Social Science*, *93*, 1–20. doi:10.1016/j.erss.2022.102812

Pulselli, F. M., Bravi, M., & Tiezzi, E. (2012). Application and use of the ISEW for assessing the sustainability of a regional system: A case study in Italy. *Journal of Economic Behavior & Organization, 81*(3), 766–778. doi:10.1016/j.jebo.2010.12.021

Pulselli, F. M., Ciampalini, F., Tiezzi, E., & Zappia, C. (2006). The Index of Sustainable Economic Welfare (ISEW) for a local authority: A case study in Italy. *Ecological Economics, 60*(1), 271–281. doi:10.1016/j.ecolecon.2005.12.004

Purvis, B., Mao, Y., & Robinson, D. (2019). Three pillars of sustainability: In search of conceptual origins. *Sustainability Science, 14*(3), 681–695. doi:10.100711625-018-0627-5

Qi, S., Peng, H., Zhang, X., & Tan, X. (2019). Is energy efficiency of Belt and Road Initiative countries catching up or falling behind? Evidence from a panel quantile regression approach. *Applied Energy, 253*, 113581. doi:10.1016/j.apenergy.2019.113581

Ranjbar, O., Chang, T., Nel, E., & Gupta, R. (2017). Energy consumption and economic growth nexus in South Africa: Asymmetric frequency domain approach. *Energy Sources. Part B, Economics, Planning, and Policy, 12*(1), 24–31. doi:10.1080/15567249.2015.1020120

Ray, S. C., Kumbhakar, S. C., & Dua, P. (2015). *Benchmarking for performance evaluation: A production frontier approach.* Springer. doi:10.1007/978-81-322-2253-8

Ray, S., & Chen, L. (2015). Data Envelopment Analysis for Performance Evaluation: A Child's Guide. In S. Ray, S. Kumbhakar, & P. Dua (Eds.), *s), Benchmarking for performance evaluation: A production frontier approach.* Springer. doi:10.1007/978-81-322-2253-8_2

Rechner. (2010). Letter to the Editor: Outdoor recreation stimulates the economy. *Washington Post.*

Rees, W. E. (1992). Ecological footprints and appropriated carrying capacity: What urban economics leaves out. *Environment and Urbanization, 4*(2), 121–130. doi:10.1177/095624789200400212

Rej, S., Bandyopadhyay, A., Murshed, M., Mahmood, H., & Razzaq, A. (2022). Pathways to decarbonization in India: The role of environmentally friendly tourism development. *Environmental Science and Pollution Research International*, 29(33), 50281-50302. doi.org/10.1007/s11356-022-19239-2.

Reliefweb. (2005). *Senator Paul Simon Water for the Poor Act: 2005 Report to Congress.* Retrieved from: https://reliefweb.int/report/world/senator-paul-simon-water-poor-act-2005-report-congress

Rettie, R., Burchell, K., & Barnham, C. (2014). Social normalisation: Using marketing to make green normal. *Journal of Consumer Behaviour, 13*(1), 9–17. doi:10.1002/cb.1439

Rettie, R., Burchell, K., & Riley, D. (2012). Normalising green behaviours: A new approach to sustainability marketing. *Journal of Marketing Management, 28*(3-4), 420–444. doi:10.1080/0267257X.2012.658840

Reynolds, L. (1974). *Labor Economics and Labor Relations* (6th ed.). Prentice-Hall, Inc.

Riskos, K., Paraskevi, D., Mylonas, N., & Tsourvakas, G. (2021). Ecolabels and the attitude-behavior relationship towards green product purchase: A multiple mediation model. *Sustainability (Basel)*, *13*(12), 6867. doi:10.3390u13126867

Robert, D. S., & James, A. R. (1999). Environmental segmentation alternatives: A look at green consumer behavior in the new millennium. *Journal of Consumer Marketing*, *16*(6), 558–575. doi:10.1108/07363769910297506

Roberts, J. A. (1995). Profiling levels of socially responsible behavior: A cluster analytic approach and its implications for marketing. *Journal of Marketing Theory and Practice*, *3*(4), 97–117. doi:10.1080/10696679.1995.11501709

Roberts, J. A. (1996). Green consumers in the 1990's: Profile and implications for advertising. *Journal of Business Research*, *36*(3), 217–231. doi:10.1016/0148-2963(95)00150-6

Rodrigues, P. M., & Taylor, A. M. R. (2012). The flexible Fourier form and local generalised least squares de-trended unit root tests. *Oxford Bulletin of Economics and Statistics*, *74*(5), 736–759. doi:10.1111/j.1468-0084.2011.00665.x

Rogers, W. H. (1993). Regression standart errors in clustered samples. *Stata Technical Bulletin*, *13*, 19–23.

Roitner-Schobesberger, B., Darnhofer, I., Somsook, S., & Vogl, C. R. (2008). Consumer perceptions of organic foods in Bangkok. *Food Policy*, *33*(2), 112–121. doi:10.1016/j.foodpol.2007.09.004

Rooney, D., Paulsen, N., Callan, V. J., Brabant, M., Gallois, C., & Jones, E. (2010). A new role for place identity in managing organizational change. *Management Communication Quarterly*, *24*(1), 44-73.

Rugani, B., Marvuglia, A., & Pulselli, F. M. (2018). Predicting sustainable economic welfare-analysis and perspectives for Luxembourg based on energy policy scenarios. *Technological Forecasting and Social Change*, *137*, 288–303. doi:10.1016/j.techfore.2018.08.005

Sadiq, M., Wen, F., & Dagestani, A. A. (2022). Environmental footprint impacts of nuclear energy consumption: The role of environmental technology and globalization in ten largest ecological footprint countries. *Nuclear Engineering and Technology*, *54*(10), 3672–3681. Advance online publication. doi:10.1016/j.net.2022.05.016

Sagner, J. S. (1974). The impact of the energy crisis on American cities based on dispersion of employment, utilization of transit, and carpooling. *Transportation Research*, *8*(4-5), 307–316. doi:10.1016/0041-1647(74)90049-5

Sahoo, B., Behera, D. K., & Rahut, D. (2022). Decarbonization: Examining the role of environmental innovation versus renewable energy use. *Environmental Science and Pollution Research International*, *29*(32), 1–16. doi:10.100711356-022-18686-1 PMID:35199265

Salahodjaev, R., Sharipov, K., Rakhmanov, N., & Khabirov, D. (2022). Tourism, renewable energy and CO_2 emissions: Evidence from Europe and Central Asia. *Environment, Development and Sustainability*, *24*(11), 1–12. doi:10.100710668-021-01993-x

Samdahl, D. M., & Robertson, R. (1989). Social determinants of environmental concern. *Environment and Behavior*, *21*(1), 57–81. doi:10.1177/0013916589211004

Sánchez, M., Toledo, E., Ordóñez, J., & Associates. (2020). The relevance of Index of Sustainable Economic Wellbeing. Case study of Ecuador. Environmental and sustainability Indicators, 6, 100037.

Santiago, R., Fuinhas, J. A., & Marques, A. C. (2020). An analysis of the energy intensity of Latin American and Caribbean countries: Empirical evidence on the role of public and private capital stock. *Energy*, *211*, 118925. doi:10.1016/j.energy.2020.118925

Sarıtaş, H., Genç, A., & Avcı, T. (2018). Türkiye'de Enerji İthalatı, Cari Açık ve Büyüme İlişkisi: VAR ve Granger Nedensellik Analizi. *Ekonomik ve Sosyal Araştırmalar Dergisi*, *14*, 181–200.

Sarkodie, S. A. (2018). The invisible hand and EKC hypothesis: What are the drivers of environmental degradation and pollution in Africa? *Environmental Science and Pollution Research International*, *25*(22), 21993–22022. doi:10.100711356-018-2347-x PMID:29797200

Saxena, R., Rath, R., Gupta, S., & Sood, N. (2021). A review on ecological degradation, its causes and sustainable development in Delhi, India. *Journal of Applied and Natural Science*, *13*(4), 1294–1304. doi:10.31018/jans.v13i4.2978

Scardigli, V. (1975). Lifestyles, inequalities and the process of social change in France. *Futures*, *7*(3), 197–208. doi:10.1016/0016-3287(75)90064-6

Schlegelmilch, B. B., Bohlen, G. M., & Diamantopoulos, A. (1996). The link between green purchasing decisions and measures of environmental consciousness. *European Journal of Marketing*, *30*(5), 35–55. doi:10.1108/03090569610118740

Schmidt, P. (1985). Frontier production functions. *Econometric Reviews*, *4*(2), 289–328. doi:10.1080/07474938608800089

Schneider, T., Maier, M., & Michelon, G. (2017). Environmental liabilities and diversity in practice under international financial reporting standards. *Accounting, Auditing & Accountability Journal*, *30*(2), 378–403. doi:10.1108/AAAJ-01-2014-1585

Schuh, G. E. (1978). The case of Brazil: Import substitution revisited. *Journal of Comparative Economics*, *2*(2), 97–110. doi:10.1016/0147-5967(78)90042-2

Schwartz, M. (2021). *How technology can provide a more sustainable future for the industrial sector.* Retrieved May 30, 2022, from https://forbes.com/2021/08/31

Schwartz, K. B., & Soo, B. S. (1996). Evidence of regulatory noncompliance with SEC disclosure rules on auditor changes. *The Accounting Review*, *71*(4), 555–572.

Schwepker, C. H. Jr, & Cornwell, T. B. (1991). An examination of ecologically concerned consumers and their intention to purchase ecologically packaged products. *Journal of Public Policy & Marketing*, *10*(2), 77–101. doi:10.1177/074391569101000205

Securities and Exchange Commission (SEC). (2010). *Commission guidance regarding disclosure related to climate change.* Release Nos. 33-9106; 34-61469; FR-82 February 2 2010, Securities and Exchange Commission.

Securities and Exchange Commission (SEC). (2019). Modernization of Regulation S-K Items 101 103 and 105. Release Nos. 33-10668; 34-86614; File S7-11-19; RIN 3235-AL78; 17 CFR 229 239 and 240. August 8 2019 Securities and Exchange Commission.

Şengül, Ü., Eslemian, S., & Eren, M. (2013). Türkiye'de İstatistikî Bölge Birimleri Sınıflamasına göre Düzey 2 Bölgelerinin Ekonomik Etkinlikle-rinin VZA Yöntemi ile Belirlenmesi ve Tobit Model Uygulaması. *Yönetim Bilimleri Dergisi, 11*(21), 75–99.

Sengupta, J. K. (1992). A fuzzy systems approach in data envelopment analysis. *Computers & Mathematics with Applications (Oxford, England), 24*(8-9), 259–266. doi:10.1016/0898-1221(92)90203-T

Senn, J., & Giordano-Spring, S. (2020). The limits of environmental accounting disclosure: Enforcement of regulations standards and interpretative strategies. *Accounting, Auditing & Accountability Journal, 33*(6), 1367–1393. doi:10.1108/AAAJ-04-2018-3461

Shabani, Z. D., & Shahnazi, R. (2021). Energy intensity convergence in Iranian provinces: Evidence from energy carriers' consumption intensity. *Environmental Science and Pollution Research International*, 1–20. PMID:33495948

Shakouri, B., Khoshnevis Yazdi, S., & Ghorchebigi, E. (2017). Does tourism development promote CO_2 emissions? *Anatolia, 28*(3), 444–452. doi:10.1080/13032917.2017.1335648

Sharma, A. P. (2021). Consumers' purchase behavior and green marketing: A synthesis, review, and agenda. *International Journal of Consumer Studies, 45*(6), 1217–1238. doi:10.1111/ijcs.12722

Sherafatian-Jahromi, R., Othman, M. S., Law, S. H., & Ismail, N. W. (2017). Tourism and CO_2 emissions nexus in Southeast Asia: New evidence from panel estimation. *Environment, Development and Sustainability, 19*(4), 1407–1423. doi:10.100710668-016-9811-x

Shinnar, R. (1975). The effect of the energy crisis on the private car in the U.S. *Transportation Research, 9*(2), 87–95. doi:10.1016/0041-1647(75)90043-X

Shiu, A., & Lam, P. (2004). Electricity consumption and economic growth in China. *Energy Policy, 32*(1), 47–54. doi:10.1016/S0301-4215(02)00250-1

Shorrocks, A. (1978). Income inequality and income mobility. *Journal of Economic Theory, 19*(2), 376–393. doi:10.1016/0022-0531(78)90101-1

Shrum, L. J., McCarty, J. A., & Lowrey, T. M. (1995). Buyer characteristics of the green consumer and their implications for advertising strategy. *Journal of Advertising, 24*(2), 71–82. doi:10.1080/00913367.1995.10673477

Singh, A., & Verma, P. (2017). Factors influencing Indian consumers' actual buying behavior towards organic food products. *Journal of Cleaner Production, 167*, 473–483. doi:10.1016/j.jclepro.2017.08.106

Sinha, J. N. (1967). Dynamics of Female Participation in Economic Activity in a Developing Economy. In *United Nations, Department of Economic and Social Affairs. Proceedings of the World Population Conference*. UN Publications.

Sinha, A., Lorente, D. B., Zafar, M. W., & Saleem, M. M. (2022). Analyzing global inequality in access to energy: Developing policy framework by inequality decomposition. *Journal of Environmental Management, 304*, 1–17. doi:10.1016/j.jenvman.2021.114299 PMID:34923413

Skinner, D. (1994). Why firms untarily disclose bad news. *Journal of Accounting Research, 32*(1), 38–60. doi:10.2307/2491386

Skorov, G. (1973). The green revolution and social progress. *World Development, 1*(11), 13–21. doi:10.1016/0305-750X(73)90252-0

Smith, A. E., & Humphreys, M. S. (2006). Evaluation of unsupervised semantic mapping of natural language with Leximancer concept mapping. *Behavior Research Methods, 38*(2), 262–279. doi:10.3758/BF03192778 PMID:16956103

Smith, L. V., Leybourne, S., Kim, T. H., & Newbold, P. (2004). More powerful panel data unit root tests with an application to mean reversion in real exchange rates. *Journal of Applied Econometrics, 19*(2), 147–170. doi:10.1002/jae.723

Soba, M., Kumcu, S, & Akıncı, Ö. (2018). Bretton Woods ikizler: aşırı yoksulluk ve derinleşen eşitsizlik. *Süleyman Demirel Üniversitesi İktisadi ve İdari Bilimler Fakültesi Dergisi, 23*, 669-687.

Solarin, S. A. (2014). Tourist arrivals and macroeconomic determinants of CO_2 emissions in Malaysia. *Anatolia, 25*(2), 228–241. doi:10.1080/13032917.2013.868364

Solarin, S. A. (2019). Convergence in CO_2 emissions, carbon footprint and ecological footprint: Evidence from OECD countries. *Environmental Science and Pollution Research International, 26*(6), 6167–6181. doi:10.100711356-018-3993-8 PMID:30617875

Sonone, S. S., Jadhav, S., Sankhla, M. S., & Kumar, R. (2020). Water contamination by heavy metals and their toxic effect on aquaculture and human health through food Chain. *Lett. Appl. NanoBioScience, 10*(2), 2148–2166.

Sosyo Politik Saha Araştırmaları Merkezi. (2019). Kadının Genel İstihdamdaki Yeri Ve İstihdam Edildiği İş Kollarındaki Durumu Üzerine Çalışma. *Kadın Çalışmaları/Slider Category*. Retrieved February 13, 2022, from https://tinyurl.com/sahamerkezi-kadin-istihdami

Squalli, J. (2007). Electricity consumption and economic growth: Bounds and causality analyses of OPEC members. *Energy Economics, 29*(6), 1192–1205. doi:10.1016/j.eneco.2006.10.001

Sreen, N., Purbey, S., & Sadarangani, P. (2018). Impact of culture, behavior and gender on green purchase intention. *Journal of Retailing and Consumer Services*, *41*, 177–189. doi:10.1016/j.jretconser.2017.12.002

Stadelmann, M., & Schubert, R. (2018). How do different designs of energy labels influence purchases of household appliances? A field study in Switzerland. *Ecological Economics*, *144*, 112–123. doi:10.1016/j.ecolecon.2017.07.031

Stavi, I., Paschalidou, A., Kyriazopoulos, A. P., Halbac-Cotoara-Zamfir, R., Siad, S. M., Suska-Malawska, M., Savic, D., Roque de Pinho, J., Thalheimer, L., Williams, D. S., Hashimshony-Yaffe, N., van der Geest, K., Cordovil, C. M. S., & Ficko, A. (2021). Multidimensional food security nexus in drylands under the slow onset effects of climate change. *Land (Basel)*, *10*(12), 1350. doi:10.3390/land10121350

Stern, D. I. (1993). Energy and economic growth in the USA: A multivariate approach. *Energy Economics*, *15*(2), 137–150. doi:10.1016/0140-9883(93)90033-N

Stockhammer, E., Hochreiter, H., Obermayr, B., & Steiner, K. (1997). The Index of Sustainable Economic Welfare (ISEW) as an alternative to GDP in measuring economic welfare. The results of the Austrian (revised) ISEW calculation 1955-1992. *Ecological Economics*, *21*(1), 19–34. doi:10.1016/S0921-8009(96)00088-2

Strateji ve Bütçe Başkanlığı. (2022). *Kamu sektörü istihdam sayıları (2007-2022).* Retrieved August 31, 2022, from https://www.sbb.gov.tr/kamu-istihdami/

Talas, C. (1997). Toplumsal Ekonomi Çalışma Ekonomis (7th ed.). İmge Kitabevi.

Tanner, C., & Kast, S. W. (2003). Promoting sustainable consumption: Determinants of green purchases by Swiss consumers. *Psychology and Marketing*, *20*(10), 883–902. doi:10.1002/mar.10101

Tatlıdil, E., & Xanthocou, Y. (2002). Türk İşgücünün Yapısı ve Avrupa Birliği İstihdam Politikaları. *Ege Academic Review*, *2*(2), 1–14.

TCMB. (2002). *TÜFE Bazlı Reel Efektif Döviz Kuru (2003=100) (Temmuz 2022).* Retrieved July 31, 2022, from https://tinyurl.com/TCMB-ReelKur-Temmuz2022

Tennison, I., Roschnik, S., Ashby, B., Boyd, R., Hamilton, I., Oreszczyn, T., Owen, A., Romanello, M., Ruyssevelt, P., Sherman, J. D., Smith, A. Z. P., Steele, K., Watts, N., & Eckelman, M. J. (2021). Health care's response to climate change: A carbon footprint assessment of the NHS in England. *The Lancet. Planetary Health*, *5*(2), e84–e92. doi:10.1016/S2542-5196(20)30271-0 PMID:33581070

Testa, F., Pretner, G., Iovino, R., Bianchi, G., Tessitore, S., & Iraldo, F. (2021). Drivers to green consumption: A systematic review. *Environment, Development and Sustainability*, *23*(4), 4826-4880.

Testa, F., Iraldo, F., Vaccari, A., & Ferrari, E. (2015). Why eco-labels can be effective marketing tools: Evidence from a study on Italian consumers. *Business Strategy and the Environment*, *24*(4), 252–265. doi:10.1002/bse.1821

Tilikidou, I., & Delistavrou, A. (2012). The components of pro-environmental purchasing behavior. In *Proceedings of the International Conference on Contemporary Marketing Issues* (pp. 549-555). Academic Press.

Timur, M. C., & Doğan Çalışkan, Z. (2015). İstihdam Yaratmayan Büyüme: Türkiye Analizi. Jobless Growth: Analysis of Turkey. *Ardahan Üniversitesi İİBF Dergisi*, *1*, 231–247.

Tisdell, C. (1987). Tourism, the environment and profit. *Economic Analysis and Policy*, *17*(1), 13–30. doi:10.1016/S0313-5926(87)50009-1

Tiwari, A. K., Ozturk, I., & Aruna, M. (2013). Tourism, energy consumption and climate change in OECD countries. *International Journal of Energy Economics and Policy*, *3*(3), 247–261.

Tltaudit. (2023). *Examples of rational use of natural resources*. Retrieved from: https://tltaudit.ru/ka/cardiovascular-disease/racionalnoe-ispolzovanie-prirodnyh-resursov-primery-prirodopolzovanie-3---referat/

Toda, H. Y., & Yamamoto, T. (1995). Statistical inference in vector autoregressions with possibly integrated processes. *Journal of Econometrics*, *66*(1 2), 225 250. doi:10.1016/0304-4076(94)01616-8

Tsalavoutas, I., Tsoligkas, F., & Evans, L. (2020). Compliance with IFRS Mandatory Disclosure Requirements: A Structured Literature Review. *Journal of International Accounting, Auditing & Taxation*, *40*, 100338. Advance online publication. doi:10.1016/j.intaccaudtax.2020.100338

Tsani, S., & Menegaki, A. N. (2018). The Energy-Growth Nexus (EGN) Checklist for Authors. *The Economics and Econometrics of the Energy-Growth Nexus*, 347-376. . doi:10.1016/B978-0-12-812746-9.00011-0

Tseng, C., Wu, B., Morrison, A. M., Zhang, J., & Chen, Y. C. (2015). Travel blogs on China as a destination image formation agent: A qualitative analysis using Leximancer. *Tourism Management*, *46*, 347–358. doi:10.1016/j.tourman.2014.07.012

Tugcu, C. T., Ozturk, I., & Aslan, A. (2012). Renewable and non-renewable energy consumption and economic growth relationship revisited: Evidence from G7 countries. *Energy Economics*, *34*(6), 1942–1950. doi:10.1016/j.eneco.2012.08.021

TUİK. (2021). *Doğum İstatistikleri, 2020. Sayı 37229*. Retrieved February 15, 2022, from https://data.tuik.gov.tr/Bulten/Index?p=Dogum-Istatistikleri-2020-37229

TUİK. (2022a). *İşgücü İstatistikleri, Haziran 2022*. Retrieved August 12, 2022, from https://data.tuik.gov.tr/Bulten/Index?p=Isgucu-Istatistikleri-Haziran-2022-45651&dil=1

TUİK. (2022b). *İşgücü İstatistikleri (1988-2013)*. Retrieved February 15, 2022, from https://biruni.tuik.gov.tr/isgucuapp/isgucu.zul

TUİK. (2022c). *İşgücü İstatistikleri (2014 ve Sonrası).* Retrieved February 15, 2022, from https://biruni.tuik.gov.tr/medas/?kn=72&locale=tr

TUİK. (2022d). *Dış Ticaret İstatistikleri, Temmuz 2022.* Retrieved August 31, 2022, from https://biruni.tuik.gov.tr/medas/?kn=72&locale=tr

Tunalı, İ. (2003). Background Study on the Labour Market and Employment in Turkey. *European Training Foundation Final Report*, 125.

Tung, J., Shih, C., Wei, S., & Chen, H. (2012). Attitudinal inconsistency toward organic food in relation to purchase intention and behavior: An illustration of Taiwan consumers. *British Food Journal*, *114*, 997–1015. doi:10.1108/00070701211241581

U.S. Bureau of Labor Statistics. (2022a). *Household Data. Table A-15. Alternative measures of labor underutilization.* Retrieved July 31, 2022, from https://www.bls.gov/news.release/empsit.t15.htm

U.S. Bureau of Labor Statistics. (2022b). *Concepts and Definitions. Labor Force Statistics from the Current Population Survey.* Retrieved July 31, 2022, from https://www.bls.gov/cps/definitions.htm

Uddin, G. A., Salahuddin, M., Alam, K., & Gow, J. (2017). Ecological footprint and real income: Panel data evidence from the 27 highest emitting countries. *Ecological Indicators*, *77*, 166–175. doi:10.1016/j.ecolind.2017.01.003

Ukaogo, P. O., Ewuzie, U., & Onwuka, C. V. (2020). Environmental pollution: causes, effects, and the remedies. In *Microorganisms for sustainable environment and health* (pp. 419–429). Elsevier. doi:10.1016/B978-0-12-819001-2.00021-8

Ülgener, S. F. (1980). *Milli Gelir, İstihdam ve İktisadi Büyüm* (6th ed.). Der Yayınevi.

Ullrich, H. D. (1974). Energy crisis today, food crisis tomorrow? *Journal of Nutrition Education*, *6*(1), 4. doi:10.1016/S0022-3182(74)80171-8

Ulucak, R., & Lin, D. (2017). Persistence of policy shocks to ecological footprint of the USA. *Ecological Indicators*, *80*, 337–343. doi:10.1016/j.ecolind.2017.05.020

Ulucak, R., Sari, R., Erdogan, S., & Alexandre Castanho, R. (2021). Bibliometric literature analysis of a multi-dimensional sustainable development issue: Energy poverty. *Sustainability (Basel)*, *13*(17), 9780. doi:10.3390u13179780

Umwelt and Bundesamt. (2019). *Resource use and its consequences.* Retrieved from: https://www.umweltbundesamt.de/en/topics/waste-resources/resource-use-its-consequences

UN. (2022). *International Day for Biological Diversity.* Retrieved from: https://www.un.org/en/observances/biological-diversity-day

UNEP. (2023). *Annual Report, 2022.* Retrieved from: https://www.unep.org/resources/annual-report-2022

Unerman, J., Bebbington, J., & O'Dwyer, B. (2018). Corporate reporting and accounting for externalities. *Accounting and Business Research*, *48*(5), 497–522. doi:10.1080/00014788.2018.1470155

United Nations. (2005). *Basic Facts about the United Nations*. UN Department of Public Information.

United Nations. (2015). *Resolution adopted by the General Assembly on 25 September 2015, Transforming our world: The 2030 Agenda for Sustainable Development (A/RES/70/1)*. The New York Declaration.

University of Oxford. (2022). *Science Encyclopedia*. Oxford University Press.

Ünsal, E. (2000). Makro İktisat (3rd ed.). İmaj Yayıncılık.

UNWTO. (2020). *International tourism highlights*. Retrieved June 11, 2022, from https://www.e-unwto.org/doi/epdf/10.18111/9789284422456

UNWTO. (2021). *World tourism barometer-statistical annex*. Retrieved June 7, 2022, from https://www.wto.org/english/tratop_e/envir_e/unwto_barom21.pdf

UNWTO. (2022). *UN tourism news*. Retrieved June 4, 2022, from https://www.unwto.org/un-tourism-news-fitur-edition-2022

Uri, N. D. (1979). Energy demand and interfuel substitution in India. *European Economic Review*, *12*(2), 181–190. doi:10.1016/0014-2921(79)90021-7

Uslu, H. (2020). İstihdam Yaratmayan Ekonomik Büyüme: Türkiye İçin Okun Yasası Çerçevesinde Ekonometrik Bir Analiz. *Optimum Ekonomi ve Yönetim Bilimleri Dergisi*, *7*(1), 101–126. doi:10.17541/optimum.578251

Usman, M., Kousar, R., & Makhdum, M. S. A. (2020). The role of financial development, tourism, and energy utilization in environmental deficit: Evidence from 20 highest emitting economies. *Environmental Science and Pollution Research International*, *27*(34), 42980–42995. doi:10.100711356-020-10197-1 PMID:32725560

Uzar, U. (2020). Is income inequality a driver for renewable energy consumption? *Journal of Cleaner Production*, *255*, 1–11. doi:10.1016/j.jclepro.2020.120287

van Riper, C. J., Lum, C., Kyle, G. T., Wallen, K. E., Absher, J., & Landon, A. C. (2020). Values, motivations, and intentions to engage in pro-environmental behavior. *Environment and Behavior*, *52*(4), 437–462. doi:10.1177/0013916518807963

van Tonder, E., Fullerton, S., & de Beer, L. T. (2020). Cognitive and emotional factors contributing to green customer citizenship behaviours: A moderated mediation model. *Journal of Consumer Marketing*, *37*(6), 639–650. doi:10.1108/JCM-06-2019-3268

Vashishtha, R. (2014). The role of bank monitoring in borrowers- discretionary disclosure: Evidence from covenant violations. *Journal of Accounting and Economics*, *57*(2-3), 176–195. doi:10.1016/j.jacceco.2014.04.002

Vaughan, D. (2022). *What is the most widely practiced religion in the world?* Retrieved June 17, 2022, from https://www.britannica.com/story/what-is-the-most-widely-practiced-religion-in-the-world

Verick, S. (2014). Female labor force participation in developing countries. *IZA World of Labor: Evidence-Based Policy Making*, *87*. Advance online publication. doi:10.15185/izawol.87

Verrecchia, R. E. (1983). Discretionary Disclosure. *Journal of Accounting and Economics*, *5*, 179–194. doi:10.1016/0165-4101(83)90011-3

Vieira, L. C., Longo, M., & Mura, M. (2021). Are the European manufacturing and energy sectors on track for achieving net-zero emissions in 2050? An empirical analysis. *Energy Policy*, *156*, 112464. doi:10.1016/j.enpol.2021.112464

Von Lyncker, K., & Thoennessen, R. (2017). Regional club convergence in the EU: Evidence from a panel data analysis. *Empirical Economics*, *52*(2), 525–553. doi:10.100700181-016-1096-2

Voutsa, M. E., & Borovas, G. (2015). The operation of the Bretton Woods institutions in the modern globalized environment: Problems and challenges. *Procedia Economics and Finance*, *33*, 168–180. doi:10.1016/S2212-5671(15)01702-5

Wackernagel, M., & Rees, W. (1996). Our ecological footprint: Reducing human impact on the earth. New Society Publishers, 1-29.

Wang, H. (2022). Exploring the influencing factors of environmental deterioration: Evidence from China employing ARDL–VECM method with structural breaks. *International Journal of Climate Change Strategies and Management*. doi:10.1108/IJCCSM-10-2021-0114

Wang, Y. F., & Wang, C. J. (2016). Do psychological factors affect green food and beverage behavior? An application of the theory of planned behavior. *British Food Journal*, *118*(9), 2171–2199. doi:10.1108/BFJ-12-2015-0469

Wassie, S. B. (2020). Natural resource degradation tendencies in Ethiopia: A review. *Environmental Systems Research*, *9*(1), 1–29. doi:10.118640068-020-00194-1

Water Footprint Network. (2014). *Water Footprint Network partners with WWF to create the "Living Planet Report 2014"*. Retrieved from: https://waterfootprint.org/en/about-us/news/news/living-planet-report/

Watkins, L., Aitken, R., & Mather, D. (2016). Conscientious consumers: A relationship between moral foundations, political orientation and sustainable consumption. *Journal of Cleaner Production*, *134*, 137–146. doi:10.1016/j.jclepro.2015.06.009

WCED. (1987). Our common future. Oxford University Press.

WDI. (2022). *World development indicators online database*. Retrieved May 21, 2022, from https://databank.worldbank.org/source/world-development-indicators

Webster, F. E. Jr. (1975). Determining the characteristics of the socially conscious consumer. *The Journal of Consumer Research*, *2*(3), 188–196. doi:10.1086/208631

Westerlund, J. (2008). Panel cointegration tests of the Fisher effect. *Journal of Applied Econometrics*, *23*(2), 193–233. doi:10.1002/jae.967

Westerlund, J., & Edgerton, D. L. (2007). A panel bootstrap cointegration test. *Economics Letters*, *97*(3), 185–190. doi:10.1016/j.econlet.2007.03.003

White, D. C. (1973). The U.S. energy crisis: A scientist's view. *Energy Policy*, *1*(2), 130–135. doi:10.1016/0301-4215(73)90035-9

White, K., Habib, R., & Hardisty, D. J. (2019). How to SHIFT consumer behaviors to be more sustainable: A literature review and guiding framework. *Journal of Marketing*, *83*(3), 22–49. doi:10.1177/0022242919825649

Wiedman, C., & Wier, H. (1999). Management of note disclosures: The case of unconsolidated subsidiaries prior to FAS 94. *Journal of Accounting, Auditing & Finance*, *14*(1), 73–94. doi:10.1177/0148558X9901400104

Wijekoon, R., & Sabri, M. F. (2021). Determinants that influence green product purchase intention and behavior: A literature review and guiding framework. *Sustainability (Basel)*, *13*(11), 6219. doi:10.3390u13116219

Williams, C. A., & Nagy, D. M. (2021). ESG and climate change blind spots: Turning the corner on SEC disclosure. *Texas Law Review*, *99*(7), 1453–1485.

Winsten, C. B. (1957). Discussion on Mr. harrells paper. *J. Royal Statist. Soc. Sen A. Statist. Soc.*, *120*(3), 282–284.

Witek, L., & Kuzniar, W. (2021). Green purchase behavior: The effectiveness of sociodemographic variables for explaining green purchases in an emerging market. *Sustainability (Basel)*, *13*(1), 209. doi:10.3390u13010209

Wolkomir, M. J., Futreal, M., Woodrum, E., & Hoban, T. (1997). Denominational subcultures of environmentalism. *Review of Religious Research*, *38*(4), 325–343. doi:10.2307/3512194

World Ocean Review. (2010). *Marine Minerals*. Retrieved from: https://worldoceanreview.com/en/wor-1/energy/marine-minerals/

World Resources Institute. (2010). *Forests*. Sustaining Forests for People and Planet. Retrieved from: https://www.wri.org/our-work/topics/forests

WTTC. (2022). *World travel & tourism council*. Retrieved April 18, 2022, from https://wttc.org/Research/Economic-Impact

Wymer, W., & Polonsky, M. J. (2015). The limitations and potentialities of green marketing. *Journal of Nonprofit & Public Sector Marketing*, *27*(3), 239–262. doi:10.1080/10495142.2015.1053341

Xie, H., Zhang, Y., Wu, Z., & Lv, T. (2020). A bibliometric analysis on land degradation: Current status, development, and future directions. *Land (Basel)*, *9*(1), 28. doi:10.3390/land9010028

Yan, C. (1975). Income distribution effect on higher energy costs. *The Journal of Behavioral Economics*, *4*(2), 45–53. doi:10.1016/0090-5720(75)90062-5

Yang, J., & Malone, E. (2008). Working to shape what society's expectations of us should be: Philip Morris's societal alignment strategy. *Tobacco Control*, *17*(6), 391–398. doi:10.1136/tc.2008.026476 PMID:18845623

Yan, Z., Du, K., Yang, Z., & Deng, M. (2017). Convergence or divergence? Understanding the global development trend of low-carbon technologies. *Energy Policy*, *109*, 499–509. doi:10.1016/j.enpol.2017.07.024

Yatchew, A. (2003). *Semiparametric Regression for the Applied Econometrician*. Cambridge University Press. doi:10.1017/CBO9780511615887

Yatchew, A. J. (1998). Nonparametric Regression Techniques in Economics. *Journal of Economic Literature*, *36*, 669–721.

Yilanci, V., Eryuzlu, H., & Hopoglu, S. (2020). Convergence of military burdens in the MENA region. *International Relations*, *17*(66), 41–59.

Yilanci, V., & Gorus, M. S. (2021). Considering nonlinearity and structural changes in the convergence of clean energy consumption: The case of OECD countries. *Management of Environmental Quality*, *32*(2), 243–259. doi:10.1108/MEQ-03-2020-0047

Yilanci, V., Gorus, M. S., & Aydin, M. (2019). Are shocks to ecological footprint in OECD countries permanent or temporary? *Journal of Cleaner Production*, *212*, 270–301. doi:10.1016/j.jclepro.2018.11.299

Yilanci, V., Gorus, M. S., & Solarin, S. A. (2022). Convergence in per capita carbon footprint and ecological footprint for G7 countries: Evidence from panel Fourier threshold unit root test. *Energy & Environment*, *33*(3), 527–545. doi:10.1177/0958305X211011461

Yilanci, V., & Pata, U. K. (2020). Investigating the EKC hypothesis for China: The role of economic complexity on ecological footprint. *Environmental Science and Pollution Research International*, *27*(26), 32683–32694. doi:10.100711356-020-09434-4 PMID:32514916

Yin, J., Zheng, M., & Chen, J. (2015). The effects of environmental regulation and technical progress on CO2 Kuznets curve: An evidence from China. *Energy Policy*, *77*, 97–108. doi:10.1016/j.enpol.2014.11.008

YÖK. (2022). *Öğrenci Sayıları Özet Tablosu. 2020-2021 Öğretim Yılı*. Yüksek Öğretim Bilgi Yönetim Sistemi. Retrieved Feburuary 2, 2022, from https://istatistik.yok.gov.tr/

Yu, E. S. H., & Hwang, B. K. (1984). The relationship between energy and GNP: Further results. *Energy Economics*, *6*(3), 186–190. doi:10.1016/0140-9883(84)90015-X

Yu, Y., Zhang, Y., & Song, F. (2015). World energy intensity revisited: A cluster analysis. *Applied Economics Letters*, *22*(14), 1158–1169. doi:10.1080/13504851.2015.1013603

Zhang, L., Godil, D. I., Bibi, M., Khan, M. K., Sarwat, S., & Anser, M. K. (2021). Caring for the environment: How human capital, natural resources, and economic growth interact with environmental degradation in Pakistan? A dynamic ARDL approach. *The Science of the Total Environment*, *774*, 145553. doi:10.1016/j.scitotenv.2021.145553 PMID:33611006

Zheng, S., Wang, R., Mak, T. M., Hsu, S. C., & Tsang, D. C. (2021). How energy service companies moderate the impact of industrialization and urbanization on carbon emissions in China? *The Science of the Total Environment*, *751*, 141610. doi:10.1016/j.scitotenv.2020.141610 PMID:32877794

Zhu, J., & Lin, B. (2020). Convergence analysis of city-level energy intensity in China. *Energy Policy*, *139*, 111357. doi:10.1016/j.enpol.2020.111357

Zivot, E., & Andrews, D. (1992). Further evidence on the great crash, the oil price shock, and the unit root hypothesis. *Journal of Business & Economic Statistics*, *10*(3), 251–270.

Zwolak, A., Sarzyńska, M., Szpyrka, E., & Stawarczyk, K. (2019). Sources of soil pollution by heavy metals and their accumulation in vegetables: A review. *Water, Air, and Soil Pollution*, *230*(7), 1–9. doi:10.100711270-019-4221-y

About the Contributors

Veli Yilanci is a Professor of Economics at Canakkale Onsekiz Mart University - Türkiye. He completed his Ph.D. in Econometrics at the Istanbul University in 2012. He earned a Master's degree in the same department in 2007 and Bachelor's degree in Econometrics at Istanbul University in 2005. He is interested in Applied Economics and Time Series Econometrics. His current research focuses on the Fourier unit root and cointegration tests.

* * *

Rashid Ameer is a lecturer in the postgraduate programme at IPU and is highly regarded for his research expertise in corporate finance, corporate governance, corporate social responsibility, and sustainability. His research in these areas has been published in leading international journals such as the Journal of Business Ethics, Journal of Multinational Management, Managerial Finance, Asian Economic Journal, and Journal of Asian Economics. He received from Emerald publishing the outstanding reviewer award in 2006 and the highly commended award for two papers published in 2013.

Murat Aslan is a Professor of Economics at Ankara Yıldırım Beyazıt University - Türkiye. He completed his Ph.D. in Economics at George Mason University in 2004. He earned a Master's degree at the American University in 1997 and Bachelor's degree in Public Finance at Hacettepe University in 1993. He has been working as the head of the Department of Economics at Ankara Yıldırım Beyazıt University, Faculty of Political Sciences, Department of Economics, since 2017.

F. Gül Aygen-Ispahi is an independent researcher. She worked at Istanbul Bilgi University as an assistant professor of marketing from 1998 to 2000. She later worked at Telsim (former Vodafone), a mobile telecommunications company, as a marketing researcher for three and a half years. In 2003, she began working at Doğuş University, Istanbul, Turkey, both as the Dean of Students and an assistant professor, till the second half of 2014. She taught at Boğaziçi University as a part-time lecturer, in the second half of 2014 and the first half of 2023. She presently concentrates on conducting research with regard to consumers. Her research interests are in the fields of consumer behavior, food safety, green marketing, sustainable development, and environmental sustainability.

Can Bekaroglu received his (English) Economics bachelor in 2006 from Anadolu University Faculty of Economics and Administrative Sciences; his Master of Arts in Economics in 2010 as well as his Phd in Economics in 2015 from University of Connecticut, USA, where he also served, first as a teaching assistant (2008 thru 2015), then as an adjunct professor (2015- 2016); after which, he continued to work in Roger Williams University in Rhode Island as an adjunct professor (2016), and a visiting professor in Central Connecticut State University (2016-2017). From 2018 on, he has been working as an assistant professor in Canakkale Onsekiz Mart University, Faculty of Political Sciences, Department of Economics.

Muhammed Sehid Gorus holds a PhD in economics from Ankara Yıldırım Beyazıt University, Turkey. Currently, he serves as an assistant professor in the department of economics at Ankara Yildirim Beyazit University, Turkey. His research interests include energy economics, environmental economics, and applied macroeconomics. He published lots of academic papers and book chapters.

Erdal Tanas Karagol holds a PhD in economics from the University of York, UK. Currently, he is serving in Germany as a Berlin education attache of Turkey. His research interests include energy economics, defense economics, and applied macroeconomics. He published more than 50 international papers/book chapters/books.

Radiah Othman is currently an Associate Professor at Massey University, New Zealand. Her Masters in Accountancy (with distinction) is from Curtin University of Technology (Australia), and her Ph.D. is from Aston University (UK). She has more than 26 years of teaching experience in auditing and accounting information systems. She is an active researcher and has published papers in the area of forensic accounting, accounting education, and sustainability in reputable journals such as the Journal of Business Ethics, Accounting Education, and Accounting and Finance. She has successfully supervised 7 Masters and 2 Ph.D. students and examined 3 Ph.D. theses in the areas of accounting, auditing, and accounting information systems. She received a Special Award for Corporate Social Responsibility (CSR) at the

Malaysian Outstanding Research Paper Awards (2011), Pearson Best Author Award (2013), and Highly Commended Award Winner at the Literati Network Awards for Excellence from Emerald (2013).

Ece Özer Çizer is a Ph.D. candidate and Research Assistant at the Department of Business Administration, Faculty of Economics and Administrative Sciences, Yildiz Technical University. Her research interests lie in the areas of digital marketing, circular economy, and sharing economy. She has authored numerous international academic articles and book chapters in these fields.

Onder Ozgur is an Associate Professor of Economics at Ankara Yıldırım Beyazıt University - Türkiye. He completed his Ph.D. in Economics at the Ankara Yildirim Beyazit University in 2020. He earned a Master's degree in Economics at the Ankara Yildirim Beyazit University in 2017 and a Bachelor's degree in Anadolu University, Faculty of Economics and Administrative Sciences, Department of Economics, Turkey, in 2010. He has been working as deputy head of the Department of Economics at Ankara Yıldırım Beyazıt University, Faculty of Political Sciences, Department of Economics since 2021.

Muhlis Selman Saglam completed his undergraduate education at Istanbul University, Faculty of Business Administration, in 2018. He received his master's degree in 2020 with his thesis titled "A Comparative Evaluation of Islamic Economics and Liberal Economics Approaches in the Context of Market Intervention based on Morality" from the same university. In the same year, he started his doctorate education at Istanbul University, Department of Economic History. He writes his thesis on theoretical approaches to the history of Ottoman economics. He works in economic history, the history of economic thought, the Enlightenment Period, economic philosophy, economic sociology and economic theory.

Index

A

B

C

D

E

F

www.ingramcontent.com/pod-product-compliance
Lightning Source LLC
LaVergne TN
LVHW081345300926
849161LV00008B/2065